D0919539

STATISTICS

A first course

Second edition

JOHN E. FREUND

Arizona State University

Prentice-Hall, Inc., Englewood Cliffs, New Jersey

Library of Congress Cataloging in Publication Data

Freund, John E.
Statistics: a first course.

Bibliography: p.
Includes index.
1.–Statistics. I.–Title.
HA29.F687–1976 519.5 75-22037
ISBN 0-13-846055-8

Printed in the United States of America

10 9 8 7 6 5 4 3 2

Prentice-Hall International, Inc., London
Prentice-Hall of Australia, Pty. Ltd., Sydney
Prentice-Hall of Canada, Ltd., Toronto
Prentice-Hall of India Private Limited, New Delhi
Prentice-Hall of Japan, Inc., Tokyo
Prentice-Hall of Southeast Asia (Pte) Ltd., Singapore

Affectionately dedicated to my sons
Doug and John

Contents

Preface

In the last few decades there have been great changes in the teaching of statistics. Not only has there been a pronounced shift in emphasis from descriptive methods to more and more inference, but there has also been a change in the level at which first courses in statistics are being taught. Whereas these courses used to be taught mostly to college Juniors and Seniors, they are nowadays taught also to Freshmen and Sophomores, and attempts have been made and are being made to introduce probability and statistics into advanced high school programs. The purpose of this book, like that of the first edition, is to reach the student at this somewhat earlier level, as will be apparent from the organization of the material, the language of the book, its format, its notation, and above all, the exercises and the illustrations. Following the suggestions of many colleagues and students, the author has made extensive changes in the format which should make the book easier to use. There is less cross-referencing among the exercises, and the numbers in many of the exercises are easier to work with, thus requiring less attention to arithmetical details. Also, the vast majority of the illustrations and exercises are new. There is some new material, for example, on nonparametric methods and the Poisson distribution; also, there is a much better square root table.

As in the first edition, controversial material (say, about the meaning of probability) has not been avoided. The reader is exposed to the strengths of statistical techniques as well as their weaknesses, and it is hoped that this honest approach will provide a stimulus as well as a challenge. To attain a measure of flexibility and to accommodate some of the differences of opinion as to what should be taught in a first course in statistics, a number of topics are included among the exercises *with detailed explanations.* These exercises, which are clearly marked, have made it possible to present special material, extra details, and some more advanced work without cluttering up the main body of the text.

The author is indebted to the Literary Executor of the late Sir Ronald Fisher, F.R.S., Cambridge, to Dr. Frank Yates, F.R.S., Rothamsted, and also to Oliver and Boyd, Edinburgh, for permission to reprint parts of Tables IV and VI, respectively, from their books *Statistical Methods for Research Workers and Statistical Tables for Biological, Agricultural, and Medical Research*; and to Professor E. S. Pearson and the Biometrika trustees for permission to reproduce Tables 8, 18, and 41 from their *Biometrika Tables for Statisticians.*

Finally, the author would like to express his appreciation to his many colleagues and students whose helpful criticisms and suggestions contributed greatly to the previous edition of this book and also to this second edition. In particular, he would like to thank his nephew Ray for checking the calculations in the text and working out the answers to the exercises, and his son Doug for helping with the proofreading. The author would also like to express his appreciation to the staff of Prentice-Hall, Inc., for their courteous cooperation in the production of this book and, above all, to his wife for her cheerful encouragement in spite of the demands made by this project on her husband's time.

John E. Freund
Scottsdale, Arizona

1

Statistical thinking: introduction

Introduction

Each chapter of this book begins with an introduction which is designed to set the tone for the remainder of the chapter. Some of these introductions outline the contents of the respective chapters, some specify their objectives, and some contain a discussion of preliminary concepts or definitions. Since this section is an *introduction* to an introduction, let us merely point out that the objectives of this chapter are to explain what statistics is all about—its history, its current state, its scope, and its limitations.

Statistics, past and present

The origin of modern statistics can be traced to two areas of interest which, on the surface, have very little in common: *games of chance and what we now call political science.* Mid-eighteenth-century studies in probability (motivated largely by interest in games of chance) led to the mathematical treatment of errors of measurement and the theory which now forms the foundation of statistics. In the same century, interest in the numerical description of political units (cities, provinces, counties, etc.) led to the development of methods which, until the last few decades, constituted all there was to the subject of statistics. This included any treatment designed to summarize, or describe, important features of numerical data, mainly by means of tables and charts. Nowadays, we refer to this as **descriptive statistics**, and it includes any kind of data processing which does not attempt to infer anything that goes beyond the data themselves.

Example 1. If the government reports on the basis of census counts that the population of the United States was 151,325,798 in 1950 and 203,211,926 in 1970, they are giving us facts which belong to the field of descriptive statistics. This would also be the case if we calculated the corresponding percentage growth, which, as can easily be verified, was 34.3%, but *not* if we use these data to predict, say, the population of the United States in the year 2000.

There are essentially two reasons why the scope of statistics and the need to study statistics have grown enormously in the last few decades. One reason is that the amount of data that is collected, processed, and disseminated to the public for one reason or another has increased almost beyond comprehension. To act as watchdogs, more and more persons with some knowledge of statistics are needed to take an active part in the collection of the data, in the analysis of the data, and, what is equally important, they must take part in all of the preliminary planning It is really frightening when one starts to think of all the things that can go wrong in the compilation of statistical data. The results of very costly surveys can be completely useless if questions are ambiguous or asked in the wrong way, if they are asked of the wrong persons, in the wrong place, or at the wrong time.

Example 2. A famous example in which questions were asked of the wrong persons is the poll conducted by the (no longer existing) *Literary Digest* to predict the 1936 presidential election. The *Literary Digest* predicted a Republican victory on the basis of a tremendously large sample of voters chosen from telephone directories and lists of automobile registrations, but as things turned out, President Roosevelt's support came mainly from the lower-income groups, which were not adequately represented among the owners of telephones and automobiles. Thus, the poll was highly **biased**, and this error turned out to be disastrous to the magazine.

There are many subtle reasons why it is often difficult to get honest answers and, hence, "usable" data. For one thing, many people are understandably reluctant to talk about personal matters, say, how often they take a shower or brush their teeth; they are unlikely to return a mail questionnaire inquiring about their success in life unless they happen to be doing rather well; and they may refer to their refrigerator as a "Frigidaire" even though it may have been made by Hotpoint or RCA. There is also the human element, which may lead a poll taker to interview mostly curvaceous blondes, cut his expenses by interviewing only persons living in the same city block, or turn in reports on interviews which never took place.

All the biases that we have discussed so far pertain to surveys, but it is just as easy to run into difficulties when it comes to laboratory experiments. Not only can instrument readings be off, or affected by changes in temperature or humidity, but there is always the possibility of human error, even in this age of automation. There are also those annoying hidden biases, which may be very difficult to detect. For instance, if a quality inspection procedure calls for a check of every tenth bottle of catsup filled and sealed by a machine, the results could be disastrously misleading if it so happened that due to a worn gear every fifth bottle is improperly sealed. Similarly, a missile may well fail in flight if its stability was tested for vibrations of the wrong frequency, and all sorts of things can go wrong with manufactured products unless there is a very careful and conscientious check of all raw materials. Needless to say, it can also happen that computers are programmed incorrectly, that information is transmitted or recorded incorrectly, and there is literally no end to the list of things that can conceivably cause trouble when it comes to the collection, recording, and processing of data. All these things must be watched very carefully, and a competent statistician will never commit himself to any results or conclusions unless he is pretty sure that his data are free of *unintentional* biases. (Of course, there are also such things as *intentional* biases and outright fraudulent misrepresentations, but this is a question of ethics rather than statistics.)

The second, and even more important, reason why the scope of statistics and the need to study statistics has grown so tremendously in recent years is the increasingly *quantitative approach* employed in all the sciences, as well as in business and many other activities which directly affect our lives. Since most of the information required by this approach comes from *samples* (namely, from observations made on only part of a large set of items), its analysis requires generalizations which go beyond the data, and this is why there has been a pronounced shift in emphasis from descriptive statistics to **statistical inference**, or **inductive statistics**. In other words,

> **Statistics has grown from the art of constructing charts and tables to the science of making generalizations on the basis of numerical data, or even more generally the science of decision making in the face of uncertainty.**

To mention a few examples, generalizations (that is, methods of statistical inference) are needed to estimate the 1985 assessed value of all property in San Diego County (on the basis of business trends, population projections, and other factors); to compare the effectiveness of two or more computer-aided methods of instruction (on the basis of the performance of samples of students thus taught); to determine the most effective dose of a new medicine (on the basis of experiments conducted with volunteer patients from selected hospitals); to evaluate conflicting and uncertain legal evidence; to predict the traffic over a bridge which has not yet been built; to judge the average fuel consumption of a new engine; to determine the age of excavated bones; and so forth. In each of these examples there are uncertainties, only partial or incomplete information, and it is the job of statistics to judge the merits of all possible alternatives and, perhaps, suggest a "most profitable" choice, a "most promising" prediction, or a "most reasonable" course of action.

All this would be impossible unless statistics concerned itself with the question of how data are obtained and how experiments are conducted. As elsewhere, we get "nothing for nothing," and unless great care is exercised in all phases of an investigation, it may be impossible to reach any valid (or useful) conclusions whatsoever. Generally speaking, *no amount of fancy mathematics or statistical manipulation can salvage poorly planned studies, surveys, or experiments.*

Example 3. A classical example of this is the following "experiment" concerning the merits of a new remedy for seasickness: All the passengers on one ocean liner were given the new remedy while all the passengers on another ocean liner were given a placebo containing nothing but sugar (since there might conceivably be a psychological effect from the mere knowledge of having taken a new kind of pill, or even the knowledge of being part of an "experiment"). Things turned out beautifully—hardly any of the passengers on the first ship got seasick, most of the passengers of the other one did, but, alas, the second ship had run into a severe storm while the first one's sailing had been as smooth as silk.

To avoid situations like this, statistics must encompass all questions concerning the collection of data, including the design (or planning) of all phases of a survey or experiment. Only then will it be possible to distinguish between "good" statistics and "bad" statistics, namely, between statistical techniques that are correctly (and usually profitably) applied, and those that are unintentionally or intentionally perverted.

The study of statistics

If the reader had taken a first course in statistics thirty or so years ago, the demands on his ability in mathematics would have been extremely modest. He would have spent most of his time constructing charts and tables, or calculating some of the very simple and straightforward statistical descriptions which we shall discuss in Chapter 2. This would not have required much more than basic arithmetic, the ability to count, and, say, the calculation of percentages. Now all this has changed: The principles which we use in the collection of data as well as the techniques which we use in their analysis are based on the theory of probability. This theory, which we shall study in Chapter 4, was originally developed in connection with games of chance, and, as such, it dates back to the seventeenth century. Its growth was rapid, but it was not until the work of Karl Gauss (1777–1855) and Pierre Laplace (1749–1827) that the theory was applied to other fields. Noting that the mathematical theory which had been developed for "heads or tails," or "red or black" in roulette, applied also to situations where the alternatives are "life or death," "male or female," "young or old," . . . , probability theory was applied to actuarial science and to some phases of social science. Later, it was also introduced into physics and biology, and most recently probability theory has found applications in practically all aspects of human endeavor which in some way involve an element of uncertainty or risk.

So far as modern statistics is concerned, we might well refer to it as *applied probability theory*. To begin with, we have to worry about probabilities when it comes to the collection of most kinds of data; practically all of the statistical techniques which we shall study in this book are based on the assumption that the data constitute **random samples**. Although the dictionary definition of "random" says that it means "haphazard" or "without aim or purpose," in practice this can be assured only by leaving the ultimate choice of the data (at least partially) to chance—for instance, by drawing lots or by employing tables of **random numbers**, which will be introduced in Chapter 7. Virtually all of the techniques treated in Chapters 7 through 11 require that our data are obtained in accordance with well-specified rules of probability *which will control every subsequent step*—they will enable us to assess the magnitude of possible errors in problems of estimation; they will enable us to decide whether discrepancies between theory and practice (namely, differences between what we might expect and what we get) can reasonably be attributed to chance; and they will enable us to judge the relative merits of predictions which are based on what has happened in the past. For instance, the rules of probability which we observe in getting the necessary data will enable us to assess the error we might make in trying to estimate the average time it takes to learn twenty French verbs, the age of a fossilized skeleton, or the average amount of instant coffee that is actually contained in a "6-ounce" jar; they will enable us to judge whether we should get suspicious if we get only 38 heads in 100 flips of a coin instead of the 50 which we might expect; and they will enable us to specify limits (that is, two numbers) about which we can be "pretty sure" that their range will contain, say, the 1982 demand for oranges in New York City, or the rush hour traffic on the Santa Ana freeway in 1984.

Much of the theory which underlies these statistical techniques requires a good deal of mathematics (including calculus), but most of it can be understood and intelligently applied with a modest understanding of what probability theory is all about. Of course, there is the temptation to make things easy by assuming that everyone has some intuitive notion about probability and let it go at that; experience has shown, however, that this would leave many readers hanging very uncomfortably in midair. In fact, the concept of probability is by no means "intuitively obvious"; it is highly controversial and the study of its meaning (or meanings) presents a fascinating challenge.

In all fairness, we must admit that most students find the material on probability the hardest part of a beginning course in statistics. However, unless the reader is prejudiced against mathematics before he starts, he will find the study of probability stimulating and rewarding. In this book we shall introduce some of the preliminary ideas relating to probabilities in Chapter 3, and we shall study some of the basic rules and formulas in Chapter 4. References to more advanced treatments of probability theory and mathematical statistics (namely, the part of probability theory which relates directly to statistics) may be found in the Bibliography at the end of the book.

In contrast to this text, which presents a general introduction to the subject of statistics, numerous books have been written on business statistics, educational statistics, engineering statistics, medical statistics, psychological statistics, and so on. Although it is true that different fields of application require some (though very few) special techniques, the beauty of it is that *most statistical methods are the same regardless of the field of application.*

The approach which we shall use in this book is in our opinion a "healthy" balance between theory and application. There is just enough theory to give meaning to fundamental ideas and to provide an understanding of the merits as well as the shortcomings of the most widely used statistical techniques. So far as applications are concerned, there is ample material pertaining to various fields (among the exercises and the illustrations), and it is hoped that this will give the reader an appreciation of the scope and the importance of the entire subject of statistics. If we had to list a single objective which we hope to accomplish, it would be that the reader develop a critical attitude toward anything connected with the collection, treatment, and analysis of data—an attitude which may well be referred to as *statistical thinking.*

Statistics, what lies ahead

Earlier in this chapter we indicated how statistics has evolved from its original job of summarizing data by means of charts and tables to its role in inference, namely, in making generalizations on the basis of samples. This is not meant to imply, however, that the subject of statistics has now become stable and inflexible, and that it has ceased to grow. Aside from the fact that new statistical techniques are constantly being developed to meet particular needs, the whole philosophy of statistics continues to be in a state of change. Most recently, attempts have been made to treat all problems of statistical inference within the framework of a unified theory called **decision theory**, which, so to speak, covers everything "from cradle to grave." One of the main features of this theory is that we must account for *all of the consequences* which can arise when we base decisions on statistical data. This poses tremendous practical difficulties, and we must say in all fairness that very little of these developments has "filtered through" into elementary texts. To understand some of the difficulties posed by such an all-encompassing theory, we must appreciate the fact that

> **No matter how objective one tries to be in the planning and in the performance of an investigation (survey or experiment), it is virtually impossible to eliminate all elements of subjectivity; furthermore, it is generally difficult, if not impossible, to put "cash values" on all possible consequences of one's acts.**

So far as the first point is concerned, subjective decisions generally affect the hiring of research personnel, the purchase of equipment, the choice of a particular statistical technique, the number of measurements or observations one decides to make, and the conditions under which they are to be performed. Needless to say, perhaps, it is quite a problem to account for all these things in one general theory. So far as the second point is concerned, let us merely ask the reader how he would put a "cash value" on the consequences of the decision whether or not to market a new medicine—especially if the wrong decision may involve the loss of human lives.

We have mentioned all this primarily to impress upon the reader that statistics, like most other fields of learning, is not static. Indeed, it is difficult to picture what a beginning course in statistics may be like twenty years hence, although it is a pretty fair bet that its contents will differ considerably from the material covered in this book. Certain aspects will probably still be the same, and that includes the role of probability theory in the foundations of statistics as well as certain "bread and butter" techniques which have been very useful in the past and will undoubtedly continue to be widely used in the future. On the other hand, there may well be much more emphasis on some of the ideas introduced in the final sections of Chapters 3 and 4. Since all this should make more sense to the reader after he has studied this book, we shall return to it briefly in Chapter 11, the concluding chapter, in which we shall summarize what we have learned and preview briefly what lies ahead.

Exercises

1. On three consecutive days, Fred had to wait 8, 13, and 9 minutes for the bus he takes to work, and 4, 9, and 11 minutes for the bus he takes home from work. Which of the following conclusions can be obtained from these figures by means of purely descriptive methods and which require a statistical inference, namely, a generalization? Explain your answers.

a. On these days, Fred had to wait on the average 10 minutes for the bus he takes to work.

b. On two of the three days, Fred had to wait longer for the bus he takes to work than for the bus he takes home from work.

c. The most Fred will ever have to wait for the bus he takes home from work is 11 minutes.

d. For the three days, the number of minutes Fred had to wait for the bus he takes home from work increased from day to day.

e. Probably, he had to wait 13 minutes for the bus he takes to work because it rained on the second day.

2. A technician working for a consumers' rating service found that four brand *A* pillows withstood, respectively, 65, 72, 57, and 46 hours of continuous torture tests, while four brand *B* pillows withstood 70, 82, 53, and 55 hours of continuous torture tests. Which of the following conclusions can be obtained from these figures by purely descriptive methods and which require a statistical inference, namely, a generalization? Explain your answers.

a. The pillow which lasted the shortest period of time was a brand *A* pillow.

b. All brand *B* pillows will withstand the continuous torture test for at least 50 hours.

c. The four brand *A* pillows lasted 60 hours on the average, while the four brand *B* pillows lasted 65 hours on the average.

d. The difference between the two averages obtained in part (c) is so small that it is impossible to judge whether brand B pillows are really better than brand A pillows.

e. Since the second figure obtained for brand B is much larger than all the others, it was probably recorded incorrectly and should have been 72 instead of 82.

3. In five French tests, Harry received grades of 42, 66, 70, 74, and 83. Which of the following conclusions can be obtained from these figures by purely descriptive methods and which require a statistical inference, namely, a generalization? Explain your answer.

a. Only one of Harry's scores was above 75.

b. Harry was probably sick on the day he took the first test.

c. Harry's score increased from each test to the next.

d. Harry studied harder for each successive test.

e. Two of Harry's grades were below 70 and two were above 70.

4. On page 2 we said that the results of costly surveys can be completely useless if questions are ambiguous or asked in the *wrong way*, if they are asked of the *wrong persons*, in the *wrong place*, or at the *wrong time*, and the first of these *biases* was illustrated in Example 2. Which of the other biases would affect the following studies?

a. To determine housewives' opinions about certain convenience foods, a house-to-house survey is conducted during the morning hours on weekdays.

b. In a study of art appreciation, persons are asked whether they like Indian art.

c. To determine public sentiment about certain foreign trade restrictions, an interviewer asks voters: "Do you feel that this unfair practice should be stopped?"

d. To predict an election, a poll taker interviews a random sample of persons coming out of a building which houses the national headquarters of a political party.

2

Summarizing data

Introduction

In recent years the collection of statistical data has grown at such a rate that it would be impossible to keep up even with a small part of the things which directly affect one's life *unless this information were disseminated in "predigested" or summarized form.* The whole matter of putting large masses of data into a usable form has always been important, and it has multiplied greatly in the last few decades. This has been due partly to the development of electronic computers which have made it possible to accomplish in minutes what previously had to be left undone because it would have taken months or even years, and partly to the increasingly quantitative approach of the sciences, especially the social sciences, where nearly every aspect of human life is now measured in one way or another.

The most common method of summarizing data is to present them in condensed form in tables or charts, and this used to take up the better part of elementary courses in statistics. Nowadays, the scope of statistics has expanded to such an extent that much less time is devoted to this kind of work—in fact, we shall talk about it only in the next few pages of this chapter. In the remainder of the chapter we shall see how data can be summarized in other ways, namely, by means of well-chosen statistical descriptions. Since one of the main objectives of this chapter is to impress upon the reader that there are situations in which given data do not require extensive statistical treatment, namely, situations in which the data can "almost" speak for themselves, we shall discuss here only the simplest kinds of statistical descriptions. More complicated kinds of descriptions (the standard deviation, the coefficient of correlation, etc.) will be taken up later, *when needed.*

Frequency distributions

When one deals with large sets of data, a good overall picture and sufficient information can often be conveyed by grouping the data into a number of classes.

Example 1. To make information about the size of U.S. farms available to whoever needs such data, the Bureau of the Census summarizes the results of its 1969 Census of Agriculture as in the following table:

Size of Farm (acres)	*Number of Farms (thousands)*
Under 10	162
10–49	473
50–99	460
100–179	542
180–259	307
260–499	419
500–999	216
1,000–1,999	91
2,000 and over	60
Total	2,730

This kind of table is called a **frequency distribution**—it shows how the sizes of the farms are *distributed* among the chosen classes. Since the data are grouped according to their numerical size, we also refer to such a table as a **quantitative distribution**. In contrast, tables such as the one in Example 2, in which data are grouped into nonnumerical categories, are called **qualitative distributions**.

Example 2. To summarize information about the frequency with which newspapers were published in the United States in 1972, the *Ayer Directory of Newspapers, Magazines, and Trade Publications* presents the following qualitative distribution:

	Number of Newspapers
Semiweekly	398
Weekly	8,682
Daily	1,809
Other	410
Total	11,299

Although frequency distributions present data in a relatively compact form, give a good overall picture, and contain information which is adequate for many purposes, there are evidently some things which can be obtained from the original data that cannot be obtained from a distribution. With reference to Example 1, for instance, we cannot determine the exact size of the smallest and largest farms from the distribution, nor can we find the exact average size of all the farms in the 50–99 acre group. Similarly, with reference to Example 2, we cannot tell how many of the newspapers are published monthly or only four times a year, and we cannot determine how often any given paper, say, the *Phoenix Gazette*, is published. Nevertheless, frequency distributions present **raw** (unprocessed) data in a more usable form, and the price we must pay for this, the loss of certain details, is usually a fair exchange.

The construction of a numerical, or quantitative, distribution consists essentially of the following *four* steps:

First we must choose the classes into which the data are to be grouped; then we sort (tally) the data into the appropriate classes; then we count the number of items in each class; and finally we display the results in a chart or table.

Since the second and third steps are purely mechanical and the fourth step is a matter of craftsmanship and taste, we shall concentrate here on the first step, namely, that of choosing suitable classifications. This involves *choosing the number of classes and deciding from where to where each class is to go*; these are both, essentially, arbitrary decisions. Nevertheless, the following are rules that are generally considered "sound practice":

We seldom use fewer than 6 or more than 15 classes, and in actual practice this choice will depend mostly on the number of measurements or observations that we have to group.

Clearly, we would lose rather than gain anything if we grouped 6 observations into 12 classes, and we would probably give away too much information if we grouped 10,000 observations into 3 or 4 classes.

We always make sure that each item will go into one and only one class.

To this end we must make sure that the smallest and largest values fall within the classification, that none of the values can fall into possible gaps between successive classes, and that the classes do not overlap.

Whenever feasible, we make the classes cover equal ranges of values.

Indeed, it is generally desirable to make these ranges (or intervals) multiples of 5, 10, 100, . . . , or other numbers which facilitate the tally (especially, if the sorting is done by machine) and make the resulting chart or table easy to read.

Note that only the last of these rules was violated in the construction of the farm-acreage distribution of Example 1, provided the original data were given to the nearest acre. (Had the original figures been rounded to the nearest tenth of an acre, a farm of, say, 49.6 acres would not have been accommodated, as it would have fallen between the second and the third class.) Not only do the intervals from 10 to 49 acres, from 100 to 179 acres, and from 260 to 499 acres, among others, cover *unequal* ranges of values, but the first and last classes are **open**. For all we know, the last class might include a farm of 100,000 acres or more, and if we had grouped profits and losses instead of acreages, the first class might even have included neagtive values. **Open classes** labeled ". . . or less," "less than . . . ," ". . . or more," or "more than . . ." usually serve to simplify the overall picture presented by a distribution when a few of the values are much greater (or much smaller) than the rest.

So far as the second of the above rules is concerned, an important thing to watch is whether the data are given to the nearest dollar or to the nearest cent, whether they are given to the nearest inch or the nearest tenth of an inch, whether they are given to the nearest ounce or the nearest hundredth of an ounce, and so forth. In other words, *we must watch the extent to which the numbers are rounded.*

Example 3. If we wanted to group the heights of children, we would use the first of the following three classifications if the heights are given to the nearest inch, the second if the heights are given to the nearest tenth of an inch, and the third if the heights are given to the nearest hundredth of an inch:

Height (inches)	*Height (inches)*	*Height (inches)*
25–29	25.0–29.9	25.00–29.99
30–34	30.0–34.9	30.00–34.99
35–39	35.0–39.9	35.00–39.99
40–44	40.0–44.9	40.00–44.99
45–49	45.0–49.9	45.00–49.99
etc.	etc.	etc.

To illustrate what we have discussed so far in this section, let us now go through the actual steps of constructing a frequency distribution for a given set of data.

Example 4. Let us consider the following data on the number of students who were absent from an elementary school on 80 school days:

11	20	16	10	22	7	13	14	23	11
15	5	10	12	14	18	15	13	11	6
26	16	14	18	27	29	23	33	17	24
17	18	21	11	10	12	13	18	15	16
22	14	15	8	9	17	15	14	10	11
10	13	12	17	16	20	19	10	7	12
16	18	21	19	19	15	9	13	19	18
6	11	15	20	13	24	24	17	12	25

The fact that the smallest number of absences is 5 and the largest is 33 suggests that we might use the *four* classes 0–9, 10–19, 20–29, and 30–39, or, perhaps, the *six* classes 5–9, 10–14, 15–19, 20–24, 25–29, and 30–34. However, if we use as few as *four* classes (and violate the first rule on page 12), we will often find that the bulk of the data falls into one class and, hence, that we lose too much information. Indeed, if we chose the four classes for our example, we would find that almost 70% of the data fall into the second class. Thus, choosing the six classes, performing the actual tally, and counting the number of values in each class, we obtain the results shown in the following **frequency table**:

Number of Absences	*Tally*	*Frequency*
5–9	~~////~~ ///	8
10–14	~~////~~ ~~////~~ ~~////~~ ~~////~~ ~~////~~ ///	28
15–19	~~////~~ ~~////~~ ~~////~~ ~~////~~ ~~////~~ //	27
20–24	~~////~~ ~~////~~ //	12
25–29	////	4
30–34	/	1
	Total	80

The numbers in the right-hand column of this frequency table are called the **class frequencies**, and they simply give the number of items in each class. Also, the smallest and largest values that can go into any given class are referred to as its **class limits**, the **lower class limit** and the **upper class limit**.

Example 4 (Continued). In our example, the lower class limits are 5, 10, 15, 20, 25, and 30, and the upper class limits are 9, 14, 19, 24, 29, and 34.

The number of absences of Example 4 were all whole numbers, but if they had been weights rounded to the nearest pound, the first class would have covered the interval from 4.5 pounds to 9.5 pounds, the second class would have covered the interval from 9.5 pounds to 14.5 pounds, the third class would have covered the interval from 14.5 pounds to 19.5 pounds, and so on. It is customary to refer to these numbers as the **class boundaries** or the **"real" class limits**. We also speak of **upper** and **lower class boundaries**, and it should be observed that the upper boundary of one class is the lower boundary of the next. This will not lead to ambiguities (as to whether an item should go into one class or another), so long as we are careful in giving the class limits to a sufficient number of decimals (see, for instance, Example 3 on page 13). In actual practice, class limits are used much more widely than class boundaries, and we have mentioned the latter here only because they will be needed later in this chapter for calculating a certain description of a distribution, called the *median*.

Two other terms used in connection with frequency distributions are "class mark" and "class interval." **Class marks** are simply the midpoints of the classes, and they are obtained by averaging the respective class limits (or boundaries), namely, by adding the class limits (or boundaries) and dividing by 2. A **class interval** is simply the length or width of a class (namely, the range of values that it contains), and it is of interest mainly when each class of a distribution has the same interval. In that case we speak of it as the **interval** of the distribution, and it should be noted that it is given by the difference between successive class boundaries or class marks and *not* by the difference between the respective class limits.

Example 4 (Continued). For the distribution of the absences, the class marks are $\frac{5+9}{2} = 7$, $\frac{10+14}{2} = 12$, 17, 22, 27, and 32; also, the interval of the distribution is 5.

There are essentially two ways in which frequency distributions can be modified to suit particular needs. One way is to convert the distribution into a **percentage distribution** by dividing each class frequency by the number of items grouped, and then multiplying by 100.

Example 4 (Continued). Dividing each of the class frequencies of the distribution of the absence figures by 80, and then multiplying by 100, we get

Number of Absences	*Frequency*	*Percentage*
5–9	8	10
10–14	28	35
15–19	27	33.75
20–24	12	15
25–29	4	5
30–34	1	1.25
	Total	100

Percentage distributions are often used when it is desired to compare two or more distributions. For instance, if we wanted to compare absenteeism at the school of Example 4 with corresponding figures obtained during another school year, we might easily be misled if we did not account for the fact that in the other school year absences were recorded, say, for 40 days instead of 80 days, or perhaps for 150 days.

The other way of modifying a frequency distribution is to convert it into an "or more," "more than," "or less," or "less than" **cumulative distribution**. To this end we simply add the class frequencies, starting either at the top or at the bottom of the distribution.

Example 4 (Continued). For the data on absenteeism, we thus obtain the following "less than" cumulative distribution:

Number of Absences	*Cumulative Frequency*
Less than 5	0
Less than 10	8
Less than 15	36
Less than 20	63
Less than 25	75
Less than 30	79
Less than 35	80

Note that instead of "less than 5" we could have written "4 or less," instead of "less than 10" we could have written "9 or less," instead of "less than 15" we could have written "14 or less," and so on.

The problem of constructing *qualitative distributions* is very much the same as the problem of constructing quantitative distributions. We must decide how many categories (classes) to use and what kind of items each category is to contain, making sure that there are no ambiguities and that all the data are accommodated. Since the categories are usually chosen before any data are actually collected, it is a sound practice to include a category labeled "others," or "miscellaneous."

When dealing with categorical distributions we do not have to worry about such mathematical details as class limits or class boundaries, but we can run into serious difficulties in trying to avoid ambiguities. For instance, if we tried to classify items sold at a supermarket into "vegetables," "meats," "baked goods," "frozen foods," etc., it would be difficult to decide where to put frozen vegetables or meat pies. To avoid difficulties like this, it is advisable (when possible) to use standard categories developed by the Bureau of the Census and other government agencies. References to lists of such categories may be found in the book by P. M. Hauser and W. R. Leonard listed in the Bibliography at the end of the book.

Graphs and charts

When frequency distributions are constructed primarily to condense large sets of data into an "easy to digest" form, it is usually advisable to present them graphically, that is, in a form which has visual appeal. The most common kind of graphical presentation of a frequency distribution is the **histogram**, an example of which (pertaining to the data on absenteeism of Example 4) is shown in Figure 2.1. Histograms are constructed by representing the quantities which are grouped on a horizontal scale, the class frequencies on a vertical scale, and drawing rectangles whose bases equal the class interval and whose heights are determined by the respective class frequencies. (The markings on the horizontal scale can be the class limits, as in Figure 2.1; the class boundaries; or arbitrary key values.) All this assumes that the class intervals are all equal; otherwise, it is best to think of the class frequencies as represented by the *areas* of the rectangles instead of their heights. Unless we do this we can run into all sorts of trouble [see, for example, part (h) of Exercise 10 on page 24]; also, the practice of representing class frequencies by means of areas is essential if we want to approximate histograms with smooth curves. For instance, if we want to approximate the distribution of absences with a smooth curve as in Figure 2.2, we can say that the tinted region under the curve represents the 17 days on which there were 20 or more absences. Clearly, this area is just about equal to the *sum* of the areas of the corresponding three rectangles.

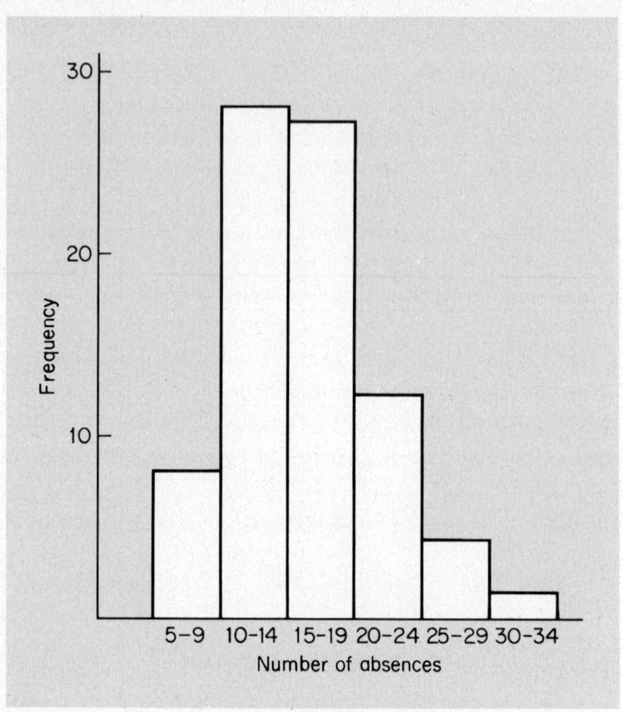

Figure 2.1 Histogram of distribution of absences.

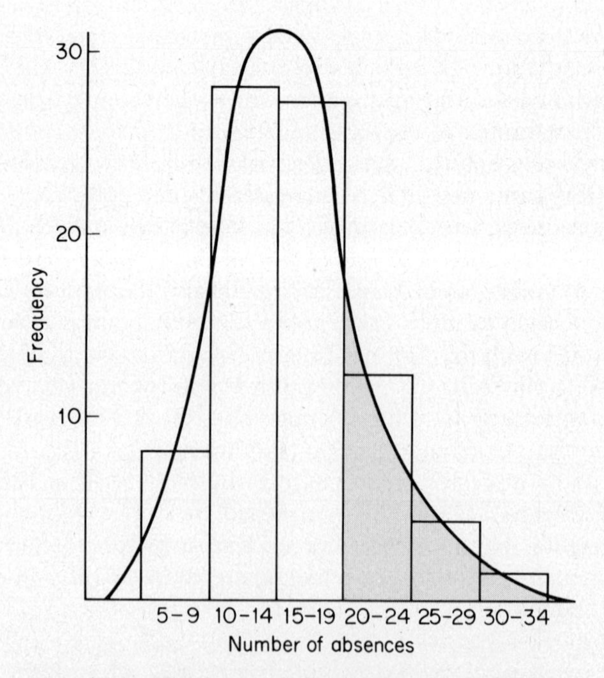

Figure 2.2 Histogram of distribution of absences approximated by means of smooth curve.

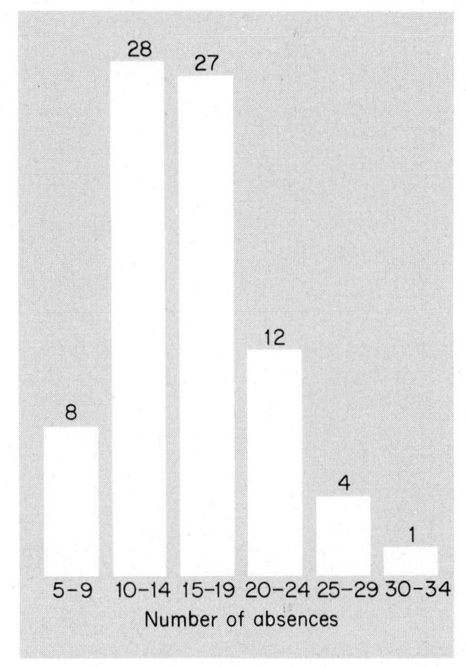

Figure 2.3 Bar chart of distribution of absences.

Similar to histograms are **bar charts**, such as the one shown in Figure 2.3; the lengths of the bars are proportional to the class frequencies, but there is no pretence of having a continuous (horizontal) scale.

Another kind of graphical presentation of a frequency distribution is the **frequency polygon**, an example of which is shown in Figure 2.4. Here the class frequencies are plotted at the class marks and the successive points are joined by means of straight lines. Note that we added a class with a zero frequency at each end of the distribution to "tie down" the graph to the horizontal scale. If we apply the same technique to the cumulative distribution on page 16, as in Figure 2.5, we obtain what is called an **ogive** (rhymes with "alive"). Observe, however, that in an ogive the cumulative frequencies are *not* plotted at the class marks; it stands to reason that the cumulative frequency corresponding to "less than 5 absences" should be plotted at 5, the one corresponding to "less than 10 absences" should be plotted at 10, and so forth. When we deal with such things as heights or weights, which are measured on *continuous scales*, the cumulative frequencies must be plotted at the dividing lines between the classes, namely at the class boundaries.

Although the visual appeal of histograms, bar charts, frequency polygons, and ogives exceeds that of the corresponding tables, there are various ways in which frequency distributions can be presented more dramatically

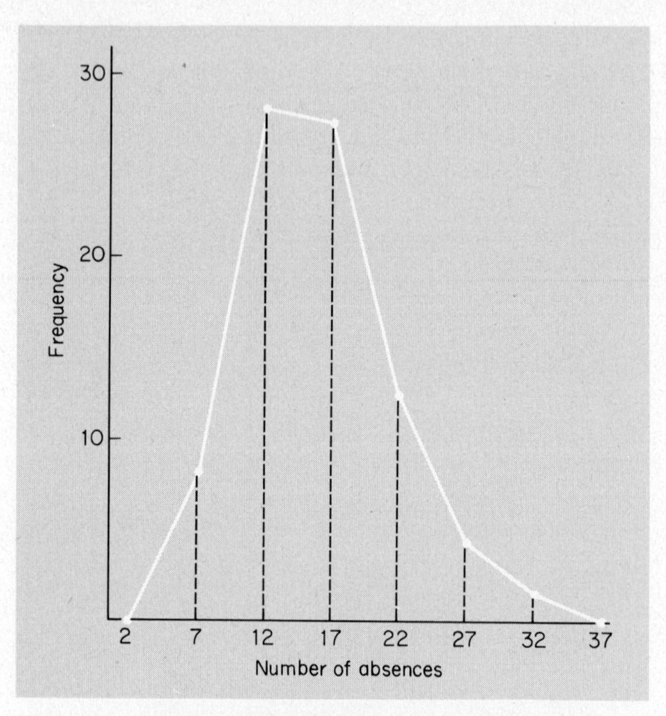

Figure 2.4 Frequency polygon of distribution of absences.

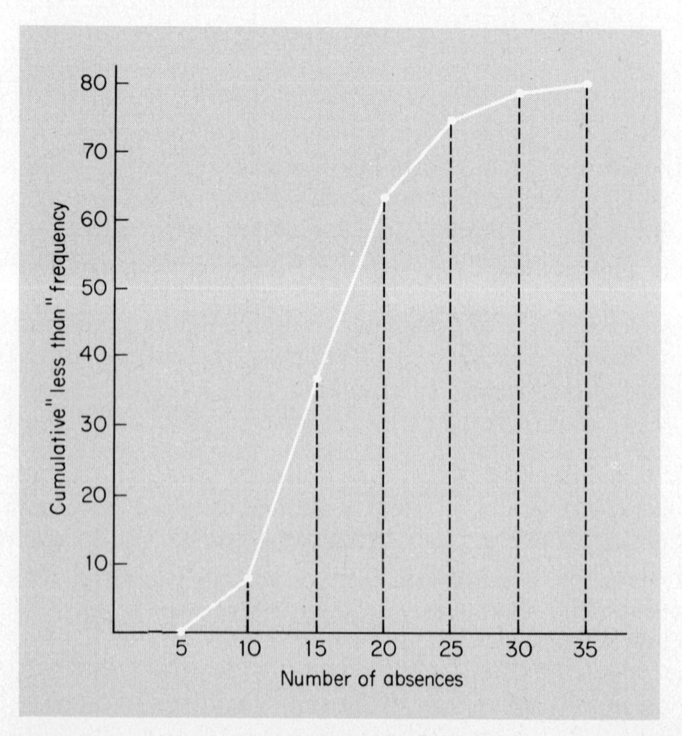

Figure 2.5 Ogive of distribution of absences.

and, it is hoped, more effectively. We are referring here to the various kinds of pictorial presentations, such as the **pictogram** of Figure 2.6, with which the reader is surely familiar through newspapers, magazines, and various forms of advertising. *The number of ways in which frequency distributions can be displayed pictorially is practically unlimited, and it depends only on the imagination and the artistic talent of the person who is preparing the presentation.*

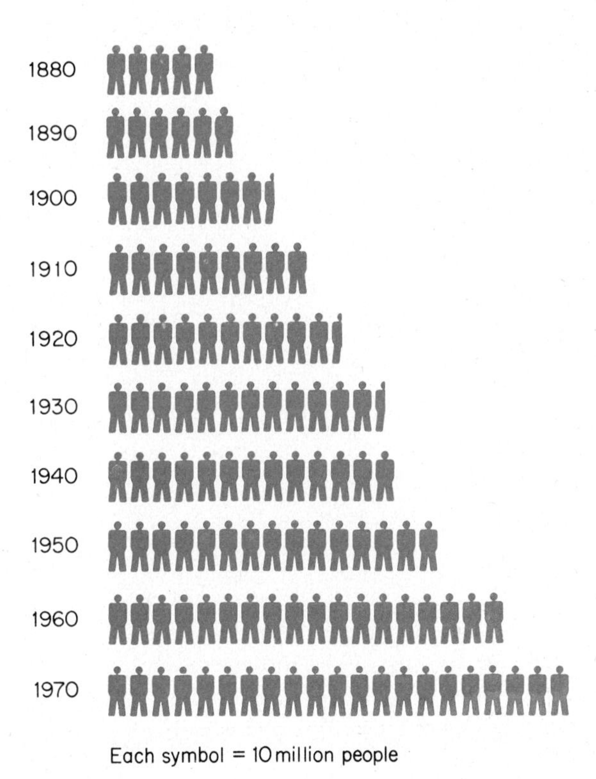

Figure 2.6 Pictogram of the population of the United States.

Qualitative distributions are often presented graphically as **pie charts** such as the one shown in Figure 2.7, where a circle is divided into sectors (pie-shaped pieces) which are proportional in size to the frequencies of the corresponding categories. To construct a pie chart we first convert the distribution into a percentage distribution (by dividing each frequency by the total number of items grouped and multiplying by 100); then we make use of the fact that 1 percent of the data is represented by a sector (pie-shaped piece) with a *central angle* of one hundredth of 360 degrees, namely, an angle of 3.6 degrees.

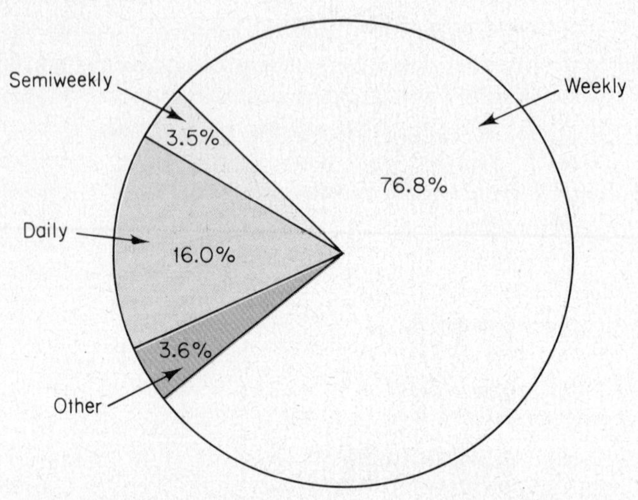

Figure 2.7 Pie chart of the frequency of publication of newspapers.

Example 2 (Continued). For the categorical distribution on page 11 we find that $\frac{398}{11,299} \cdot 100 = 3.5\%$ of the newspapers are published semiweekly, $\frac{8,682}{11,299} \cdot 199 = 76.8\%$ are published weekly, $\frac{1,809}{11,299} \cdot 100 = 16.0\%$ are published daily, and the others constitute $\frac{410}{11,299} \cdot 100 = 3.6\%$ of the total. Then, if we multiply each of these percentages by 3.6, we get 13, 276, 58, and 13 (rounded to the nearest degree), and these are the central angles of the four sectors into which we divided the circle in the pie chart of Figure 2.7.

Like histograms, pie charts can be improved by using various artistic devices; for instance, by shading or coloring the sectors as in Figure 2.7, by giving the whole diagram a three-dimensional effect, or by actually cutting out a piece of the "pie" to draw attention to a particular category.

Exercises

1. Check whether the following quantities can be determined on the basis of the distribution of absences on page 14; if possible, give a numerical answer.

a. The number of days on which less than 12 students were absent.
b. The number of days on which 15 or more students were absent.
c. The number of days on which more than 15 students were absent.
d. The number of days on which 19 or less students were absent.
e. The number of days on which less than 19 students were absent.
f. The number of days on which 33 or more students were absent.

2. If the number of empty seats on commuter buses in Vancouver, B.C., are grouped into a table having the classes 0–3, 4–7, 8–11, 12–15, 16–19, and 20 *or more*, check whether the following quantities can be determined on the basis of the resulting distribution.
a. The number of times there were at least 12 empty seats.
b. The number of times there were more than 12 empty seats.
c. The number of times there were exactly 8 empty seats.
d. The number of times there were at most 7 empty seats.
e. The number of times there were anywhere from 4 through 11 empty seats.
f. The number of times there were at most 22 empty seats.

3. To summarize data on how many rainy days there were at a particular weather station in the month of May during the last thirty years, a meteorologist uses the classes 0–5, 6–11, 11–16, 17–22, and 24–31. Explain where the rules on pages 12 and 13 are violated by this classification.

4. If the prices paid for furniture at an auction varied from \$22.35 to \$94.60, indicate the limits of eight classes into which these prices might reasonably be grouped.

5. The number of speeding tickets issued per day in a certain city are grouped into a table having the classes 2–7, 8–13, 14–19, 20–25, 26–31, and 32–37. Find
a. the lower class limits;
b. the upper class limits;
c. the class boundaries;
d. the class marks;
e. the class interval.

6. The class marks of a distribution of temperature readings, given to the nearest degree Fahrenheit, are 84, 93, 102, 111, and 120. Find the class limits.

7. The weights of certain mineral specimens, given to the nearest tenth of an ounce, are grouped into a table having the classes 11.5–12.4, 12.5–13.4, 13.5–14.4, 14.5–15.4, 15.5–16.4, and 16.5–17.4 ounces. Find
a. the class marks;
b. the class boundaries;
c. the class interval.

8. Class limits and class boundaries have to be interpreted very carefully when we are dealing with ages, for the age group from 5 to 9, for example, includes all those who have passed their fifth birthday but have not yet reached their tenth. Taking this into account, what are the boundaries and the class marks of the following age groups: 10–19, 20–29, 30–39, and 40–49?

9. If the amounts which tourists spend at a giftshop are grouped into a table having the classes $0.00–$9.99, $10.00–$19.99, $20.00–29.99, . . . , and $60.00–$69.99, find

a. the class marks;
b. the class boundaries;
c. the class interval.

10. The following is a distribution of the final examination grades which 200 students obtained in a freshman course in psychology:

Grades	*Frequency*
0–19	24
20–39	55
40–59	76
60–79	32
80–99	13
Total	200

a. Convert this distribution into a percentage distribution.
b. Convert this distribution into a cumulative "less than" distribution.
c. Convert this distribution into a cumulative "more than" distribution.
d. Convert this distribution into a cumulative "or more" percentage distribution.
e. Draw a histogram of the original distribution.
f. Draw a frequency polygon of the original distribution.
g. Draw an ogive of the cumulative "less than" distribution of part (b).
h. Draw a histogram of the distribution which is obtained after putting the grades from 20 to 59, inclusive, into *one* class. (*Hint*: If one class interval is twice as wide as the others, we compensate for this by dividing the height of its rectangle by 2.)

11. The following are the amounts (in dollars) which 40 students attending various four-year colleges paid for room and board during the academic year 1972–1973:

1,225	912	1,116	1,033	1,172	1,016	933	1,135
1,010	1,190	975	1,125	843	1,285	1,140	1,077
967	1,045	1,263	1,048	1,171	1,105	1,055	960
1,073	892	1,075	1,059	1,166	1,325	1,083	980
1,267	1,085	1,120	984	952	1,021	1,293	1,190

a. Group these amounts into a distribution having the classes 800–899, 900–999, 1,000–1,099, 1,100–1,199, 1,200–1,299, and 1,300–1,399.
b. Draw a histogram of the distribution obtained in part (a).
c. Convert the distribution obtained in part (a) into a cumulative "or more" distribution and draw its ogive.

12. A study of air pollution in a city yielded the following daily readings of the concentration of sulfur dioxide (in parts per million):

0.04	0.14	0.17	0.11	0.18	0.20	0.13	0.17	0.10	0.07
0.15	0.09	0.10	0.16	0.12	0.27	0.10	0.12	0.05	0.06
0.08	0.15	0.05	0.09	0.11	0.13	0.07	0.14	0.09	0.17
0.10	0.08	0.14	0.02	0.14	0.08	0.11	0.08	0.01	0.18
0.03	0.13	0.06	0.19	0.12	0.15	0.01	0.05	0.11	0.04

a. Group these data into a table having the classes 0.00–0.04, 0.05–0.09, 0.10–0.14, 0.15–0.19, 0.20–0.24, and 0.25–0.29.
b. Convert the distribution obtained in part (a) into a cumulative "less than" percentage distribution.
c. Construct a histogram of the distribution obtained in part (a).
d. Construct an ogive of the cumulative percentage distribution obtained in part (b).

13. The following are the body weights (in grams) of 48 rats used in a study of vitamin deficiencies:

103	125	91	115	109	137	148	127
112	153	133	135	88	128	115	119
124	106	120	136	124	115	93	134
141	117	118	125	116	104	123	125
121	115	105	117	122	113	100	155
130	95	125	110	106	112	120	138

Group these data into a table having the classes 80–89, 90–99, 100–109, . . . , 140–149, and 150–159, and draw a bar chart of this distribution.

14. The following are the burning times of certain solid-fuel rockets given to the nearest tenth of a second:

4.1	5.0	4.8	4.3	4.2	5.3	4.2	3.6	4.5	4.4
4.5	3.2	4.0	3.8	3.8	5.3	4.5	4.6	4.0	5.2
5.2	4.4	4.7	4.1	4.6	4.9	4.1	5.8	4.2	4.2
4.8	4.1	5.6	4.5	5.1	4.6	4.3	5.2	4.7	3.2
4.0	4.6	4.0	4.2	4.5	3.5	4.7	4.9	3.9	4.8
3.7	5.4	4.9	4.6	4.3	5.4	5.0	4.5	4.7	4.3

a. Group these data into a table having the classes 3.0–3.4, 3.5–3.9, 4.0–4.4, . . . , and 5.5–5.9, and draw a histogram of this distribution.
b. Convert the distribution obtained in part (a) into a cumulative distribution showing how many of the burning times were "2.95 seconds or less," "3.45 seconds or less," "3.95 seconds or less," . . . , and "5.95 seconds or less."
c. Draw an ogive of the cumulative distribution obtained in part (b) and use it to read off (roughly) the value below which we should find the lowest 25% of the data.

15. Use the entertainment section of the newspaper of a fairly large city to construct a table showing how many of the movies that are playing are rated G, PG, R, X, or are unrated.

16. Use the program listings of a local television station to construct a table showing how many of the half-hour periods between 1 P.M. and 5 P.M. during one week are situation comedies, game shows, serious dramas, educational programs, sports events, news, and so forth.

17. Use the American Stock Exchange listings in a Sunday paper to construct a table showing how many of the R, S, and T stocks showed a net increase, a net decrease, or no change in price for the preceding week.

18. The following table shows how the dogs entered in a dog show are distributed according to A.K.C. classifications:

Group	*Number*
Sporting dogs	26
Hounds	34
Working dogs	76
Terriers	17
Toys	24
Nonsporting dogs	23
Total	200

Construct a pie chart of this distribution.

19. Among the engineers employed by U.S. colleges and universities in 1971, 7.5% were aeronautical engineers, 9.4% were chemical engineers, 21.0% were civil engineers, 34.9% were electrical engineers, and 27.2% were mechanical engineers. Construct a pie chart of this percentage distribution.

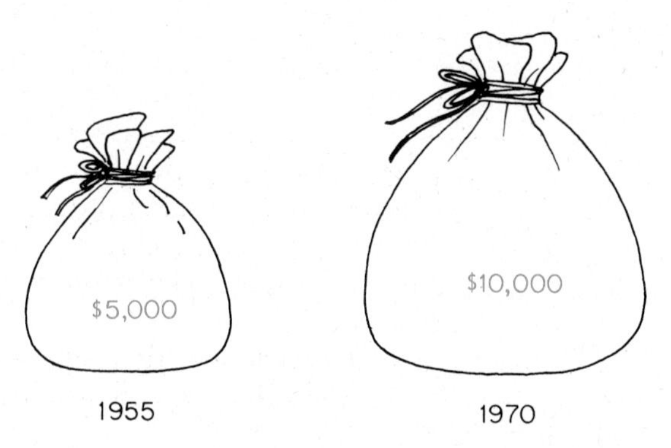

Figure 2.8 Average family income in a certain part of the United States.

20. The pictogram of Figure 2.8 is intended to illustrate the fact that, in a certain part of the country, average family income has *doubled* from $5,000 in 1955 to $10,000 in 1970. Explain why this pictogram does *not* convey a fair impression of the actual change, and suggest how it might be improved.

Statistics

As we are using it here, the word "statistic" is intended to have an entirely different meaning than in the title of this book. In the title of the book we are referring to the subject of statistics, namely, the totality of methods employed in the collection, processing, and analysis of numerical data (or, more generally, to the science of decision making in the face of uncertainty). In the heading of this section, statistics is meant to be the plural of **statistic**, namely, a particular numerical description of a set of data, say, its average. In this sense, a statistic is also referred to as a **statistical measure** or simply as a **statistical description**.

It is customary to classify statistics (that is, statistical measures) according to the particular features of a set of data which they are supposed to describe. Thus, we speak of **measures of location**, which may be described crudely as "averages" in the sense that they are generally indicative of the "center," "middle," or the "most typical" of a set of data. Then there are **measures of variation**, which are indicative of the variability, spread, or dispersion of a set of data. Some of these will be introduced in Chapter 7, where we shall also mention **measures of skewness** in an exercise on page 203. In Chapters 9 and 10 we shall study some **measures of association** which are meant to be indicative of the strength of the relationship (dependence) between two variables, say, a person's income and the amount of his or her schooling, air pollution and the incidence of respiratory diseases, or gasoline consumption and the speed at which one drives one's car. In addition to these general kinds of descriptions, there are numerous other ways in which statistical data can be summarized. So far as the work of this chapter is concerned, it will be devoted mostly to measures of location.

The mean

The most popular measure of location is what the layman calls an "average" and what the statistician calls a **mean**. The reason we put the word "average" in quotes is that it has all sorts of connotations in everyday language —we speak about a baseball player's batting average, we talk about the average suburban family, we describe a bank robber's appearance as average, and so forth. The mean of a set of n numbers is simply *their sum divided by n*, and, hence, it is very easy to calculate.

Example 5. If the ages of six teenagers attending a party are 16, 18, 17, 15, 17, and 16, the mean of their ages is

$$\frac{16 + 18 + 17 + 15 + 17 + 16}{6} = \frac{99}{6} = 16.5 \text{ years}$$

Example 6. If ten cars are timed passing a check point going 53, 48, 50, 47, 43, 58, 42, 46, 51, and 38 miles per hour, the mean of these speeds is

$$\frac{53 + 48 + 50 + 47 + 43 + 58 + 42 + 46 + 51 + 38}{10} = 47.6$$

miles per hour.

To give a formula for the mean which is applicable to any kind of data, it will be necessary to represent the measurements or observations by means of symbols such as x, y, or z. In the last example we might have used the letter x, for instance, and denoted the ten values x_1 (*x sub-one*), x_2 (*x sub-*

two), x_3 (*x sub-three*), . . . , and x_{10} (*x sub-ten*). More generally, if we have n measurements or observations which we denote $x_1, x_2, x_3, \ldots$, and x_n, their mean can be written as

$$\frac{x_1 + x_2 + x_3 + \cdots + x_n}{n}$$

To simplify our notation, we refer to this expression, namely, the mean of the x's, as $\bar{x}$ (*x bar*). (Of course, if we refer to the measurements as y's or z's, we would correspondingly write their mean as $\bar{y}$ or $\bar{z}$.) Furthermore, let us introduce the symbol $\sum$ (capital *sigma*, the Greek letter for S), which is simply a mathematical shorthand symbol indicating the process of summation, or addition. If we write $\sum x$, this means literally "the sum of the x's," and we can thus give the formula for the mean as*

$$\bar{x} = \frac{\sum x}{n}$$

The popularity of the mean as a measure of the "center" or "middle" of a set of data is not just accidental. Any time we use a single number to describe a whole set of data, there are certain requirements (namely, certain desirable properties) that we must keep in mind. Some of the noteworthy properties of the mean are:

It is familiar to most persons, although it may not be called by that name.

It always exists, that is, it can be calculated for any kind of numerical data.

It is always unique; in other words, a set of numerical data has one and only one mean.

It takes into account each individual item.

It lends itself to further statistical treatment (for instance, the means of several sets of data can always be combined into an overall mean for all the data; see Exercise 15 on page 37).

It is relatively reliable in the sense that for sample data it is generally not as strongly affected by chance as some of the other measures of location.

*See also Exercise 20 on page 41.

This question of **reliability** is of fundamental importance when it comes to problems of estimation, hypothesis testing, and making predictions, and we shall study it in quite some detail in Chapter 7.

Whether the fourth property which we have listed is actually desirable is open to some doubt; a single extreme (very small or very large) value can affect the mean to such an extent that it is debatable whether it is really "representative" or "typical" of the data it is supposed to describe.

Example 5 (Continued). To illustrate this point, suppose that the party is chaperoned by the 48-year-old mother of one of the teenagers. This would make the mean age

$$\frac{16 + 18 + 17 + 15 + 17 + 16 + 48}{7} = \frac{147}{7} = 21$$

and the statement that "the average age of the persons at the party is 21" is apt to be misinterpreted by most people.

Example 7. To give another illustration, suppose that the attendance at the six home games of a high school's basketball team was 903, 867, 789, 840, 827, and 832. This makes the *average attendance* $\bar{x} = 843$, as can easily be checked, but if the person calculating this mean had misread 832 as 238, he would have obtained

$$\bar{x} = \frac{903 + 867 + 789 + 840 + 827 + 238}{6} = 744$$

This shows how one careless mistake can have a pronounced effect on the mean.

Since the calculation of means is very easy, involving only addition and a division, there is really no need to look for shortcuts or simplifications. However, if the numbers we may want to average have many digits and no adding machine is available, it can be advantageous to group the data first and then calculate the mean on the basis of the resulting distribution. Another reason we shall devote some time to the calculation of the mean of grouped data is that published data (to which we may have to refer in a study) are generally available only in grouped form.

Results thus obtained will generally not equal the *actual* means of the data. These can no longer be determined once the data have been grouped —each item, so to speak, loses its identity, and we know only how many items there are in each class. Nevertheless, *a good approximation can generally be obtained by treating each value in a class as if it equaled the class mark, namely, the midpoint of the class.* Although some of the values in a class will usually exceed the class mark, others will be less than the class mark, and all this will more or less "average out." Thus, the 8 values in the 5–9 class of the distribution of absences on page 14 will be treated as if they all equaled 7, the 28 values in the 10–14 class will be treated as if they all equaled 12, . . . , and the single value in the 30–34 class will be treated as if it equaled 32.

To obtain a formula for the mean of a frequency distribution with k classes, let us write the successive class marks as $x_1, x_2, \ldots$, and x_k, and the corresponding frequencies as $f_1, f_2, \ldots$, and f_k. The total that goes into the numerator for the formula for the mean is thus obtained by adding f_1 times the class mark x_1, f_2 times the class mark x_2, . . . , and f_k times the class mark x_k; in other words, it is given by

$$x_1f_1 + x_2f_2 + \cdots + x_kf_k$$

Using the $\sum$ notation introduced on page 29, we can thus write the formula for the mean of a distribution as*

$$\bar{x} = \frac{\sum x \cdot f}{n}$$

where n equals the sum of the class frequencies, namely, $n = \sum f$. In words, to find the mean of a distribution *we add the products obtained by multiplying each class mark by the corresponding frequency, and then divide by n (the total number of items grouped).*

Example 4 (Continued). To illustrate the calculation of the mean of a frequency distribution, let us refer again to the distribution of absences on page 14. Writing the class marks in the second column and calculating the necessary products $x \cdot f$, we get

Number of Absences	*Class Mark* x	*Frequency* f	$x \cdot f$
5–9	7	8	56
10–14	12	28	336
15–19	17	27	459
20–24	22	12	264
25–29	27	4	108
30–34	32	1	32
	Totals	80	1,255

*See also Exercise 20 on page 41.

and the mean of the distribution is

$$\bar{x} = \frac{1{,}255}{80} = 15.7$$

absences (rounded to the nearest tenth). It is of interest to note that the actual mean of the ungrouped data on page 14 is $\frac{1{,}249}{80} = 15.6$, and, hence, that the error introduced by calculating the mean on the basis of the distribution is very small.

The calculation of the mean of the distribution of Example 4 was quite easy, but it could have been made even easier by replacing the class marks with consecutive integers, namely, with numbers which are easier to handle. This process is referred to as **coding**, and we shall illustrate how it is done in Exercise 11 on page 35.

The weighted mean

There are many situations in which it would be quite misleading to average quantities without accounting in some way for their relative importance in the overall picture we are trying to describe.

Example 8. Given that in 1972 the average salaries of elementary school teachers in Oregon, Washington, and Alaska were, respectively, \$9,309, \$9,824, and \$14,154, we find that the mean of these three figures is

$$\frac{9{,}309 + 9{,}824 + 14{,}154}{3} = \$11{,}096$$

to the nearest dollar, but we *cannot* conclude that this is the average salary paid in 1972 to elementary school teachers in these three states. To find this figure, we would have to know how many elementary school teachers were employed that year in each of the states.

Similarly, it would be meaningless to calculate the mean of the prices of various food items without accounting in some way for the respective roles which they play in the average family's budget, and it would be misleading if a student averaged all the grades he got in a course without taking into account how much each test counts.

Example 8 (Continued). Let us now add the information that in 1972 there were, respectively, 11,870, 17,249, and 2,591 elementary school teachers in Oregon, Washington, and Alaska. Then, simple calculations show that the *total earnings* of these 11,870 + 17,249 + 2,591 = 31,710 teachers were

$$11{,}870(\$9{,}309) + 17{,}249(\$9{,}824) + 2{,}591(\$14{,}154) = \$316{,}625{,}020$$

and, hence, that on the average they earned

$$\frac{316{,}625{,}020}{31{,}710} = \$9{,}985$$

to the nearest dollar.

The average we calculated in this example is called a **weighted mean**—we averaged the salary figures for the three states, giving due weight to their relative importance. This provides a more realistic answer than the "ordinary" or "unweighted" mean calculated on page 32, which was much too high.

In general, the weighted mean of a set of n numbers $x_1, x_2, \ldots,$ and x_n, whose relative importance is measured by a corresponding set of numbers $w_1, w_2, \ldots,$ and w_n called the **weights**, is given by the formula*

$$\bar{x}_w = \frac{\sum w \cdot x}{\sum w}$$

Here $\sum w \cdot x$ stands for the sum of the products obtained by multiplying each x by the corresponding weight, while $\sum w$ is simply the sum of the weights. Note that when the weights are all equal, the formula for the weighted mean reduces to that of the (ordinary) mean. Also, the mean of grouped data (as given by the formula on page 31) can be looked upon as a weighted mean, with the weights of the class marks being the corresponding class frequencies.

Example 9. To give another example, suppose that someone invests \$2,000 at 5% in a savings account, \$5,000 at 7% in a certificate of deposit, and \$12,000 at 9% in second mortgages. To average these

*See also Exercise 20 on page 41.

three percentage rates, we must weight them with the respective amounts which the person invests, and, hence, substitution into the formula yields

$$\bar{x}_w = \frac{2{,}000 \cdot 5 + 5{,}000 \cdot 7 + 12{,}000 \cdot 9}{2{,}000 + 5{,}000 + 12{,}000} = 8.05\%$$

The choice of the weights did not pose any difficulties in Examples 8 and 9, but there are situations in which the selection of the weights is far from obvious. For instance, if we wanted to compare figures on the cost of living in different cities (or in different years), it would be difficult to account for the *relative importance* of such items as food, rent, entertainment, medical care, and so on.

Exercises

1. In a political science survey, ten persons were able to name, respectively, 2, 1, 5, 3, 0, 1, 2, 1, 4, and 2 of a state's eight congressmen. Does the mean support the claim that *on the average* a person in this state can name at least two of the state's eight congressmen?

2. The records of 15 persons convicted of various crimes showed that, respectively, 4, 3, 0, 0, 2, 4, 4, 3, 1, 0, 2, 0, 2, 1, and 4 of their grandparents were born in the United States. Find the mean and discuss whether it can be used to support the contention that the "average criminal" has two foreign-born grandparents.

3. The following are the I.Q.'s of 16 persons selected for jury duty by a court: 112, 107, 103, 96, 118, 94, 110, 88, 92, 103, 127, 111, 89, 97, 101, and 108.

a. Find the mean and comment on the argument that "the average juror has an I.Q. over 100."

b. Recalculate the mean of the 16 I.Q.'s by first subtracting 100 from each value, finding the mean of the numbers thus obtained, and then adding 100 to the result. (What general simplification does this suggest for the calculation of a mean?)

4. The following are the number of seconds which ten insects survived after being sprayed with a certain insecticide: 64, 55, 59, 73, 65, 50, 58, 76, 63, and 54.

a. Calculate the mean of these figures.

b. Recalculate the mean of these figures by first subtracting 60 from each value, finding the mean of the numbers thus obtained, and then adding 60 to the result. (What general simplification does this suggest for the calculation of a mean?)

5. By mistake, an instructor has erased the grade which one of the ten students in his class received in a final examination. However, he knows that the ten students averaged (had a mean grade of) 74 in the examination, and that the other nine students received grades of 47, 50, 59, 73, 77, 81, 85, 97, and 99. What must have been the erased grade?

6. A bridge is designed to carry a maximum load of 120,000 pounds. If at a given moment it is loaded with 32 vehicles having an average weight of 3,600 pounds, is there any real danger that it might collapse?

7. In Example 6 on page 28 we gave the speeds of ten cars passing a check point as 53, 48, 50, 47, 43, 58, 42, 46, 51, and 38 miles per hour. As we showed, the mean of these figures in 47.6, but what value would we have obtained if we had erroneously recorded the speed of the first car as 35 instead of 53?

8. The following is a distribution of the amount of time it took sales persons of a department store to serve 150 customers:

Time Required (minutes)	*Number of Customers*
1–2	8
3–4	40
5–6	46
7–8	41
9–10	12
11–12	3

Find the mean of this distribution.

9. Find the mean of the grade distribution of Exercise 10 on page 24.

10. Find the mean of whichever data you grouped among those of Exercises 11, 12, 13, and 14 on pages 25 through 26, using

a. the raw (ungrouped) data;
b. the distribution obtained in that exercise.

11. CODING As we indicated on page 32, the calculation of the mean of a distribution can usually be simplified by replacing the class marks with consecutive integers. For the distribution of Example 4 on page 14, the one dealing with the number of absences, we might thus get

Class Mark x	u	f	$u \cdot f$
7	−2	8	−16
12	−1	28	−28
17	0	27	0
22	1	12	12
27	2	4	8
32	3	1	3
	Totals	80	−21

where we put the 0 of the new scale (which we referred to as the u-scale) near the middle of the table *to keep the numbers conveniently small.* Of course, if we use this kind of coding for a distribution with equal class intervals, we must compensate for it in the formula for the mean, which becomes

$$\bar{x} = x_0 + c \cdot \frac{\sum u \cdot f}{n}$$

where x_0 is the class mark (in the original scale) to which we assign the number 0 in the u-scale, c is the class interval, and n is the total number of items grouped. So far as the absences are concerned, we thus get

$$\bar{x} = 17 + 5 \cdot \frac{-21}{80} = 17 - 1.3 = 15.7$$

absences (rounded to the nearest tenth). This result is *identical* (as it should be) with the result obtained on page 32. Use this shortcut technique to rework

a. Exercise 8;
b. Exercise 9;
c. part (b) of Exercise 10.

12. If an instructor counts the midterm examination in a chemistry course twice as much as each hour examination, and the final examination four times as much as each hour examination, what is the weighted average grade of a student who received grades of 86, 59, 73, and 82 in four one-hour examinations, a midterm grade of 67, and a final examination grade of 44?

13. The four investments which someone made five years ago increased, respectively, by 8.1%, 5.3%, 6.5%, and 12.9%.

a. Find the mean of these four percentages and explain under what conditions it would represent the person's overall yield on his investments.

b. Find the actual overall percentage increase in the person's investments, if the corresponding amounts invested were, respectively, \$1,000, \$5,000, \$2,000, and \$2,000.

14. A butcher sells three grades of beef, respectively, for \$1.69, \$1.89, and \$2.09. If in a certain week he sells 800 pounds of the cheapest grade, 200 pounds of the medium-priced grade, and 200 pounds of the most expensive grade, what is the average price he gets per pound?

15. THE OVERALL MEAN OF SEVERAL SETS OF DATA If k sets of data consisting of $n_1, n_2, \ldots,$ and n_k observations have the respective means $\bar{x}_1, \bar{x}_2, \ldots, \bar{x}_k$, the overall mean of all the data is given by

$$\frac{n_1\bar{x}_1 + n_2\bar{x}_2 + \ldots + n_k\bar{x}_k}{n_1 + n_2 + \ldots + n_k} = \frac{\sum n \cdot \bar{x}}{\sum n}$$

where the numerator represents the actual sum of all the observations, while the denominator represents the total number of observations. *Thus, the overall mean is the weighted mean of the $\bar{x}$'s, with the weights being the number of observations in the corresponding sets of data.* For instance, if the 20 students in one section of a psychology course received an average (mean) grade of 67 in an examination, the 18 students in a second section of the course averaged 57, and the 12 students in a third section of the course averaged 62, then the overall mean of the grades of these $20 + 18 + 12 = 50$ students is

$$\bar{x} = \frac{20 \cdot 67 + 18 \cdot 57 + 12 \cdot 62}{20 + 18 + 12} = \frac{3{,}110}{50} = 62.2$$

a. If the mean weight of the four offensive starting backs of a professional football team is 214 pounds and the mean weight of the seven offensive starting linemen is 258 pounds, what is the mean weight of this starting eleven?

b. In a primary election, the expenses of the three candidates of one party averaged (had a mean of) \$12,450, and those of the two candidates of the other party averaged \$18,620. Find the mean of the expenses of these five candidates.

c. In 1973, a college paid its 42 instructors an average salary of \$8,400, its 121 assistant professors an average salary of \$10,600, its 88 associate professors an average salary of \$12,700, and its 49 full professors an average salary of \$16,500. What was the mean salary paid to the 300 members of this faculty?

16. THE GEOMETRIC MEAN* For any set of n positive numbers, *the geometric mean is given by the* nth *root of their product.* For instance, the geometric mean of 3 and 12 is

$$\sqrt[2]{3 \cdot 12} = \sqrt[2]{36} = 6$$

and the geometric mean of $\frac{1}{3}$, 1, and 81 is

$$\sqrt[3]{\tfrac{1}{3} \cdot 1 \cdot 81} = \sqrt[3]{27} = 3$$

a. Find the geometric mean of 10 and 40.
b. Find the geometric mean of 1, 1, and 8.
c. Find the geometric mean of 1, 2, 8, and 16.
d. In his first season as a professional, a hockey player scored 6 goals, in his second season he scored 9 goals, and in his third season he scored 24 goals. Thus, from the first season to the second his "output" was multiplied by $\frac{9}{6} = \frac{3}{2}$, and from the second season to the third his "output" was multiplied by $\frac{24}{9} = \frac{8}{3}$. Find the geometric mean of these two "growth rates," and apply it to the number of goals he scored in his third season to predict the number of goals he will score in his fourth season.

In actual practice, the geometric mean is rarely used, and then mainly to average rates of change.

17. THE HARMONIC MEAN For any set of n positive numbers, *the harmonic mean is given by* n *divided by the sum of the reciprocals of the* n *numbers*, namely, by $\dfrac{n}{\sum 1/x}$. For instance, if \$12 is spent on vitamin pills costing 40 cents a dozen and another \$12 is spent on vitamin pills costing 60 cents a dozen, the average price is *not* $\dfrac{40 + 60}{2} = 50$ cents a dozen; since a total of \$24 is spent on a total of $\dfrac{1{,}200}{40} + \dfrac{1{,}200}{60} = 50$ dozen vitamin pills, the actual cost is $\dfrac{2{,}400}{50} = 48$ cents a dozen. Note that this result *is* the harmonic mean of 40 and 60, that is,

$$\frac{2}{\frac{1}{40} + \frac{1}{60}} = \frac{2}{\frac{3}{120} + \frac{2}{120}} = \frac{2}{\frac{5}{120}} = \frac{2 \cdot 120}{5} = 48$$

*To distinguish the "ordinary" mean from the geometric mean of this exercise and the *harmonic mean* of Exercise 17, the mean is often referred to more explicitly as the arithmetic mean.

a. If an investor buys \$5,400 worth of capital stock at \$45 a share and \$5,400 worth at \$36 a share, calculate the average price he or she pays per share, and verify that this is the harmonic mean of \$45 and \$36.

b. If a jet travels the first third of a trip at 300 miles per hour, the next third at 450 miles per hour, and the final third at 360 miles per hour, use the harmonic mean to determine its average speed for the whole trip.

In actual practice, the harmonic mean is used only when dictated by special circumstances, as in the above examples.

18. AGGREGATIVE INDEX NUMBERS (RATIOS OF MEANS)

In business and economics there are many problems in which we are interested in **index numbers**, namely, *in measures of the overall changes that have taken place in the prices (quantities, or values) of various commodities.* In general, the year or period which we want to compare by means of an index number is called the **given year** or **given period**, while the year or period relative to which the comparison is to be made is called the **base year** or **base period**. Furthermore, given-year prices are denoted p_n and base-year prices are denoted p_0. Among the many ways in which we can measure overall changes in prices, a simple and straightforward method consists of *averaging (calculating the mean of) the two sets of prices separately, taking the ratio of these two means, and then multiplying by 100 to express the index as a percentage.* After we cancel the common denominator k (the number of commodities being compared), the formula for this kind of **simple aggregative index** can be written as

$$I = \frac{\sum p_n}{\sum p_0} \cdot 100$$

where $\sum p_n$ is the sum of the given-year prices and $\sum p_0$ is the sum of the base-year prices. To illustrate the use of this formula, consider the following prices of three metals in cents per pound:

	1960	*1970*
Copper	32.5	58.2
Lead	11.9	15.6
Zinc	12.9	15.3

Substituting into the formula, we get

$$I = \frac{58.2 + 15.6 + 15.3}{32.5 + 11.9 + 12.9} \cdot 100 = \frac{89.1}{57.3} \cdot 100 = 155.5$$

and this tells us that the average price of the three metals in the given year 1970 was 55.5% higher than it was in the base year 1960.

a. Construct a simple aggregative index to compare the 1970 (given year) prices received by farmers for the following four major crops (in dollars per bushel) with the corresponding 1965 (base year) prices:

	1965 Prices	*1970 Prices*
Wheat	1.35	1.33
Corn	1.16	1.33
Oats	0.62	0.62
Soybeans	2.54	2.85

b. The following are the average prices (per hundred pounds) received by farmers in the United States for selected livestock and poultry items:

	1970	*1971*
Hogs	$22.70	$17.50
Cattle	34.50	36.30
Chickens	9.10	7.70
Turkeys	22.60	22.10

Calculate the simple aggregative index which compares the 1971 prices with those of 1970.

19. WEIGHTED AGGREGATIVE INDEX NUMBERS (RATIOS OF WEIGHTED MEANS) Aggregative index numbers (see Exercise 18) can often be greatly improved by weighting the various prices by the corresponding quantities (produced, sold, or consumed) in the given year, the base year, or some other fixed period of time. If we use the given-year quantities as weights, we obtain the **weighted aggregative index,**

$$I = \frac{\sum p_n q_n}{\sum p_0 q_n} \cdot 100$$

and if we use the base-year quantities as weights we obtain the **weighted aggregative index,**

$$I = \frac{\sum p_n q_0}{\sum p_0 q_0} \cdot 100$$

where the notation is the same as in Exercise 18, with the addition that q_n and q_0 denote given-year and base-year quantities. *Note that these index numbers are ratios of weighted means, multiplied by 100, with the denominators, the sums of the weights, canceled.* To illustrate, let us refer again to the example of Exercise 18, and add the information that in 1970 the production of copper, lead, and zinc in the United States totaled, respectively, 1,719, 571, and 534 thousands of short tons. Substituting into the first of the above formulas, we thus get

$$I = \frac{58.2(1{,}719) + 15.6(571) + 15.3(534)}{32.5(1{,}719) + 11.9(571) + 12.9(534)} \cdot 100 = 168.4$$

The difference between this value, indicating an increase of 68.4%, and the 55.5% obtained in Exercise 18 can be attributed to the greater attention paid here to the change in the price of copper.

a. Refer to part (a) of Exercise 18, and using base-year quantities as weights, construct a weighted aggregative index comparing the 1970 prices of the four crops with those of 1965. The 1965 production of wheat, corn, oats, and soybeans in the United States totaled, respectively, 1,316, 4,084, 927, and 846 million bushels.

b. The following table gives the prices (in cents per pound) and production totals (in millions of pounds) of three vegetable oils:

	Prices		*Quantities*	
	1968	*1971*	*1968*	*1971*
Cottonseed oil	16.3	19.0	1,115	1,209
Linseed oil	12.7	8.9	307	412
Soybean oil	10.3	15.1	6,150	8,082

Construct a weighted aggregative index comparing the 1971 prices of these vegetable oils with those of 1968, using

(1) the 1968 quantities as weights;

(2) the 1971 quantities as weights.

20. SUMMATIONS The Σ notation which we introduced on page 29 is highly abbreviated, and many mathematicians prefer to write the sum $x_1 + x_2 + \ldots + x_n$ more explicitly as $\sum_{i=1}^{n} x_i$. This is meant to indicate that we are adding the x's whose subscript i is 1, 2, . . . , and n. Similarly, $\sum_{i=1}^{n} x_i^2$ denotes the sum of the squares of the x's with the subscripts 1, 2, . . . , and n, namely $x_1^2 + x_2^2 + \ldots + x_n^2$, and $\sum_{i=1}^{n} x_i y_i$ denotes the sum of the products $x \cdot y$ with the subscripts of x and y both equal to 1, 2, . . . , and n, namely, $x_1y_1 + x_2y_2 + \ldots + x_ny_n$. Using this more explicit summation notation, rewrite

a. the formula for the mean on page 29;
b. the formula for the mean of a distribution on page 31;
c. the formula for the weighted mean on page 33.

Also write each of the following expressions without the use of summation signs:

d. $\sum_{i=1}^{6} x_i$; **f.** $\sum_{i=1}^{5} x_if_i$; **h.** $\sum_{i=2}^{4} x_i$;

e. $\sum_{i=1}^{8} y_i$; **g.** $\sum_{i=1}^{4} x_i^2$; **i.** $\sum_{i=3}^{6} (y_i - k)$.

Note that as in parts (h) and (i), sums expressed in this notation need not start with the subscript 1.

The median

If we have to work with data given to us in the form of a distribution with an *open class* or if we do not want to risk getting a misleading picture due to an *extreme value* or a *mistake* (see Example 7 on page 30), we will have to describe the "middle" or "center" of a set of data with a statistical measure other than the mean. One possibility is the **median**, which, for ungrouped data, is obtained by *first arranging the measurements according to size and then choosing the one in the middle, or the mean of the two that are nearest to the middle.* The symbol which we use for the median of a set of x's is $\tilde{x}$ (and, hence, $\tilde{y}$ or $\tilde{z}$ when we refer to the measurements as y's or z's).

When an *odd* number of measurements are arranged according to size, there is always a middle one whose value is the median.

Example 10. For instance, 42, 39, 31, 35, and 38 bighorn sheep were killed by hunters in Arizona in the years 1969 through 1973. If we arrange these figures according to size, we get

$$\begin{array}{ccccc} 31 & 35 & 38 & 39 & 42 \\ & & \uparrow & & \end{array}$$

and it can be seen that the median is 38.

Example 11. Also, if the low-temperature readings at a weather station were 38, 43, 40, 44, 40, 39, 42, 41, 37, 38, and 41 on eleven consecutive days, we get

$$\begin{array}{ccccccccccc} 37 & 38 & 38 & 39 & 40 & 40 & 41 & 41 & 42 & 43 & 44 \\ & & & & & \uparrow & & & & & \end{array}$$

arranging them according to size, and it can be seen that the median is 40. Note that there are two 40's in this example and that we do not refer to either of them as *the* median—*the median is a number and not necessarily a particular measurement or observation.*

Generally speaking, if there are n measurements and n is *odd*, the median is the value of the $\frac{n+1}{2}$th *largest*; for instance, the median of 35 measurements is given by the value of the $\frac{35+1}{2} = 18$th largest, the median of 65 measurements is given by the value of the $\frac{65+1}{2} = 33$rd largest, and the median of 113 numbers is given by the value of the $\frac{113+1}{2} =$ 57th largest.

When an *even* number of measurements are arranged according to size, none of them can be exactly in the middle, and the median is defined as the *mean* of the values of the two measurements that are nearest to the middle.

Example 6 (Continued). On page 28 we gave the speeds of ten cars passing a check point as 53, 48, 50, 47, 43, 58, 42, 46, 51, and 38 miles per hour. If we arrange these figures according to size, we get

$$\begin{array}{cccccccccc} 38 & 42 & 43 & 46 & 47 & & 48 & 50 & 51 & 53 & 58 \\ & & & & & \uparrow & & & & & \end{array}$$

and it can be seen that the median is $\frac{47+48}{2} = 47.5$.

Example 7 (Continued). If we arrange the attendance figures given on page 30 according to size, we get

$$\begin{array}{ccccccc} 789 & 827 & 832 & & 840 & 867 & 903 \\ & & & \uparrow & & & \end{array}$$

and it can be seen that the median is $\frac{832+840}{2} = 836$. Then, if we make the same mistake as on page 30 and misread 832 as 238, the array becomes

$$238 \quad 789 \quad 827 \quad 840 \quad 867 \quad 903$$

↑

and the median is $\frac{827 + 840}{2} = 833.5$. Note that in this case the error is much smaller than it was on page 30, where the mean was reduced from 843 to 744. This illustrates the fact that *the median is generally not as strongly affected by an extreme value or a careless mistake as is the mean.*

In general, if there are n measurements and n is *even*, the formula $\frac{n+1}{2}$ will again give the *position* of the median, provided that it is interpreted correctly. For the first of the above two examples we had $n = 10$, and

$$\frac{n+1}{2} = \frac{10+1}{2} = 5.5$$

tells us that the median is *halfway between the 5th and 6th largest values.* Similarly, for the other example we had $n = 6$, and

$$\frac{n+1}{2} = \frac{6+1}{2} = 3.5$$

tells us that the median is *halfway between the 3rd and 4th largest values.* It is important to remember that $\frac{n+1}{2}$ is *not* a formula for the median—it merely tells us where the median is located.

In the above continuation of Example 7 it should not have surprised the reader that the median of the attendance figures did not coincide with their mean; the respective values were 836 and 843. The mean and the median measure the "middle" or "center" of a set of data in a different way, and as we shall see in Exercise 17 on page 203, the very fact that the median and the mean of a set of data *are (or are not) close together* is indicative of a further property of the data, namely, whether their distribution is **symmetrical** or **skewed** (that is, lopsided).

As we have already said, *the median is generally not as strongly affected by an extreme value as is the mean.* Among the other desirable properties of the median we find that, like the mean, *it can be calculated for any kind of data and it is always unique.* Unlike the mean, on the other hand, *the median of grouped data can generally be found when there are open classes, and the median can even be used to define the middle of a number of objects, properties, or qualities, which do not permit a quantitative description.* It is possible, for example, to rank a number of college courses according to their difficulty and then describe the one in the middle of the list as being of "average difficulty"; also, we might rank samples of fudge according to their consistency and then describe the middle one as having "average consistency." Perhaps the most important distinction between the median and the mean is that in problems of inference (estimation, prediction, etc.) *the median is generally less reliable than the mean.* In other words, the median is generally subject to greater chance fluctuations than the mean; that is, it is apt to *vary more from sample to sample,* as is illustrated in Exercise 7 on page 51. Finally, let us mention that the median is easy enough to find once the data have been arranged according to size, but unless we have automatic equipment, *the process of arranging a large set of data according to size can be an extremely tedious job.*

If we want to determine the median of a set of data that has already been grouped, we find ourselves in the same position as on page 30—we can no longer find the *actual* value of the median, although we *can* find the class into which the median will have to fall. The median of a distribution will thus have to be defined in a special way, and Figure 2.9 illustrates

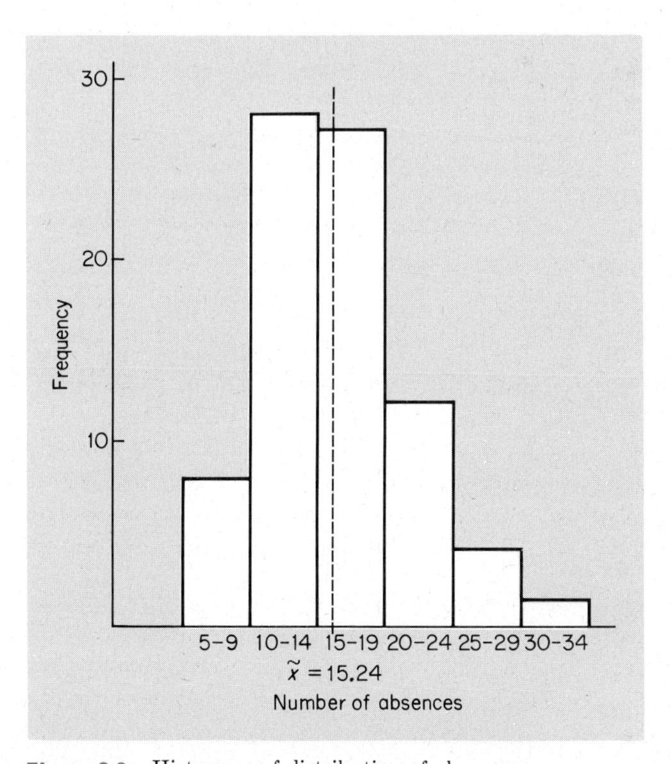

Figure 2.9 Histogram of distribution of absences.

how it is done. The median of a frequency distribution is defined as the number or point on the horizontal scale of Figure 2.9, which is such that *half the total area of the rectangles of the histogram of the distribution lies to its left and the other half lies to its right.* This means that the sum of the areas of the three rectangles to the left of the dashed line of Figure 2.9 must equal the sum of the areas of the four rectangles which lie to its right. Actually, this definition is equivalent to the assumption that the items are distributed evenly (that is, *spread out evenly*) throughout the class into which the median must fall.

To find the dividing line between the two halves of the histogram (each of which represents $n/2$ of the items grouped), we must somehow count $n/2$ items, starting at either end of the distribution.

Example 4 (Continued). To illustrate how this is done, let us refer again to the distribution of absences on page 14. Since $n = 80$ in this example, we shall have to count $\frac{80}{2} = 40$ items, starting at either end. If we begin with the smallest values, we find that $8 + 28 = 36$ of the values fall into the first two classes, $8 + 28 + 27 = 63$ fall into the first three, so that the median will have to fall into the class whose limits are 15–19. Having reached the *lower boundary* of this class, namely, 14.5, we will thus have to count another $40 - 36 = 4$ items in addition to the 36 items which fall below this class. We accomplish this by adding $\frac{4}{27}$ of the class interval of 5 to the lower boundary of the class. *Note that we add $\frac{4}{27}$ of the class interval because we want to count 4 of the 27 items which are contained in this class.* Thus, we get

$$\tilde{x} = 14.5 + \frac{4}{27} \cdot 5 = 15.24$$

or approximately 15.2.

In general, if L is the lower boundary of the class into which the median must fall, f is its frequency, c is its interval, and j is the number of items we still have to count after reaching L, then the **median of the distribution** is given by the formula

$$\tilde{x} = L + \frac{j}{f} \cdot c$$

The median of a frequency distribution can also be found by starting at the other end of the distribution and *subtracting* an appropriate fraction of the class interval from the upper boundary of the class into which the median must fall. In that case, the formula becomes

$$\tilde{x} = U - \frac{j'}{f} \cdot c$$

where j' is the number of items we still have to count after reaching U, the upper boundary of the class into which the median must fall.

Example 4 (Continued). With reference to the distribution of absences on page 14, we find that 17 of the values fall above 19.5, the *upper boundary* of the class into which the median must fall, so that $j' = 40 - 17 = 23$, and substitution into the formula yields

$$\tilde{x} = 19.5 - \tfrac{23}{27} \cdot 5 = 15.24$$

This agrees, as it should, with the answer which we obtained before.

The median belongs to a general class of statistical measures called **fractiles**, which are defined as values above or below which certain fractions (or percentages) of the data must fall. Clearly, for the median this fraction is $\frac{1}{2}$ either way. Since the formulas which we gave for the median of a distribution apply also to the calculation of other fractiles, we shall ask the reader to compute some of them in the exercises which follow.

Further measures of location

Besides the mean, the median, and the weighted mean, there are numerous other ways of describing the "middle" or "center" of a set of data. Some of these we already met in Exercises 16 and 17 on pages 38 and 39, and two others worth noting are the **midrange** and the **mode**. *The midrange of a set of data is simply the mean of the smallest and the largest values*, and its main advantage is that *it is easy to find.*

Example 12. If twelve golf professionals score 70, 68, 71, 71, 72, 69, 71, 73, 67, 71, 72, and 70 playing a certain course, the lowest score is 67, the highest score is 73, and the midrange of the scores is $\dfrac{67 + 73}{2} = 70$.
As can easily be verified, the median and the mean of the scores are, respectively, 71 and 70.4, so that *in this particular example* these three measures of location do not differ by very much.

Example 4 (Continued). For the data on absenteeism on page 14, the smallest value is 5, the largest is 33, and the midrange is $\frac{5+33}{2} = 19$. In this case, the value of the midrange differs considerably from the mean of 15.7 obtained on page 32 and the median of 15.24 obtained on page 46, and this is due to the small number of *very large* values which give the distribution of the data (see Figure 2.1 on page 18) the appearance of having a "tail" on the right-hand side.

The mode of a set of data is simply the value which occurs the most often, and its two main advantages are that (1) *it requires no calculations* and (2) *it can be determined for quantitative as well as qualitative data.*

Example 12 (Continued). Since only one golfer each scores 67, 68, 69, and 73, two golfers each score 70 and 72, and four golfers each score 71, the mode is 71, and we refer to it as the **modal score**.

Example 13. If more American tourists in Europe want to visit England than any other European country, we can say that England is their **modal choice**.

So far as modes of grouped data are concerned, it is customary to refer to the class with the highest frequency as the **modal class**.

Example 4 (Continued). For the distribution of absences, the 10–14 class has the highest frequency of 28, and it is, therefore, the modal class. Sometimes, the class mark (midpoint) of the modal class is referred to as the mode of the distribution, and for the distribution of absences we would, thus, get a mode of 12.

Two definite disadvantages of the mode are that (1) *it may not exist*, which is the case when no two values in a set of data are alike, and (2) *there may be two modes, or even three or more.*

Example 14. Consider the set of data which consists of the following ten numbers:

4 9 4 7 4 5 10 9 6 9

As can easily be seen, the numbers 5, 6, 7, and 10 each occurs once, and the numbers 4 and 9 each occurs three times, so that there are *two modes*: 4 and 9. Probably, this does not mean very much, but the presence of more than one mode is often indicative of the fact that the data are *not homogeneous*, namely, that they can be looked upon as a *combination of several sets of data.* Thus, if the above numbers are the grades which ten students received in a test, we might infer that the class is a mixture of two essentially different kinds of students—some that are very good and some that are very poor.

The question of what particular "average" should be used in a given situation is not always easily answered, and the fact that there is a good deal of arbitrariness in the selection of statistical descriptions has led some persons to believe that they can take any set of data, apply the magic of statistics, and prove almost anything they want. Indeed, a famous nineteenth-century British statesman once said that there are three kinds of lies: *lies, damned lies, and statistics.*

Example 15. To give an example where this kind of criticism might be justified, suppose that a consumer testing service obtained the following miles per gallon in five test runs performed with each of three intermediate-size cars:

Car A: 16.8, 19.3, 19.5, 20.3, 20.6

Car B: 20.1, 17.6, 20.2, 17.6, 20.0

Car C: 17.5, 18.0, 17.4, 21.0, 18.6

If the manufacturers of car A want to show on the basis of these data that their car gets the best mileage, their statistician will be delighted to find that the respective *means* are 19.3, 19.1, and 18.5—he can claim that *on the average* his employers' car got the best mileage.

Now let us take a look at a statistician working for the manufacturers of car B. He cannot base the comparison on the means—that would not prove his point—but he does not have to look very far for a measure of location that will do the trick. The respective *medians* are 19.5, 20.0, and 18.0, and this provides him with the kind of "proof" he wants. The median is a perfectly respectable measure of the "average" or "middle" of a set of data, and using the medians he can claim that his employers' car came out best in the test.

Finally, let us consider the plight of a statistician working for the manufacturers of car C. He cannot use the median or the mean, but after trying several others, he finally is lucky enough to find one that "works"—the *midrange*, which we defined on page 47. Since the midranges of the three sets of data are, respectively, 18.7, 18.9, and 19.2, he can claim that "on the average" his employers' car performed best in the test.

The moral of this example is that

Methods of describing statistical data should always be decided upon before the data are collected, or at least before they have been inspected.

Of course, in Example 15 there is also the consideration that comparisons based on such small sets of data may be far from conclusive, and this is a problem which we shall discuss in Chapters 8 and 11.

Exercises

1. On five experimental farms, experiments were conducted, respectively, with 20, 24, 16, 21, and 29 varieties of corn. Find

a. the mean;
b. the median;
c. the midrange of the number of varieties of corn with which they experimented at these farms.

2. The following are the grades which nine students from one high school obtained on the verbal part of the College Board's SAT test: 519, 647, 703, 728, 572, 486, 692, 609, and 435. Find

a. the mean;
b. the median;
c. the midrange of these scores.

3. Diving at the site of a ship wrecked during the early part of the eighteenth century, eight divers found, respectively, 12, 15, 21, 6, 42, 19, 8, and 24 gold coins. Find

a. the mean;
b. the median;
c. the midrange of the number of gold coins which the divers found.

If a person knows that there were eight divers, which of the three measures of location would enable him to determine how much he would have to pay for *all* the coins at $200 per coin?

4. The following are twelve temperature readings taken at various locations in a large kiln (in degrees Fahrenheit):

415 450 510 460 475 470 410 460 425 460 500 475

Find

a. the mean;
b. the median;
c. the midrange;
d. the mode of these temperature readings.

5. During the year 1974, ten insurance salesmen wrote, respectively, 65, 44, 30, 75, 52, 32, 50, 75, 56, and 44 new policies. Find

a. the mean;
b. the median;
c. the midrange;
d. the mode of the number of new policies which these insurance salesmen wrote.

Suppose, furthermore, that the company for which these insurance salesmen work wants to give a bonus of a watch to each of the five agents who wrote the most policies and a bonus of \$2.00 for each new policy written to each of the other salesmen.

e. Which of the above measures of location would tell them (without knowledge of the actual data) *at most* how much they may have to pay out in cash bonuses?

f. If the watches cost \$120 each, *at most* how much would all these bonuses cost? Base this figure on the answer to part (e).

6. Each of seven women taking part in a fund-raising campaign for a local charity was assigned a certain quota, and the following are the percentages of their respective quotas which they actually collected: 101, 108, 93, 381, 117, 112, and 82. Calculate the mean and the median of these percentages and indicate which of the two is a better indication of these women's "average performance."

7. To verify the claim that the mean is generally *more reliable* than the median (namely, that it is subject to smaller chance fluctuations), a student conducted an experiment consisting of twelve rolls of three dice. The following are his results: 2, 4, and 6; 5, 3, and 5; 4, 5, and 3; 5, 2, and 3; 6, 1, and 5; 3, 2, and 1; 3, 1, and 4; 5, 5, and 2; 3, 3, and 4; 1, 6, and 2; 3, 3, and 3; 4, 5, and 3.

a. Calculate the twelve medians and the twelve means.

b. Group the medians and the means obtained in (a) into separate distributions having the classes 1.5–2.5, 2.5–3.5, 3.5–4.5, and 4.5–5.5. (Note that there will be no ambiguities since the medians of three whole numbers and the means of three whole numbers cannot equal 2.5, 3.5, or 4.5.)

c. Draw histograms of the two distributions obtained in part (b) and explain how they illustrate the claim that the mean is generally more reliable than the median.

d. Repeat the entire "experiment" by repeatedly rolling three dice (or one die three times) and constructing corresponding distributions for the medians and the means. If no dice are available, *simulate* the experiment by drawing numbered slips of paper out of a hat.

8. With reference to the grade distribution of Exercise 10 on page 24, find

a. the median;
b. the modal class.

9. With reference to the distribution of service times of Exercise 8 on page 35, find

a. the median;
b. the modal class.

10. Find the median and the modal class of whichever data you grouped among those of Exercises 11, 12, 13, and 14 on pages 25 through 26.

11. The following is the distribution of the social science achievement grades of 800 eighth-graders in a certain school district:

Achievement Grade	*Frequency*
20–29	16
30–39	57
40–49	112
50–59	169
60–69	200
70–79	143
80–89	70
90–99	33

Find the median of this distribution and also indicate which class is the modal class.

12. Find the mode (if it exists) of each of the following sets of figures on the number of robberies reported in three cities in twelve weeks:

City A: 23, 26, 26, 31, 26, 26, 24, 28, 24, 22, 29, 20

City B: 12, 28, 20, 23, 22, 11, 22, 28, 17, 22, 28, 23

City C: 15, 27, 13, 24, 31, 25, 14, 21, 20, 9, 11, 28

13. Thirty registered voters were asked whether they consider themselves Democrats, Republicans, or Independents. Use the following results to determine their *modal choice*: Democrat, Republican, Independent, Independent, Democrat, Independent, Republican, Republican, Independent, Democrat, Democrat, Independent, Democrat, Independent, Republican, Independent, Independent, Independent, Democrat, Democrat, Republican, Independent, Independent, Republican, Republican, Democrat, Republican, Democrat, Independent, and Independent.

14. A survey made at a resort city showed that 50 tourists arrived by the following means of transportation: car, train, plane, plane, plane, bus, train, car, car, car, plane, car, plane, train, car, car, bus, car, plane, plane, train, train, plane, plane, car, car, train, car, car, plane, car, plane, plane, bus, plane, bus, car, plane, car, car, train, train, car, plane, bus, plane, car, car, train, and bus. What is the *modal means of transportation* by which they arrived?

15. **QUARTILES** The three *quartiles* of a distribution Q_1, Q_2, and Q_3 are values such that a fourth of the total area of the rectangles of its histogram (representing the smallest 25% of the items) falls to the left of Q_1, a fourth of the total area (representing the next 25% of the items) falls between Q_1 and Q_2, a fourth of the total area (representing the next 25% of the items after that) falls between Q_2 and Q_3, and a fourth of the total area (representing the largest 25% of the items) falls to the right of Q_3. To find the first and third quartiles (Q_2 actually *is* the median), we can use either of the two formulas on pages 46 and 47; for Q_1 we usually count $\frac{1}{4}$ of the items starting with the smallest values and for Q_3 we usually count $\frac{1}{4}$ of the items starting with the largest values. For instance, for the distribution of absences on page 14 we must count the $\frac{80}{4} = 20$ smallest values to reach Q_1, and using the first of the two formulas on page 46, we get

$$Q_1 = 9.5 + \tfrac{12}{28} \cdot 5 = 11.64$$

Similarly, we must count the $\frac{80}{4} = 20$ largest values to reach Q_3, and using the second of the two formulas on page 47, we get

$$Q_3 = 19.5 - \tfrac{3}{27} \cdot 5 = 18.94$$

a. Find Q_1 and Q_3 for the grade distribution of Exercise 10 on page 24.

b. Find Q_1 and Q_3 for the distribution of service times of Exercise 8 on page 35.

c. Find Q_1 and Q_3 for the distribution of social science achievement test grades of Exercise 11.

16. QUARTILES, CONTINUED Quartiles can also be defined for *ungrouped* data, but we shall not study this problem in general in this text. What would be a logical way of *defining* the quartiles of a set of 80 measurements or observations, such as, for example, the ungrouped data of Example 4 on page 14? Compare the resulting values of Q_1 and Q_3 obtained for the ungrouped data of Example 4 on page 14 with the corresponding values determined in Exercise 15 from their distribution.

17. DECILES The nine *deciles* of a distribution are values such that a tenth of the total area of the rectangles of its histogram (representing the smallest 10% of the items) falls to the left of D_1, a tenth of the total area (representing the next 10% of the items) falls between D_1 and $D_2, \ldots,$ and a tenth of the total area (representing the largest 10% of the items) falls to the right of D_9. To find the deciles of a distribution, we can again use either of the two formulas on pages 46 and 47. For instance, to find D_1 for the distribution of absences on page 14, we must count the $\frac{80}{10} = 8$ smallest values, and using the first of the two formulas on page 46, we get

$$D_1 = 9.5 + \tfrac{0}{28} \cdot 5 = 9.5$$

Similarly, we must count the $2 \cdot \frac{80}{10} = 16$ largest values to reach D_8, and using the second of the two formulas on page 47, we get

$$D_8 = 24.5 - \tfrac{11}{12} \cdot 5 = 19.92$$

a. Find D_1 and D_9 for the grade distribution of Exercise 10 on page 24.
b. Find D_3 and D_6 for the distribution of service times of Exercise 8 on page 35.
c. Find D_2 and D_8 for the distribution of social science achievement test grades of Exercise 11.

18. PERCENTILES The ninety-nine *percentiles* of a distribution $P_1, P_2, \ldots,$ and P_{99} are defined in the same way as the quartiles and deciles in Exercises 15 and 17. For instance, to find P_5 for the distribution of absences on page 14, we must count the $5 \cdot \frac{80}{100} = 4$ smallest values, and using the first of the two formulas on page 46, we get

$$P_5 = 4.5 + \tfrac{4}{8} \cdot 5 = 7$$

Similarly, we must count the $\frac{80}{100} = 0.8$ largest value to reach P_{99}, and using the second of the two formulas on page 47, we get

$$P_{99} = 34.5 - \frac{0.8}{1} \cdot 5 = 30.5$$

Note that the quantities j and j' in the two formulas on pages 46 and 47 need not be whole numbers.

a. Find P_{15} and P_{85} for the grade distribution of Exercise 10 on page 24.

b. Find P_2 and P_{98} for the distribution of service times of Exercise 8 on page 35.

c. Find P_{97}, P_{98}, and P_{99} for the distribution of social science achievement test scores of Exercise 11.

3

Possibilities and probabilities

Introduction

We can hardly predict which television program will get the highest rating for a given week unless we know at least what shows will be on the air, and we cannot very well predict the winner of the women's singles at Wimbledon unless we know at least who is going to play. More generally, we cannot make intelligent predictions or decisions unless we know at least what is possible—in other words, *we must know what is possible before we can judge what is probable.* Thus, the next three sections will be devoted to "what is possible" in a given situation, after which the final three sections of the chapter will be devoted to "what is probable."

The sample space

In statistics, a set of all the possible outcomes of an experiment is called a **sample space**, for it usually consists of all the things that can happen when we take a sample.

Example 1. If a zoologist must choose three of his 24 guinea pigs for an experiment, the sample space consists of all the possible ways in which this can be done. As we shall see on page 74, the number of possibilities in this case is 2,024.

Example 2. If the dean of a junior college has to assign two of his 84 faculty members as advisors to a political science club, the sample space consists of all the possible ways in which the dean can make this choice. As we shall see on page 76, the number of possibilities in this case is 3,486.

To avoid misunderstandings about the words "outcome" and "experiment," let us make it clear that statisticians use these terms in a very wide sense. An **experiment** may consist simply of checking whether or not a person is wearing glasses; it may consist of measuring the speed of a car or counting the number of patients who have adverse effects from a new medication; on the other hand, an experiment may also consist of the very complicated process of determining the mass of an electron or studying environmental effects on the behavior of a child. Correspondingly, an **outcome** of an experiment may be a simple "yes or no" answer; it may be the result of a measurement or a count; or it may be an answer obtained after extensive measurements and calculations.

When we study the outcomes of an experiment, it is usually advantageous to identify the different possibilities with numbers or points, for we can then treat all questions concerning the outcomes *mathematically*, without having to go through long verbal descriptions of what has taken place, is taking place, can take place, or will take place. Actually, this is precisely what we do in sports when we refer to football players by their numbers; it is what the Internal Revenue Service does when it refers to taxpayers by their social security numbers; and it is what the post office does when it uses ZIP codes to identify geographical regions.

The use of points rather than numbers has the added advantage that it makes it easier to *visualize* the various possibilities, and perhaps discover some of the special features which several of the outcomes may have in common.

Example 3. If a psychologist measures a person's reaction time to a visual stimulus, the sample space would consist of the points on a line (coordinated in the usual way), and an outcome of 2.43 seconds, for instance, is represented by the corresponding point shown in Figure 3.1.

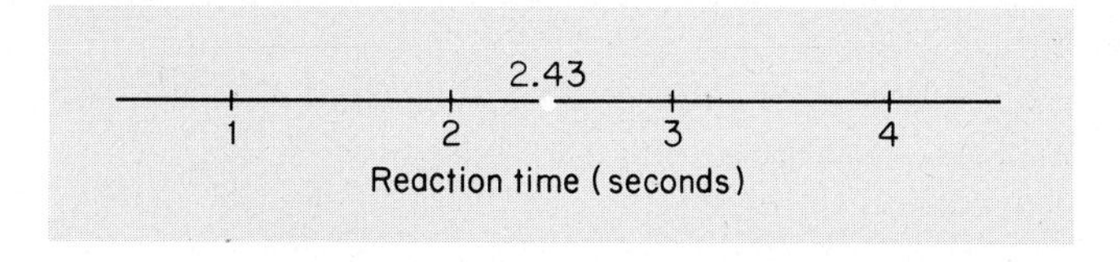

Figure 3.1 Sample space for measurement of reaction time.

Example 4. If a meteorologist measures temperature as well as humidity, he might use the sample space pictured in Figure 3.2, where the temperature is represented by the x-coordinate and the humidity by the y-coordinate. Thus, if the temperature is 73° and the humidity is

23%, this outcome would be represented by the point (72, 23) as in Figure 3.2.

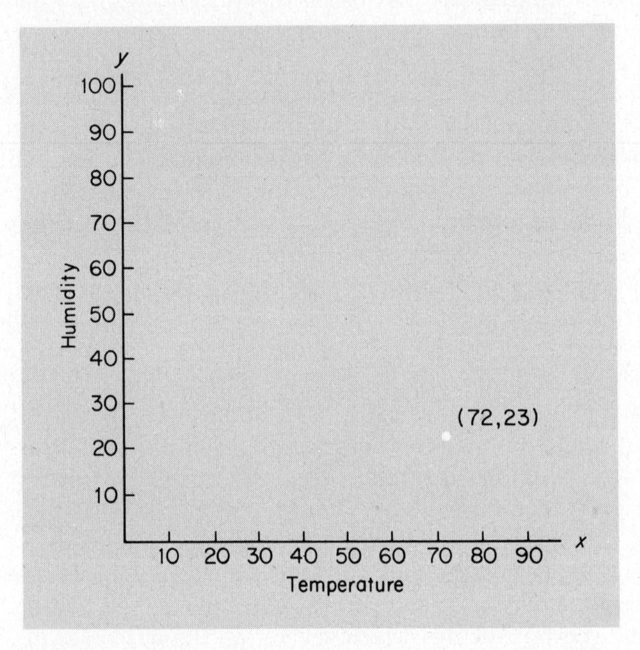

Figure 3.2 Sample space for temperature and humidity readings.

To study problems relating to sample spaces, outcomes, and the combinations of outcomes into more complicated kinds of outcomes, let us consider a sample space containing only a very small number of points.

Example 5. Suppose that two archaeological expeditions are looking for the ruins of two ancient cities, and that we are interested here only in *how many of these ancient cities each expedition will discover*. If we use two coordinates so that (0, 1) represents the outcome that the first expedition will discover none of the ancient cities and the second expedition will discover one, and (1, 1) represents the outcome that each of the expeditions will discover one of the ancient cities, the six possible outcomes are (0, 0), (1, 0), (0, 1), (2, 0), (1, 1), and (0, 2), and the corresponding points are shown in Figure 3.3. Actually, the six possible outcomes could have been represented by means of six points on a line, or six points *in any kind of pattern*, but the use of coordinates has the advantage that it makes it easy to identify each outcome with the corresponding point.

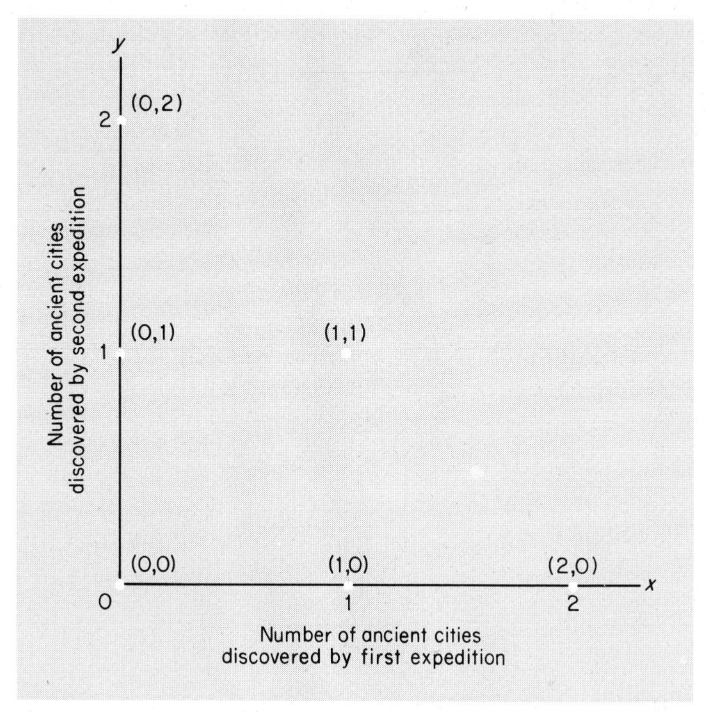

Figure 3.3 Sample space for Example 5.

Generally, we classify sample spaces according to the number of points which they contain. The ones we studied in Examples 1, 2, and 5 contained, respectively, 2,024, 3,486, and 6 points (possible outcomes), and we refer to them as **finite**. This means that they contain a limited, or fixed, number of points. In contrast, Examples 3 and 4 dealt with quantities measured with continuous scales, and we refer to the corresponding sample spaces (the line of Figure 3.1 and the plane region of Figure 3.2) as **continuous**. In this chapter and in Chapter 4, we shall consider only sample spaces that are finite; others will be taken up in Chapter 6.

Outcomes and events

In statistics, we refer to sets of outcomes of experiments (and the corresponding points of the sample space) as **events**.

Example 5 (Continued). With reference to Figure 3.3, the point (0, 0) constitutes the event that *neither expedition will discover either of the* *ancient cities*, the points (1, 0) and (0, 1) together constitute the event that *only one of the two ancient cities will be discovered by the two expeditions*, and the points (0, 0), (1, 0), (0, 1), and (1, 1) together constitute the event that *neither expedition will discover both of the ancient cities.*

To make it easy to refer to particular events and to show how two or more events can be combined to form other events, it will be convenient to denote events with capital letters.

Example 5 (Continued). With reference to Figure 3.4, which shows the same sample space as Figure 3.3, we shall let D denote the event that *the first expedition will discover one of the ancient cities*, E denote the

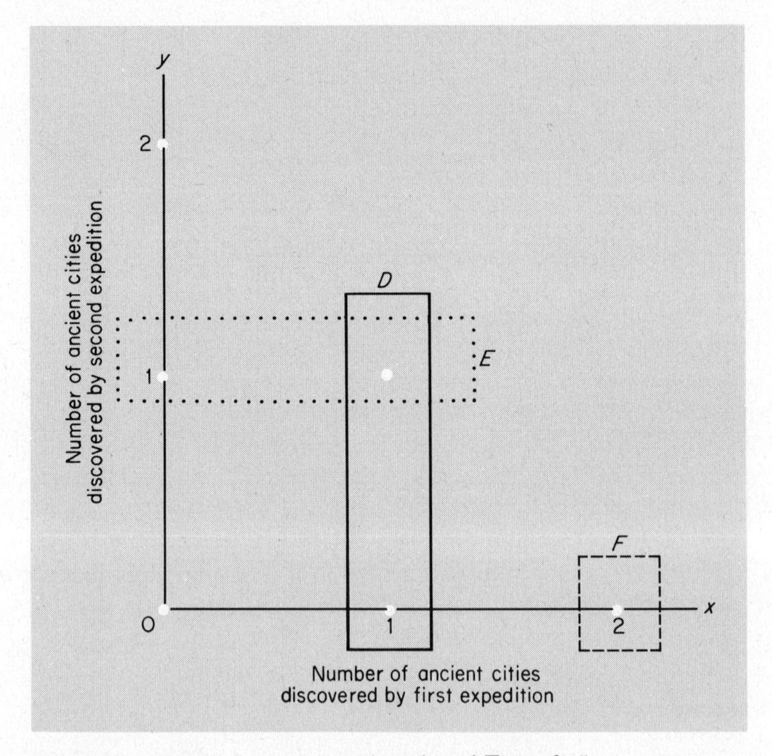

Figure 3.4 Sample space for continuation of Example 5.

event that *the second expedition will discover one of the ancient cities*, and F denote the event that *the first expedition will discover both of the ancient cities.* In the diagram, the two points which constitute D are inside the solid line, the two points which constitute E are inside the dotted line, and the point which constitutes F is inside the dashed line. Observe that events E and F have no points in common; they are referred to as **mutually exclusive events**, which means that *they cannot both occur at the same time.* Evidently, if the second expedition discovers one of the two ancient cities, the first expedition cannot very well discover both. It is also apparent from Figure 3.4 that events D and F are mutually exclusive and that events D and E are *not* mutually exclusive.

As the reader may recall from elementary mathematics, sets are usually combined by forming **unions** and **intersections**, and we shall do so now with events. Formally,

The union of two events A and B, denoted $A \cup B$, is the event which consists of all the outcomes (points) contained in event A, event B, or both. The intersection of two events A and B, denoted $A \cap B$, is the event which consists of all the outcomes (points) contained in both A and B.*

To illustrate these definitions and/or refresh the reader's memory, let us consider the following examples:

Example 6. If A is the event that Mr. Jones is out of work and B is the event that his wife is out of work, then $A \cup B$ is the event that at least one of the two (Mr. Jones, Mrs. Jones, or both) is out of work and $A \cap B$ is the event that they are both out of work.

Example 5 (Continued). With reference to Figure 3.4, $D \cup E$ is the event that *either or both of the expeditions will discover one of the ancient cities* and it contains the points $(1, 0)$, $(0, 1)$, and $(1, 1)$; $D \cap E$ is the event that *each expedition will discover one of the two ancient cities* and it contains the point $(1, 1)$. Also, as we pointed out before, $D \cap F$ is an event which cannot possibly occur since D and F are *mutually exclusive*, and we refer to it as an **empty set**, denoted with the symbol $\varnothing$.

*It is common practice to read $A \cup B$ as "A *or* B" and $A \cap B$ as "A *and* B."

Another important concept from the mathematics of sets, which we shall now apply to events, is that of a **complement**. Formally,

> **The complement of an event A, denoted A' (A-prime), is the event which consists of all the outcomes (points) of the sample space that are not contained in A.**

In some books, the complement of A is denoted $\bar{A}$ or $\sim A$.

Example 7. If W is the event that a community college's basketball team will win a certain game, then W' is the event that it will not win the game. Also, if M is the event that a student will make at most five mistakes on a test, then M' is the event that he or she will make more than five mistakes on the test.

Example 5 (Continued). With reference to Figure 3.4, E' is the event that *the second expedition will not discover exactly one of the ancient cities*, namely, the event that *it will discover neither or both.* Also, F' is the event that *the first expedition will not discover both of the ancient cities.*

Sample spaces and events, particularly relationships among events, are often pictured by means of **Venn diagrams** such as those of Figures 3.5 and 3.6. In each case the sample space is represented by a rectangle, while events are represented by regions within the rectangle, usually by circles or parts of circles. Thus, the shaded regions of the four Venn diagrams of Figure 3.5 represent, respectively, event X, the complement of X, the union of two events X and Y, and the intersection of X and Y.

Example 8. If X is the event that a certain high school senior applies for admission to Syracuse University and Y is the event that he applies for admission to Cornell, then the region shaded in the first diagram of Figure 3.5 represents the event that the student applies to Syracuse University, the region shaded in the second diagram represents the event that he does *not* apply to Syracuse University, the region shaded in the third diagram represents the event that he applies to Syracuse University and/or Cornell, and the region shaded in the fourth diagram represents the event that he applies to both of these universities.

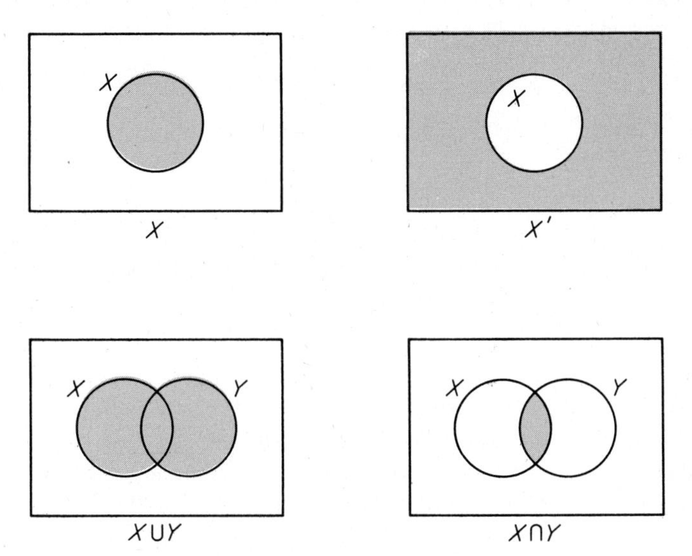

Figure 3.5 Venn diagrams.

When we deal with three events, A, B, and C, it is customary to draw the circles as in Figure 3.6. As can be seen from this diagram, the circles divide

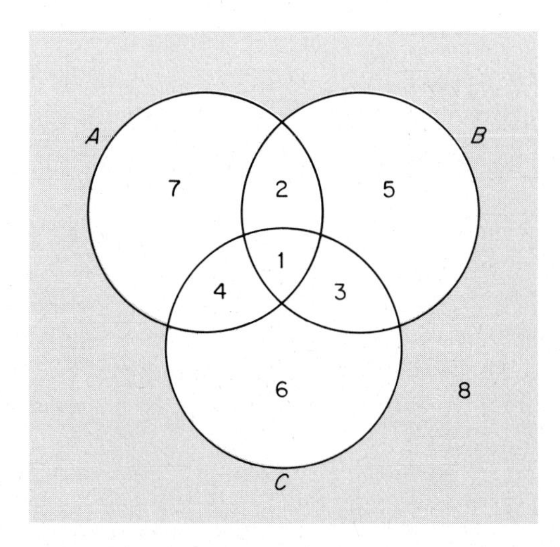

Figure 3.6 Venn diagram.

the sample space into eight regions (which we numbered 1 through 8), and it is quite easy to check for each of the corresponding events whether or not it is contained in A, whether or not it is contained in B, and whether or not it is contained in C.

Example 9. Suppose that an engineer has designed a new kind of engine and that A is the event that its gasoline consumption will be low, B is the event that its maintenance cost will be low, and C is the event that it can be produced at a profit. Looking at Figure 3.6, it can easily be seen that Region 4, for example, is contained in A and C but not in B, so that it represents the event that *the gasoline consumption of the engine will be low and it can be produced at a profit, but the maintenance cost will not be low.* Similarly, it can be seen that Region 5 represents the event that *the maintenance cost will be low, but the gasoline consumption will not be low and the engine cannot be produced at a profit.* It will be left to the reader to identify some of the other regions (and also some of their combinations) in Exercise 8 below.

Exercises

1. With reference to Example 5 and the sample space of Figure 3.3, describe *in words* the event which is represented by each of the following sets of points:

a. (0, 0) and (1, 1);
b. (0, 2);
c. (2, 0), (1, 1), and (0, 2);
d. (0, 0), (0, 1), and (0, 2).

2. Two professors and three graduate assistants are responsible for the supervision of a biology lab, and at least one professor and one graduate assistant have to be present at all times.

a. Using two coordinates so that (1, 3), for example, represents the event that one of the professors and all three of the graduate assistants are present, and (2, 1) represents the event that both professors but only one of the graduate assistants are present, draw a diagram (similar to that of Figure 3.3) showing the *six* points of the corresponding sample space.

b. Describe *in words* the event which is represented by each of the following sets of points of the sample space: the event Q which consists of the points (1, 3) and (2, 3), the event R which consists of the points (1, 1) and (2, 2), and the event S which consists of the points (1, 2) and (2, 1).

c. With reference to part (b), list the points of the sample space which represent the event $R \cup S$, and describe *in words* the corresponding event.

d. With reference to part (b), are events Q and S mutually exclusive?

3. A small taxicab company owns only three cabs, which, furthermore, are frequently in the garage for repairs.

a. Using two coordinates so that (2, 1), for example, represents the event that two of the cabs are operative and one is out on a call, and (3, 0) represents the event that all three cabs are operative but none is out on a call, draw a diagram (similar to that of Figure 3.3) showing the *ten* points of the corresponding sample space.

b. List the points which comprise the event K that at least two of the company's cabs are out on calls, the event L that only one of the company's cabs is operative, and the event M that all operative cabs are out on calls.

c. With reference to part (b), list the points of the sample space which represent M' and $M \cap L$, and describe *in words* the corresponding events.

d. With reference to part (b), which of the pairs of events K and L, K and M, and L and M are mutually exclusive?

4. A high school ranks its seniors 1, 2, 3, or 4, depending on whether they are in the top 25%, the second highest 25%, the second lowest 25%, or the lowest 25% of their class.

a. Draw a suitable sample space showing the different ways in which a senior can thus be ranked. Also encircle by means of a solid line the points which represent the event that this senior is in the top half of the class, and by means of a dotted line the points which represent the event that he or she is *not* in the top 25%. Are these two events mutually exclusive?

b. Using two coordinates to represent the ranks of two seniors, say, Doug and Ray, draw a diagram (similar to that of Figure 3.3) showing the *sixteen* points of the corresponding sample space.

c. With reference to part (b), list the points which comprise the event A that Doug and Ray receive the same rank, the event B that Doug is in the lowest half of the class, and the event C that Ray is in the top 25% of the class. Are any two of these three events mutually exclusive?

5. To construct sample spaces for experiments in which we deal with *categorical data*, we often *code* the various alternatives by assigning them numbers. For instance, if shoppers in a market are asked whether they prefer brand A coffee to brand B, whether they prefer brand B to brand A, or whether they have no preference, we might assign these three responses the codes 1, 2, and 3.

a. Draw a suitable sample space showing the different ways in which a shopper can thus respond. Also encircle by means of a solid line the event that the shopper does not prefer brand A, and by means of a dotted line the event that the shopper does not prefer brand B. Are these two events mutually exclusive?

b. Using two coordinates to represent, in order, the responses of two shoppers, draw a diagram (similar to that of Figure 3.3) showing the *nine* points of the corresponding sample space.

c. Describe *in words* the event which is represented by each of the following sets of points of the sample space of part (b): the event D, which consists of the points (1, 1), (1, 2), and (1, 3); the event E, which consists of the points (1, 2), (2, 2), and (3, 2); the event F, which consists of the points (1, 1), (1, 2), (2, 1) and (2, 2); and the event G, which consists of the points (1, 3) and (3, 1).

d. With reference to part (c), describe *in words* the events which are denoted F', $D \cup E$, and $F \cap E$. Also list the points which comprise each of these three events.

e. If we use three coordinates to represent, in order, the responses of three shoppers, what event is represented by the point (2, 3, 1), what event is represented by the set of points (1, 1, 1), (2, 2, 2), and (3, 3, 3), and what event is represented by the set of points (2, 2, 1), (2, 2, 2), and (2, 2, 3)? [If we drew suitable x-, y-, and z-axes, these points could be pictured as points in three-dimensional space. However, if we used (2, 3, 1, 2) to represent the response of four shoppers, or (1, 2, 1, 3, 2) the response of five shoppers, the corresponding sample spaces would be difficult to visualize.]

6. In Figure 3.7, U denotes the event that a given candidate for Congress will be elected and V denotes the event that he will keep his promises. Explain *in words* what events are represented by Regions 1, 2, 3, and 4 of the diagram.

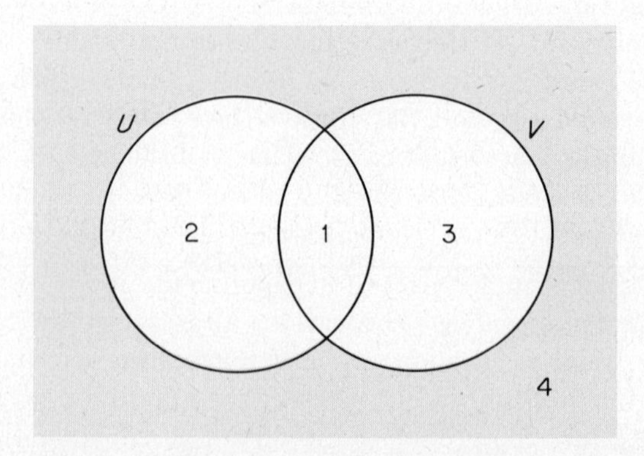

Figure 3.7 Venn diagram for Exercise 6.

7. In Figure 3.8, G is the event that a new novel will get good reviews and B is the event that it will be a best-seller. Explain *in words* what events are represented by Regions 1, 2, 3, and 4 of the diagram.

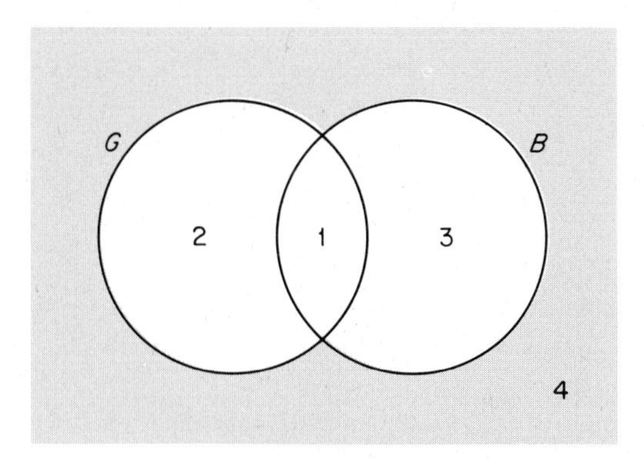

Figure 3.8 Venn diagram for Exercise 7.

8. With reference to Example 9 and Figure 3.6 on page 63, explain *in words* what events are represented by the following regions of the Venn diagram:

a. region 1;
b. region 3;
c. region 7;
d. regions 1 and 3 together;
e. regions 2 and 7 together;
f. regions 4, 6, 7, and 8 together.

9. Suppose that a group of students are planning a trip to Europe and that L is the event that they will visit London, P is the event that they will visit Paris, and G is the event that they will have a good time. With reference to the Venn diagram of Figure 3.9, list (by numbers) the regions or combinations of regions which represent the following events:

a. The event that they will visit London and Paris, but not have a good time.
b. The event that they will visit neither London nor Paris, but have a good time.
c. The event that they will visit Paris and have a good time.
d. The event that they will visit London or Paris (or both), but not have a good time.

10. With reference to Exercise 9 and the Venn diagram of Figure 3.9, express *in words* what events are represented by the following regions:

a. region 1;
b. region 5;
c. region 8;
d. regions 3 and 5 together;
e. regions 1 and 4 together;
f. regions 6 and 8 together.

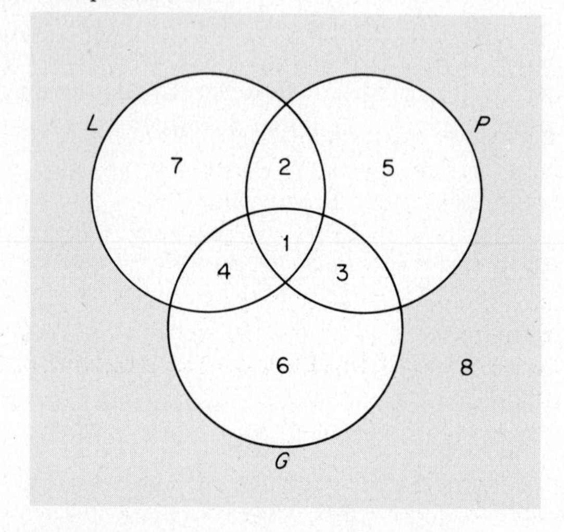

Figure 3.9 Venn diagram for Exercise 9.

11. Which of the following pairs of events are mutually exclusive? Explain your answers.

a. Having rain and sunshine on the 4th of July, 1978.

b. Being under 25 years of age and being President of the United States.

c. One and the same person wearing black shoes and green socks.

d. A driver getting a ticket for speeding and his getting a ticket for going through a red light.

e. A person leaving Los Angeles by jet at 11:45 P.M. and arriving in Washington, D.C., on the same day.

f. A baseball player getting a walk and hitting a home run in the same game.

g. A baseball player getting a walk and hitting a home run in the same time at bat.

Listing outcomes and counting events

In this section we shall discuss two kinds of problems related to sample spaces, outcomes, and events. *First, there is the problem of listing everything that can happen in a given experiment*, and although this may sound simple, it is often easier said than done. Sometimes there are so many outcomes that the job of listing them all becomes a Herculean task, and even when the number of outcomes is fairly small, it can be quite difficult to make sure that none of them has been left out. *The second kind of problem is that of determining the total number of outcomes (or the total number of events of a given kind) without actually constructing a complete list.* This is important because there are many problems in which we really do not need a complete list and, hence, can save a lot of work. For instance, it can be shown that more than 635 *billion* different 13-card bridge hands can be dealt with an ordinary deck of 52 playing cards, and it would hardly seem necessary to point out that this figure was obtained without actually listing all these hands.

Example 10. To give an example in which the listing of all possible outcomes is *not* straightforward, suppose that two students must choose a course in science to meet graduation requirements, and that, among others, they are considering chemistry, geology, and botany. The problem is to determine the number of ways in which they can make their choice, caring only *how many of them* (not which ones) will decide to take chemistry, *how many of them* will decide to take geology, and *how many of them* will decide to take botany. Clearly, there are many possibilities: both of them might decide to take chemistry; one of them might decide to take geology while the other decides to take botany; one of them might decide to take chemistry while the other decides to take a course in science other than the ones we have mentioned; and so forth. Being very careful, we may be able to complete this list, but the chances are that we will leave out some of the possibilities.

To handle this kind of problem systematically, it is helpful to refer to a diagram such as that of Figure 3.10, which is called a **tree diagram**. This diagram shows that first there are three possibilities (three branches) corresponding to 0, 1, or 2 of the students deciding to take chemistry. Then, for geology there are three branches emanating from the top branch, two from the middle branch, and none from the bottom branch. Clearly, there are again three possibilities (0, 1, or 2) for geology when neither of the students has decided to take chemistry, two possibilities (0 or 1) when one of the two students has decided to take chemistry, and there is no need to go any further once both students have decided to take chemistry. For botany the reasoning is the same, and we thus find that (going from left to right) there are altogether *ten* different paths along the "branches" of the tree; in other words, *the "experiment" has ten possible outcomes.*

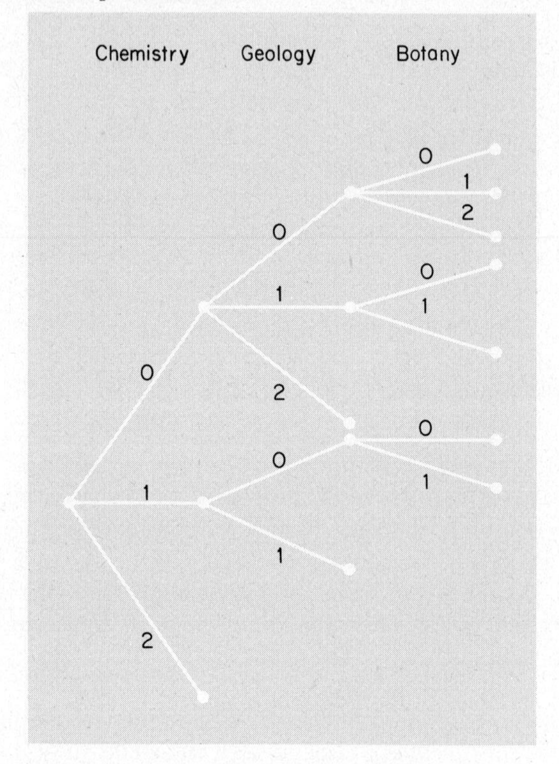

Figure 3.10 Tree diagram for Example 10.

Example 11. To consider another problem in which a tree diagram will help (at least, until we shall have studied some other techniques), suppose that in a medical study persons are classified according to whether they have blood type A, B, AB, or O, and also according to whether their blood pressure is low, normal, or high. If we want to know in how many different ways a person can, thus, be classified, we have only to look at the tree diagram of Figure 3.11, from which it is apparent that the answer is 12. Starting at the top, the first path along the "branches" of the tree corresponds to a person having blood type A and low blood pressure, the second path corresponds to a person having blood type A and normal blood pressure, . . . , and the twelfth path corresponds to a person having blood type O and high blood pressure.

Note that the answer which we obtained in Example 11 is the *product* of 4 and 3, namely, the *product* of the number of ways in which a person can be classified according to blood type and the number of ways in which he

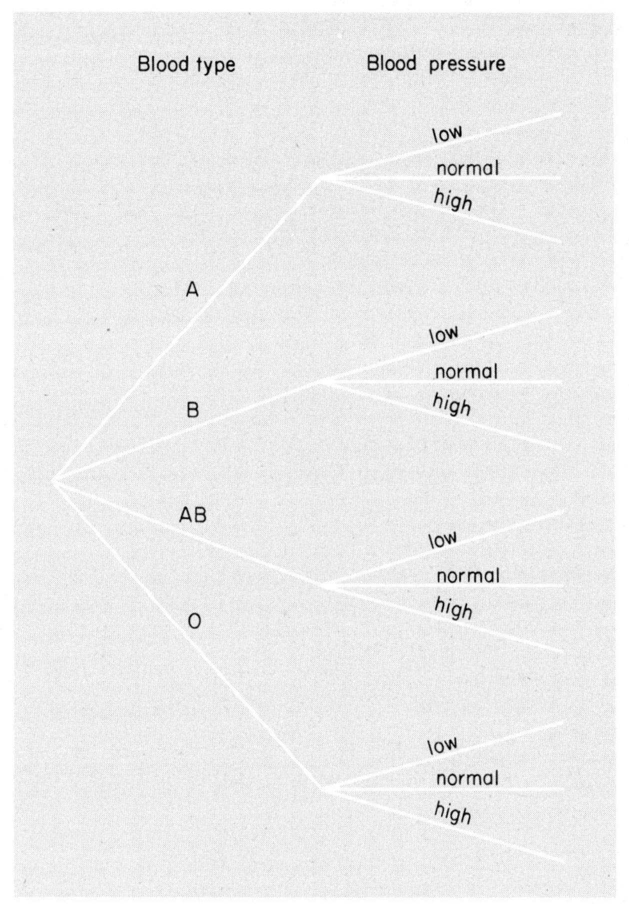

Figure 3.11 Tree diagram for Example 11.

or she can be classified according to blood pressure. In fact, the example illustrates the following general rule:

> **If a selection consists of two separate steps, of which the first can be made in m ways and for each of these the second can be made in n ways, then the whole selection can be made in $m \cdot n$ ways.**

This expresses what is sometimes called the "multiplication of choices," and to prove it, we have only to draw a tree diagram similar to that of Figure 3.11. First there are m branches corresponding to the possibilities in the first step, and then there are n branches emanating from each of these branches to represent the possibilities in the other step. This leads to $m \cdot n$ paths along the branches of the tree diagram, and hence to $m \cdot n$ possibilities.

Example 12. If a restaurant offers 9 different desserts, which it serves with coffee, Sanka, tea, milk, or hot chocolate, there are $9 \cdot 5 = 45$ different ways in which one can order a dessert and a drink. Also, if a professor has the choice of 22 different texts for his elementary statistics class and 8 different texts for his advanced statistics class, there are altogether $22 \cdot 8 = 176$ different ways in which he can choose the two texts.

By using appropriate tree diagrams, we can easily generalize the above rule for the "multiplication of choices" so it will apply to selections involving more than two steps. For k steps, where k is a positive integer, we get

If a selection consists of k separate steps, of which the first can be made in n_1 ways, for each of these the second can be made in n_2 ways, . . . , and for each of these the kth can be made in n_k ways, then the whole selection can be made in $n_1 \cdot n_2 \cdot \ldots \cdot n_k$ ways.

Thus, we simply keep multiplying the number of ways in which each selection can be made.

Example 13. If a new-car buyer has the choice of five body styles, four different engines, and twelve colors, he can make a choice in $5 \cdot 4 \cdot 12 = 240$ different ways. Furthermore, if he has the option of choosing a car with or without air conditioning, with or without an automatic transmission, with or without bucket seats, and with or without power brakes, the total number of ways in which he can make a choice becomes $5 \cdot 4 \cdot 12 \cdot 2 \cdot 2 \cdot 2 \cdot 2 = 3{,}840$.

Example 14. If a true–false test consists of 20 questions, there are

$$2 \cdot 2 \cdot 2 \cdot 2 \cdot \ldots \cdot 2 = 2^{20} = 1{,}048{,}576$$

different ways in which it can be marked—unfortunately, for the student, only one of these corresponds to the case where *all* the questions are answered correctly.

Although the rule for the "multiplication of choices" made it easy to find the answers in Examples 13 and 14, it must be used with some caution; it could not have been used, for instance, in Example 10.

The rule of the preceding paragraph is often applied when *several selections are made from one and the same set, and the order in which they are made is of importance.*

Example 15. This would be the case, for example, if the judges had to select the winner, the first runner-up, and the second runner-up from among the 10 finalists of a beauty contest, or if the 62 members of a union had to elect a president, a vice-president, and a secretary-treasurer. In the first case, the judges could make their selection in $10 \cdot 9 \cdot 8 = 720$ different ways, for *after each choice there is one less to choose from for the next selection.* Correspondingly, the 62 union members could select their three officers in $62 \cdot 61 \cdot 60 = 226{,}920$ different ways.

In general, if r objects are selected from a set of n objects, any particular arrangement of these r objects is referred to as a **permutation**. For instance, 4 2 1 3 is a permutation (a particular ordered arrangement) of the first four positive integers, and

Yankees, Orioles, Brewers, Red Sox

and

Tigers, Indians, Yankees, Brewers

are two different permutations (particular ordered arrangements) of four of the six baseball teams in the Eastern Division of the American League.

Example 16. If we were asked to list *all possible permutations* of two of the five vowels a, e, i, o, and u, the answer would be

ae ai ao au ei eo eu io iu ou

ea ia oa ua ie oe ue oi ui uo

So far as the counting of possible permutations is concerned, direct application of the rule on page 72 leads to the following result*:

The total number of permutations of r objects selected from a set of n objects is

$$n(n-1)(n-2) \cdot \ldots \cdot (n-r+1)$$

*The following are some of the symbols used in other books to denote the number of permutations of r objects selected from a set of n objects: ${}_nP_r$, $P(n, r)$, P_r^n, and $(n)_r$.

To prove this formula we have only to observe that the first selection is made from the whole set of n objects, the second selection is made from the $n - 1$ objects which remain after the first selection has been made, the third selection is made from the $n - 2$ objects which remain after the first two selections have been made, . . . , and the rth and final selection is made from the

$$n - (r - 1) = n - r + 1$$

objects which remain after the first $r - 1$ selections have been made.

Since products of consecutive integers arise in many problems dealing with permutations and other kinds of special arrangements, we write $5 \cdot 4 \cdot 3 \cdot 2 \cdot 1$, for instance, as $5!$ and refer to it as "5 factorial." More generally, $n!$, which we call **"n factorial,"** denotes the product

$$n(n - 1)(n - 2) \cdot \ldots \cdot 3 \cdot 2 \cdot 1$$

for any positive integer n. Thus, $1! = 1$, $2! = 2 \cdot 1 = 2$, $3! = 3 \cdot 2 \cdot 1 = 6$, $4! = 4 \cdot 3 \cdot 2 \cdot 1 = 24$, $5! = 5 \cdot 4 \cdot 3 \cdot 2 \cdot 1 = 120$, $6! = 6 \cdot 5 \cdot 4 \cdot 3 \cdot 2 \cdot 1 = 720$, and so on. To make certain formulas apply more generally (see Exercise 24 on page 80), it is customary also to let $0! = 1$ by definition. Using the factorial notation, we can now say that

> **The number of permutations of n objects taken all together is $n!$.**

This is simply a special case of the formula on page 73 with $r = n$, so that the last factor becomes $n - r + 1 = n - n + 1 = 1$.

Example 17. For instance, the number of ways in which eight instructors can be assigned to eight sections of a course in Freshman English is $8! = 40{,}320$, and the number of ways in which the starting eleven of a professional football team can be introduced to the public before a game is $11! = 39{,}916{,}800$.

In this example we did not actually calculate the values of $8!$ and $11!$, but looked them up in Table VII at the end of the book.

To illustrate the concept of a sample space in the beginning of this chapter, we referred to the one which consists of all the possible ways in which a zoologist can choose 3 of his 24 guinea pigs for an experiment. As we indicated on page 56, there are 2,024 possible ways, but this is *not* the number of permutations of 3 objects selected from a set of 24 objects. If we were interested also in the order in which the three guinea pigs are selected (say, to be given different dosages of an experimental medication), the answer would be $24 \cdot 23 \cdot 22 = 12{,}144$, but each possible set of 3 guinea pigs would then be counted *six times*. For instance, if D, L, and T are the letters painted (for purposes of identification) on 3 of the guinea pigs, then

DLT DTL LTD LDT TDL TLD

constitute *different permutations*, but they all represent the same sample of 3 guinea pigs. To go from the number of permutations to the number of different samples, we shall thus have to divide by 6, and we get $\frac{12{,}144}{6} =$ 2,024. Note that this is the answer we gave on page 56.

The number of different samples (or subsets) of r objects which can be chosen from a set of n objects is generally referred to as the number of **combinations** of r objects selected from a set of n objects—thus, *combinations differ from permutations insofar as we disregard the order in which the selection is made.* To obtain a formula for the number of combinations of r objects taken from a set of n objects, we have only to observe that any r objects can be arranged in $r!$ different ways; in other words, *the $r!$ permutations of any r objects count only as one combination.* Hence, the $n(n-1)(n-2) \cdot \ldots \cdot (n-r+1)$ permutations of r objects selected from a set of n objects contain each combination $r!$ times, and it follows that we obtain the number of combinations by dividing the number of permutations by $r!$. Thus, we arrive at the following rule:

The number of ways in which a sample of r objects can be selected from a set of n objects, namely, the number of combinations of r objects selected from a set of n objects, is given by

$$\frac{n(n-1)(n-2) \cdot \ldots \cdot (n-r+1)}{r!}$$

Symbolically, we shall let $\binom{n}{r}$ denote the number of combinations of r objects selected from a set of n objects, and, for reasons to be explained in Exercise 25 on page 80, these quantities are also referred to as **binomial coefficients.*** The values of binomial coefficients for $n = 2$ through $n = 20$

*The following are some alternative symbols used to denote the number of combinations of r objects selected from a set of n objects: ${}_nC_r$, $C(n, r)$, and C_r^n.

may be obtained from Table VIII at the end of the book; otherwise, an easy way of determining binomial coefficients is indicated in Exercise 26 on page 80. The above formula for the number of combinations and the formula for the number of permutations on page 73 do *not* hold for $r = 0$; however, to make alternative formulas (see Exercise 24 on page 80) apply in general, it is customary to let both of these quantities be equal to 1.

The following are some straightforward applications of the formula for the number of combinations:

Example 18. The number of ways in which a person can invite 4 of his 14 closest friends to a party is

$$\binom{14}{4} = \frac{14 \cdot 13 \cdot 12 \cdot 11}{4!} = 1{,}001$$

and the number of ways in which a social scientist can select 6 of the 20 largest cities in the United States to be included in a survey is

$$\binom{20}{6} = \frac{20 \cdot 19 \cdot 18 \cdot 17 \cdot 16 \cdot 15}{6!} = 38{,}760$$

Note that both of these answers could have been looked up in Table VIII.

Example 2 (Continued). The number of ways in which the dean of the junior college can choose 2 of his 84 faculty members to serve as advisors to the political science club is

$$\binom{84}{2} = \frac{84 \cdot 83}{2!} = 3{,}486$$

Unless Table VIII is used in problems like these, it will always help to cancel as many factors as possible before performing any multiplications.

Exercises

1. Suppose that in a baseball World Series (in which the winner is the first team to win four games) the National League champion leads the American League champion three games to two. Construct a tree diagram to show the number of ways in which these teams may win or lose the remaining game or games.

2. To cut down on his television viewing, a child is allowed to watch *at most* two cartoon shows per day. Construct a tree diagram to determine the number of ways the child can watch cartoon shows on two consecutive days. Also, use the tree diagram to read off the number of possibilities in which the child watches, respectively, 0, 1, 2, 3, or 4 cartoon shows.

3. There are four routes, A, B, C, and D, between a businessman's home and his office, but route B is one-way so that he cannot take it on the way to work, and route D is one-way so that he cannot take it on the way home.

a. Draw a tree diagram showing the various ways he can go to and from work.

b. Draw a tree diagram showing the various ways he can go to and from work, given that he never goes by the same route both ways.

4. A student can study 0, 1, or 2 hours for a Spanish test on any given night. Construct a tree diagram to show that there are six different ways in which he or she can study altogether 4 hours for the test on three consecutive nights.

5. In a traffic court, violators are classified according to whether or not they are properly licensed, whether their violations are major or minor, and whether or not they have committed any other violations in the preceding year.

a. Construct a tree diagram showing the various ways in which this traffic court classifies violators.

b. If there are 20 violators in each of the eight categories of (a) and the judge gives each violator who is not properly licensed a stern lecture, how many of the violators will receive a stern lecture?

c. If, furthermore, the judge gives a $50 fine to everybody who has committed a major violation and/or another violation in the preceding year, how many of the violators will receive a $50 fine?

d. How many of the violators will receive a stern lecture from the judge as well as a $50 fine?

6. In a primary election, there are three candidates for mayor and four candidates for city treasurer. In how many different ways can a person vote for one of the candidates for each office?

7. In a political science survey, voters are classified into six categories according to income and into five categories according to the extent of their education. In how many different ways can a voter thus be classified?

8. A coin dealer grades coins with regard to their *condition* as good, fine, very fine, extremely fine, and uncirculated, and with regard to their *rarity* as common, scarce, rare, very rare, and extremely rare.

a. In how many different ways can this dealer grade one of his coins with regard to condition and rarity?

b. In how many different ways can this dealer grade a coin which is at least very fine and at least very rare?

9. In an optics kit there are five concave lenses, four convex lenses, and two prisms. In how many different ways can one choose a concave lens, a convex lens, and a prism from this kit?

10. A psychologist preparing three-letter nonsense words for use in a memory test chooses the first letter from among the consonants q, w, x, and z; the second letter from among the vowels e, i, and u; and the third letter from among the consonants c, f, p, and v.

a. How many different three-letter nonsense words can he construct?

b. How many of these nonsense words will begin either with the letter x or the letter z?

c. How many of these nonsense words will end with the letter p?

11. In an ice cream parlor a customer can order a sundae with any one of 12 different kinds of ice cream, with any one of five different kinds of syrup, with or without whipped cream, and with or without nuts. In how many different ways can one order a sundae at this ice cream parlor?

12. A questionnaire sent through the mail as part of a market study consists of eight questions, each of which can be answered in three different ways. In how many different ways can a person answer the eight questions on this questionnaire?

13. In a biology laboratory there are 12 microscopes, and an assistant keeps a chart from which it can be seen whether or not each microscope is being used. In how many different ways can this chart be marked?

14. THE NUMBER OF SUBSETS OF A SET When we speak of the number of subsets of a set, we always include the *empty set* which has no elements at all, and also the whole set, itself. To select a subset from any given set we must decide for each element whether or not it is to be included, and for a set with n elements there are, therefore,

$$\underbrace{2 \cdot 2 \cdot 2 \cdot 2 \cdot \ldots \cdot 2 \cdot 2}_{n \text{ factors}} = 2^n$$

possibilities, namely, 2^n different subsets.

a. List the $2^3 = 8$ different subsets of the set which consists of the three foreign car manufacturers Volvo, Fiat, and BMW.

b. List the $2^4 = 16$ different subsets of the set which consists of the following professions: medicine, law, teaching, and theology.

c. In how many ways can one select a subset from the set which consists of the eight congressmen from Alabama, if it is to contain at least one of these congressmen?

d. In how many different ways can one select a subset from the set which consists of the ten best-sellers listed in a Sunday paper?

15. If there are only three horses in a race, Beauty, Cactus, and Princess, and there are no ties (for win, place, or show), list the six possible ways in which they can finish the race.

16. On page 73 we gave two permutations of four of the six baseball teams in the Eastern Division of the American League. How many different permutations like these are there altogether?

17. The price of a European tour includes four stopovers to be selected from among ten cities. In how many ways can one plan such a tour if

a. the order of the stopovers matters;

b. the order of the stopovers does not matter?

18. In how many different ways can a student arrange his six textbooks on a shelf?

19. In how many ways can a television director schedule a sponsor's five different commercials during the five time slots allocated to commercials during the telecast of the first period of a hockey game?

20. In Example 14 we determined the number of ways in which a student can mark a true–false test which consists of 20 questions. To continue with this problem, in how many ways can a student mark this test and get

a. 3 right and 17 wrong;

b. 5 right and 15 wrong;

c. 8 right and 12 wrong.

Check your answers in Table VIII.

21. Without referring to Table VIII, find

a. the number of ways in which the IRS can choose 4 of 15 income tax returns for a special audit;

b. the number of ways in which a student can choose 3 of 18 elective subjects;

c. the number of ways in which a motel chain can choose 3 of 13 locations for the construction of new motels;

d. the number of ways in which a 4-man committee can be chosen from among the 16 teachers of a private school.

22. A carton of eggs contains two bad ones. In how many ways can one select three of the eggs so that

a. none of the bad ones is included;

b. both of the bad ones are included;

c. only one of the bad ones is included?

[*Hint:* Use the results of parts (a) and (b) to find the answer to part (c).]

23. To fill a number of vacancies, the personnel manager of a company has to choose three secretaries from among six applicants and two file clerks from among five applicants. What is the total number of ways in which he can make his choice? (*Hint:* Use the rule for the "multiplication of choices.")

24. Letting $0! = 1$, as was pointed out on page 74, show that

a. the formula for the number of permutations on page 73 can be written as $\frac{n!}{(n-r)!}$ for $r = 0, 1, 2, \ldots,$ and n;

b. the formula for the number of combinations on page 75 can be written as $\frac{n!}{(n-r)!\,r!}$ for $r = 0, 1, 2, \ldots,$ and n.

Note that for $r = 0$ both formulas yield the answer 1. Use these formulas and Table VII to rework

c. Exercise 16;

d. part (a) of Exercise 17;

e. part (a) of Exercise 21;

f. part (c) of Exercise 21.

25. The quantities $\binom{n}{r}$ are referred to as *binomial coefficients* because they are, in fact, the coefficients in the binomial expansion of $(a + b)^n$. Verify that this is true for $n = 2$, 3, and 4 by expanding $(a + b)^2$, $(a + b)^3$, and $(a + b)^4$, and comparing the coefficients with the corresponding values of $\binom{n}{r}$ in Table VIII.

26. PASCAL'S TRIANGLE The number of combinations of r objects selected from a set of n objects, namely, the quantities $\binom{n}{r}$, can be determined by means of the following arrangement called *Pascal's triangle*:

$$
\begin{array}{ccccccccc}
 & & & & 1 & & & & \\
 & & & 1 & & 1 & & & \\
 & & 1 & & 2 & & 1 & & \\
 & 1 & & 3 & & 3 & & 1 & \\
1 & & 4 & & 6 & & 4 & & 1 \\
\cdots & \cdots & \cdots & \cdots & \cdots & \cdots & \cdots & \cdots & \cdots
\end{array}
$$

where each row begins with a 1, ends with a 1, and each other entry is the *sum* of the nearest two entries in the row immediately above.

a. Use Table VIII to verify that the *fourth* row of the triangle contains the values of $\binom{3}{r}$ for $r = 0, 1, 2$, and 3.

b. Use Table VIII to verify that the *fifth* row of the triangle contains the values of $\binom{4}{r}$ for $r = 0, 1, 2, 3$, and 4.

c. Construct the *sixth* row of the triangle and verify in Table VIII that it gives the values of $\binom{5}{r}$ for $r = 0, 1, 2, 3, 4$, and 5.

d. Construct the *seventh* row of the triangle and verify in Table VIII that it gives the values of $\binom{6}{r}$ for $r = 0, 1, 2, 3, 4, 5$, and 6.

Probabilities and odds

So far we have studied only *what is possible* in a given situation. In some instances we listed all possibilities and in others we merely determined how many different possibilities there are. *Now we shall go one step further and judge also what is probable and what is improbable.* The most common way of measuring the uncertainties connected with events (say, the market for a new detergent, the outcome of a presidential election, the durability of a new exterior paint, the side effects of a new serum, or the total points we may roll with a pair of dice) is to assign them **probabilities**, or to specify the **odds** at which it would be fair to bet that the events will occur. [As we shall see on page 84, these two concepts (probabilities and odds) are very closely related—in fact, if we know one we can always calculate the other.]

Among the different theories of probability—and there are many—most widely held is the **frequency concept of probability**, according to which the probability of an event is interpreted as the *proportion of the time that such kind of event will occur in the long run.**

Example 19. If we say that there is a probability of 0.82 that a jet from New York to Chicago will arrive on time, this means that such flights arrive on time about 82% of the time. More generally, we say that an event has a probability of, say, 0.90, in the same sense in which we might say that our car will start in cold weather about 90% of the time. *We cannot guarantee what will happen on any one try, the car may start and then it may not, but it would be reasonable to bet $9.00 against $1.00 or 90 cents against a dime* (*namely, at odds of 9 to 1*) *that the car will start at any given try.* This would be "reasonable" or "fair," because we would win $1.00 (or a dime) about 90% of the time, lose $9.00 (or 90 cents) about 10% of the time, and, hence, could expect to break even in the long run.

*Another popular concept of probability, based on the assumption that all outcomes are "equally likely," will be mentioned in Chapter 4.

In accordance with the frequency concept of probability, we *estimate* the probability of an event by observing how often (what part of the time) similar events have occurred in the past.

Example 19 (Continued). If airline records show that over a certain period of time 738 of 900 jets from New York to Chicago arrived on time, we *estimate* the probability that any such flight from New York to Chicago (perhaps, the next one) will arrive on time as $\frac{738}{900} = 0.82$. Similarly, if 4,061 of 5,562 voters interviewed say that they are opposed to school financing through property taxes, we *estimate* the probability that any voter will be opposed to school financing through property taxes as $\frac{4{,}061}{5{,}562}$, or approximately 0.73.

Since we defined probabilities in terms of what happens to similar events in the long run, let us examine briefly whether it is at all meaningful to talk about the probability of an event which can happen only once.

Example 20. For instance, can we ask for the probability that a well-known football player's broken leg will heal within a month, or the probability that a certain major-party candidate will win an up-coming gubernatorial election? If we put ourselves in the position of the football player's doctor, we could check medical records, observe that 34% of the time such fractures have healed within a month, and apply this figure to the football player's leg. This may not be of much comfort to this football player or his coach, but it does give a meaning to a probability statement about his broken leg—the probability that it will heal within a month is 0.34.

It follows from this example that *when we make a probability statement about a specific (nonrepeatable) event, the frequency concept of probability leaves us no choice but to refer to a set of similar events.* As can well be imagined, however, this may easily lead to complications, since the choice of "similar" events is often neither obvious nor straightforward.

Example 20 (Continued). With reference to the football player's leg, we might count as "similar" only those cases where the fracture was in the same (left or right) leg, we might count only those cases in which the patients were of the same age as the football player, we might count only those cases in which the patients were athletes, or we might count only those cases in which the patients were also of the same height and weight.

It should be apparent from this example that the choice of "similar" cases is ultimately a matter of personal judgment, and it should be observed that *the more we narrow things down, the less information we will have for estimating the corresponding probabilities.*

Example 20 (Continued). So far as the other illustration is concerned, the one concerning the gubernatorial election, suppose that we ask a person who has conducted a poll "how sure" he is that the given candidate will actually win. If he replies that he is "99% sure," namely, that he assigns the candidate's election a probability of 0.99, this is *not* meant to imply that the candidate would win 99% of the time if he ran for office a great many times. No, he means that *his conclusion is based on a method which (in the long run) will "work" 99% of the time.*

It is in the sense of this last illustration that many of the probabilities which we use in statistics to express our faith in predictions or decisions are simply *success ratios* that apply to the methods we have employed.

An alternative point of view, which is currently gaining favor, is to interpret probabilities as **personal** or **subjective** evaluations. Such probabilities express the *strength of one's belief* with regard to the uncertainties that are involved, and they apply especially when there is *very little direct evidence*, so that there really is no choice but to consider collateral (indirect) information, "educated guesses," and perhaps intuition and other subjective factors.

Example 21. For instance, a businessman may "feel" that the **odds** for the success of a new venture, say, a new restaurant, are 3 to 2. This means that he would be willing to bet (or consider it *fair* to bet) $300 against $200, or perhaps $3,000 against $2,000, that the new restaurant will be a success. In this way he would be expressing the *strength of his belief* regarding the uncertainties connected with the success of the restaurant, and it may be based on business conditions in general, the opinion of an expert consultant, or his own subjective evaluation of the whole situation, including, perhaps, a small dose of optimism.

Regardless of how we interpret probabilities and odds, subjectively or objectively (namely, as personal judgments or in terms of frequencies or proportions), the mathematical relationship between the two is always the same. It is given by the following rule:

> **If somebody considers it fair or equitable to bet a dollars against b dollars that a given event will occur, he is, in fact, assigning the event the probability $\frac{a}{a+b}$.**

Actually, the quantities a and b need not be in dollars, but it is usually easiest to think of them as cash bets.

Example 21 (Continued). The businessman who is willing to give odds of 3 to 2 that the new restaurant will be a success is actually assigning its success a probability of $\frac{3}{3+2} = 0.60$.

Example 22. Also, if a student feels that the odds are 7 to 2 that she *will not* get accepted to the college of her choice, then the probability (her personal probability) that she *will not* get accepted is $\frac{7}{7+2} = \frac{7}{9}$; correspondingly, the odds that she *will* get accepted are 2 to 7, the probability that she *will* get accepted is $\frac{2}{2+7} = \frac{2}{9}$, and it should be noted that the sum of these two probabilities is $\frac{7}{9} + \frac{2}{9} = 1$.

To illustrate how probabilities are converted into odds, let us refer back to the example on page 81, where we dealt with the question of whether or not we could start our car. As we pointed out at the time, the probability of 0.90 implies that we should win (namely, get the car started in cold weather) about 90% of the time, and lose (fail to get it started) about 10% of the time. In other words, *we should win about nine times as often as we should lose*, and the proper odds are 9 to 1. In general,

> **If the probability of an event is p, then the odds for its occurrence are p to $1-p$ and the odds against its occurrence are $1-p$ to p.**

This result can also be obtained by showing *algebraically* that $p = \frac{a}{a+b}$ leads to $\frac{a}{b} = \frac{p}{1-p}$.

Example 23. If the probability that an item left on a bus will never be claimed is 0.15, the odds are 0.15 to 1 − 0.15 = 0.85, or 3 to 17, that such an item left on a bus will never be claimed (and they are 17 to 3 that it will be claimed). Also, if the probability that a concert will be sold out is 0.25, the odds are 0.25 to 1 − 0.25 = 0.75, or 1 to 3, that the concert will be sold out (and they are 3 to 1 that it will not be sold out).

Note that in both parts of this example we followed the common practice of quoting odds as *ratios of positive integers having no common factors.*

Exercises

1. If 779 of 1,025 visitors to a national park said that they would like to return, estimate the probability that any one visitor to this national park would like to return for another visit.

2. In a sample of 400 cans of mixed nuts (taken from a very large shipment), 128 contained no pecans. Estimate the probability that there will be no pecans in a can of mixed nuts selected from this shipment.

3. Among 858 armed robberies committed in a certain city, 143 were never solved. Estimate

a. the probability that an armed robbery committed in this city will not be solved;

b. the odds that an armed robbery committed in this city will not be solved.

4. If a tennis player has won 156 of his last 195 matches and we do not know who his next opponent is going to be, estimate

a. the probability that he will win his next match;

b. the odds that he will win his next match.

5. If 42 of 525 rockets failed to function properly, estimate

a. the probability that such a rocket will fail to function properly;

b. the probability that such a rocket will function properly;

c. the odds that such a rocket will fail to function properly.

6. In a sample of 130 lakes and streams in a certain state, 91 were found to contain excessive amounts of pollutants. Estimate

a. the probability that any one of the lakes or streams in this state will contain excessive pollutants;

b. the probability that any one lake or stream in this state will not contain excessive pollutants;

c. the odds that any one lake or stream in this state will contain excessive pollutants.

Also, if we offered to bet a fisherman that the next lake or stream to which he comes will be polluted, who would be favored if

d. we offered to bet our \$35 against his \$28?

7. If a college senior feels that 2 to 1 are fair odds that he will be admitted to law school, what is his personal probability that he will be admitted to law school?

8. If somebody claims that the odds are 31 to 19 that a certain bus will be on time, what probability does he or she assign to the bus being on time?

9. If the odds are 16 to 9 that a newly hired airline stewardess will get married before she has been on the job for two years, what is the probability that this will be the case?

10. If an oil company executive feels that the odds are 13 to 7 that a drilling operation will be successful, what is the probability that this will not be the case?

11. If a sportswriter feels that the odds are *at least* 4 to 1 that the home team will win an upcoming football game, what can we say about the probability he is thus assigning

a. to the home team's winning the game;

b. to the home team's losing or tying the game?

12. Suppose that a student is willing to bet \$4 against \$1, but not \$5 against \$1, that he will get a passing grade in a physics course. What does this tell us about the personal probability that he assigns to his getting a passing grade in the course? (*Hint*: The answer should read "at least . . . but less than. . . .")

13. A television executive is willing to bet \$400 against \$4,600, but not \$600 against \$4,400, that a new daytime serial will be a success. What does this tell us about the executive's subjective probability that the new serial will *not* be a success? (*Hint*: The answer should read "greater than . . . but at most. . . .")

14. Discuss the following assertion: When the weatherman says that the probability for rain tomorrow is 0.30 (namely, that there is a "30% chance of rain"), whatever happens tomorrow cannot prove him right or wrong.

15. Explain how one might conceivably assign a probability to the truth of testimony given at a trial, using

a. the frequency concept of probability;

b. subjective probabilities.

16. Is it possible for two persons using the frequency concept of probability to arrive at correct, yet different, probabilities regarding the potential success of a new book?

Expectations and decisions

If an insurance agent tells us that in the United States a 45-year-old woman can expect to live 33 more years, this does not mean that anyone really expects a 45-year-old woman to live until her 78th birthday and then pass away the next day. Similarly, if we read that a person living in the United States can expect to eat 10.4 pounds of cheese and 324 eggs a year, or that a child in the age group from 6 to 16 can expect to visit a dentist 1.9 times a year, it must be obvious that the word "expect" is not being used in its colloquial sense. A child cannot very well go to the dentist 1.9 times, and it would be surprising, indeed, if we found somebody who has actually eaten 10.4 pounds of cheese and 324 eggs in a given year. So far as the first statement is concerned, some 45-year-old women will live another 12 years, some will live another 25 years, some will live another 38 years, . . . , and the life expectancy of "33 more years" will have to be interpreted as an average, namely, as a **mathematical expectation.**

Originally, the concept of a mathematical expectation arose in connection with games of chance, and in its simplest form it is *the product of the probability that a player will win and the amount he stands to win.*

Example 24. If we stand to receive \$10 if a balanced coin comes up *tails* and nothing if it comes up *heads*, our mathematical expectation is $10 \cdot \frac{1}{2} = \$5$. Similarly, if we buy one of 1,000 raffle tickets issued for a prize (say, a color television set) worth \$490, the probability for each ticket is $\frac{1}{1{,}000}$ and our mathematical expectation is $490\left(\frac{1}{1{,}000}\right) =$ \$0.49, or at least the equivalent of this amount in merchandise. Thus, it would be unwise to pay more than 49 cents for the ticket, unless, of course, the proceeds of the raffle went to a worthy cause. Note that in connection with this raffle we could also have argued that 999 of the tickets will not pay anything at all, one of the tickets will pay \$490, so that altogether the 1,000 tickets will pay \$490, or *on the average* 49 cents per ticket—*this is the mathematical expectation for each ticket.*

So far, we have considered only problems in which there was a single "payoff," namely, a single prize or a single payment. To see how the concept of a mathematical expectation might be generalized, let us consider the following modification of the raffle of the last example.

Example 24 (Continued). Suppose that the raffle is changed so that there is also a second prize (say, a tape recorder) worth \$150 and a third prize (say, a clock-radio) worth \$40. We can then argue that

997 of the tickets will not pay anything at all, one will pay $490 (in merchandise), another will pay $150 (in merchandise), and a third will pay $40 (in merchandise); altogether, the 1,000 tickets will thus pay 490 + 150 + 40 = $680 (in merchandise), or *on the average* $0.68 per ticket. *As before, this is the mathematical expectation for each ticket.* Looking at the problem in a different way, we could argue that if the raffle were repeated many times, we would hold a losing ticket about 99.7% of the time and win each of the three prizes about 0.1% of the time. On the average we would thus win

$$0(0.997) + 490(0.001) + 150(0.001) + 40(0.001) = \$0.68$$

which is the sum of the products obtained by multiplying each amount by the corresponding proportion or probability.

Generalizing from this example, let us now make the following definition:

If the probabilities of obtaining the amounts $a_1, a_2, \ldots,$ or a_k are, respectively, $p_1, p_2, \ldots,$ and p_k, then the mathematical expectation is

$$a_1p_1 + a_2p_2 + \ldots + a_kp_k$$

Each amount is multiplied by the corresponding probability, and the mathematical expectation, E, is given by the sum of all these products; symbolically, we could thus write the formula for a mathematical expectation as $E = \sum a \cdot p$. So far as the a's are concerned, it is important to keep in mind that they are *positive* when they represent profits, winnings, or gains (namely, amounts which we receive), and that they are *negative* when they represent losses, penalties, or deficits (namely, amounts which we have to pay).

Example 25. If we bet $2 on the flip of a coin (that is, we either win $2 or lose $2 depending on the outcome), the amounts a_1 and a_2 are +2 and −2, the probabilities p_1 and p_2 are both equal to $\frac{1}{2}$, and the mathematical expectation is

$$E = 2 \cdot \tfrac{1}{2} + (-2) \cdot \tfrac{1}{2} = 0$$

Indeed, that is what the mathematical expectation should be in an **equitable game**, namely, in a game which does not favor either player.

Example 26. To consider another example, suppose that a businessman is interested in a piece of property for which the probabilities are 0.22, 0.36, 0.28, and 0.14 that he will sell it at a profit of $2,500, that he will sell it at a profit of $1,000, that he will break even, or that he will sell it at a loss of $1,500. If we substitute all these figures into the formula for E, we get

$$\begin{aligned} E &= 2{,}500(0.22) + 1{,}000(0.36) + 0(0.28) + (-1{,}500)(0.14) \\ &= \$700 \end{aligned}$$

and this is the *profit* he can expect to make.

Although we referred to the quantities $a_1, a_2, \ldots,$ and a_k in the formula for a mathematical expectation as "amounts," they need not be *cash* winnings, losses, penalties, or rewards. When we said on page 87 that a child in the age group from 6 to 16 goes to the dentist 1.9 times a year, we were actually referring to a result which was obtained by multiplying 0, 1, 2, 3, 4, . . . , by the respective probabilities that a child in this age group will visit a dentist that many times a year.

Example 27. To illustrate, suppose that weather bureau records show that in a certain county the probabilities for 0, 1, 2, 3, 4, 5, 6, or 7 hurricanes in any given year are, respectively, 0.09, 0.22, 0.26, 0.21, 0.13, 0.06, 0.02, and 0.01. Thus, we find that people in this county can expect

$$\begin{aligned} E &= 0(0.09) + 1(0.22) + 2(0.26) + 3(0.21) + 4(0.13) \\ &\quad + 5(0.06) + 6(0.02) + 7(0.01) \\ &= 2.38 \end{aligned}$$

hurricanes per year.

A decision problem

When we are faced with uncertainties, mathematical expectations can often be used to great advantage in making decisions. Generally speaking, if we have to make a choice among several alternatives, it is considered "rational" to select the one with the "most promising" mathematical expectation: the one which *maximizes expected profits, minimizes expected costs, maximizes tax advantages, minimizes expected losses,* and so on. Although this approach to decision making has great intuitive appeal and sounds very logical, it is not without complications.

Example 28. To illustrate some of these difficulties, consider the problem faced by the director of a research laboratory, who must decide whether to expand his facilities now or wait at least another year. It all depends on whether or not he will get a certain government contract, so let us suppose that he has the following information. If the facilities are expanded right away and he gets the government contract, his operations will show a profit of \$287,000; if the facilities are expanded right away and he does not get the government contract, his operations will show a deficit of \$70,000; if the facilities are not expanded right away and he gets the government contract, his operations will show a profit of \$140,000; and if the facilities are not expanded right away and he does not get the government contract, his operations will show a profit of only \$14,000. Schematically, all this information can be presented as in the following table:

	Gets Government Contract	*Does Not Get Government Contract*
Expand Facilities Right Away	\$287,000	−\$70,000
Delay Expansion of Facilities	\$140,000	\$14,000

Evidently, it will be advantageous to expand the facilities right away only if he gets the government contract, and his decision will therefore have to depend on the chances that this will be the case. Suppose, for instance, that the director of the research laboratory judges (on the basis of his past dealings with the government) that the odds are 2 to 1 that he will *not* get the contract. According to the rule on page 84, this means that he assigns his getting the contract a probability of $\frac{1}{1+2} = \frac{1}{3}$, and his *not* getting the contract a probability of $\frac{2}{2+1} = \frac{2}{3}$. He can then argue that *if the facilities are expanded right away, the expected profit is*

$$287{,}000 \cdot \tfrac{1}{3} + (-70{,}000) \cdot \tfrac{2}{3} = \$49{,}000$$

and *if the expansion of the facilities is delayed, the expected profit is*

$$140{,}000 \cdot \tfrac{1}{3} + 14{,}000 \cdot \tfrac{2}{3} = \$56{,}000$$

Now, since an expected profit of \$56,000 is obviously preferable to an expected profit of only \$49,000, it stands to reason that the director of the research laboratory should delay the expansion of the facilities. *Or should he?* What if he had been hasty in assessing the odds as 2 to 1 against his getting the government contract? What if the odds should have been 3 to 2 instead of 2 to 1? In that case the expected profit would be

$$287{,}000 \cdot \frac{2}{2+3} + (-70{,}000) \cdot \frac{3}{3+2} = \$72{,}800$$

if the facilities are expanded right away

$$140{,}000 \cdot \tfrac{2}{5} + 14{,}000 \cdot \tfrac{3}{5} = \$64{,}400$$

if the expansion of the facilities is delayed, and it can be seen that the decision would be reversed.

This serves to illustrate that if mathematical expectations are to be used for making decisions, *it is essential to know the values of all relevant probabilities.* Not only that, but *we must also know the correct values of the "payoffs" which are associated with the various possibilities.*

Example 28 (Continued). For instance, what should the director of the research laboratory do if the \$287,000 profit (corresponding to his expanding the facilities right away and getting the government contract) has to be changed to \$280,000 or \$350,000; or if the deficit of \$70,000 (corresponding to his expanding the facilities right away and *not* getting the government contract) has to be changed to \$52,500 or \$105,000? It will be left to the reader to answer some of these questions in Exercise 14 on page 94.

The way in which we have studied the problem faced by the director of the research laboratory is referred to as a **Bayesian analysis**. In this kind of analysis, probabilities are assigned to the alternatives about which uncertainties exist (the "states of nature," which in our example were his getting or not getting the government contract); then we choose the alternative which promises the greatest expected profit or the smallest expected loss.

Exercises

1. If a charitable organization raises funds by selling 2,000 raffle tickets for a vacation trip worth $320, what is the mathematical expectation of a person who buys one of these raffle tickets?

2. Repeat Exercise 1 with the modification that there is also a second prize of a painting worth $100.

3. As part of a promotional scheme, the manufacturers of a new breakfast food offer a prize of $75,000 to some person willing to try the new product (distributed without charge) and send in his or her name on the label. The winner is to be drawn at random from all entries in front of a large television audience.

- **a.** What is each entrant's mathematical expectation, if 1,500,000 persons send in their names?
- **b.** Was it worth the 10 cents postage it cost to send in one's name?

4. In a winner-take-all bowling tournament among four professional bowlers, the prize money is $24,000. If one of these bowlers figures that the odds against his winning are 5 to 1, what is his mathematical expectation? Would he be better off if he made a secret agreement with the other bowlers to split the prize money evenly regardless of who wins?

5. A student's parents promise him a gift of $25 if he gets an A in a course in psychology, $10 if he gets a B, and otherwise no reward. What is the student's mathematical expectation if the probability of his getting an A is 0.32 and the probability of his getting a B is 0.44?

6. A wage negotiator of a labor union feels that the odds are 3 to 1 that the members of the union will get a raise of 80 cents in their hourly wage, the odds are 17 to 3 *against* their getting a raise of 40 cents in their hourly wage, and the odds are 9 to 1 *against* their getting no raise at all.

- **a.** Find the corresponding probabilities that they will get, respectively, an 80 cents raise, a 40 cents raise, or no raise in their hourly wage.
- **b.** What is the corresponding *expected* raise in their hourly wage?

7. A recent college graduate is faced with a decision which cannot wait, namely, that of accepting or rejecting a job paying $8,400 a year. What can we say about the probability he assigns to his only other prospect, a job paying $10,500 a year, if he decides to reject the $8,400-a-year job? (*Hint*: The answer should read "the probability is greater than . . . .")

8. To handle a liability suit, a lawyer has to decide whether to charge a straight fee of $400 or a contingent fee of $1,600 which he will get only if his client wins. What does the lawyer think about his client's chances if

a. he cannot make up his mind;
b. he prefers the straight fee of $400;
c. he prefers the contingent fee of $1,600?

9. An importer is offered a shipment of jade jewelry for $5,500, and the probabilities that he will be able to sell it for $8,000, $7,500, $7,000, or $6,500, are, respectively, 0.25, 0.46, 0.19, and 0.10.

a. For how much can he expect to sell this shipment of jade jewelry?
b. What is his expected gross profit?

10. The following table gives the probabilities that a woman who enters a given dress shop will buy 0, 1, 2, 3, or 4 dresses:

Number of Dresses	0	1	2	3	4
Probability	0.11	0.41	0.33	0.12	0.03

How many dresses can a woman entering this shop be expected to buy?

11. The probabilities that on any one day the office of an airline at Sky Harbor Airport will get 0, 1, 2, 3, 4, 5, 6, 7, or 8 complaints about the handling of luggage are, respectively, 0.05, 0.17, 0.25, 0.19, 0.17, 0.10, 0.04, 0.02, and 0.01. How many complaints about the handling of luggage can they expect per day?

12. The management of a mining company must decide whether to continue an operation at a certain location. If they continue and are successful, they will make a profit of $1,500,000; if they continue and are not successful, they will lose $900,000; if they do not continue but would have been successful, they will lose $600,000 (for competitive reasons); and if they do not continue and would not have been successful anyhow, they will profit by $150,000 (because funds allocated to the operation remain unspent).

a. Present all this information in a table like that on page 90.
b. What decision would maximize the company's expected gain if the odds against success are 3 to 2?
c. Repeat part (b) when the odds against success are 4 to 1.
d. Repeat part (b) when the odds against success are 2 to 1.

13. Tom is starting out to meet his friends at the beach, but he cannot remember whether he was supposed to meet them in La Jolla or in Mission Beach, which are 4 miles apart. He lives 11 miles from the spot where he would meet them in La Jolla and 9 miles from the spot where he would meet them in Mission Beach.

a. Construct a table (similar to the one on page 90) which shows the number of miles Tom has to drive to meet his friends depending on where he goes first and where they are actually supposed to meet.

b. Where should he go first if he wants to minimize the distance he can expect to drive to meet his friends and he feels that the odds are 5 to 1 that they are in La Jolla?

c. Where should he go first if he wants to minimize the distance he can expect to drive to meet his friends and he feels that the odds are 2 to 1 that they are in La Jolla?

d. Show that it does not matter where he goes first when the odds are 3 to 1 that his friends are in La Jolla.

14. With reference to Example 28 and the table on page 90, what decision should the director of the research laboratory make if

a. the $287,000 profit is replaced by $350,000 and the probability of his getting the government contract is $\frac{1}{3}$;

b. the $70,000 deficit is replaced by a deficit of $105,000 and the probability of his getting the government contract is $\frac{2}{5}$?

4

Some rules of probability

Introduction

Some familiarity with the rules of probability can be of value to anyone, and not only to gamblers for whose "benefit" the theory was originally developed. Businessmen must buy merchandise without knowing for sure whether it will sell, military strategists must commit men and equipment to the hazards of battle, we travel by car or plane without knowing for certain whether we will reach our destination, doctors risk their lives in combatting disease, important messages are sent by mail without any assurance that they will be delivered on time, etc., and in each case the study of probability makes it possible, or at least easier, to "live with the corresponding uncertainties."

In the study of probability there are three basic questions. First, there is the question of *what we mean* when we say that the probability of an event is, say, 0.90, 0.74, or 0.05; then there is the question of *how the values of probabilities are determined* in actual practice; and finally there is the question of *how the probabilities of simple events can be used to calculate those of more complicated kinds of events.* Having discussed the first two questions briefly in Chapter 3, we shall devote most of this chapter to investigating the third. [So far as the first question is concerned, we stated that if an event is assigned a probability of, say, 0.90, many people interpret this as a *proportion,* namely, the proportion of the time that such events will occur in the long run—to others it means that they feel *subjectively* that 9 to 1 would be fair odds for betting that the event will occur. Correspondingly, if probabilities are interpreted as "long-run proportions" they are *estimated* in terms of observed proportions, and if they are interpreted subjectively their values are calculated on the basis of the odds at which a person would be willing to bet (or consider it fair to bet) on the corresponding events.]

Some basic rules

To tackle the third question which we listed in the preceding section, let us begin by investigating some of the simpler rules according to which probabilities "behave." First, since probabilities are *measures of uncertainty* and since most of the things we measure are given by nonnegative numbers, let us state the rule that

> **The probability of any event must be a positive real number or zero.**

It is important to note that this rule is in complete agreement with the frequency interpretation as well as the subjective concept of probability. Clearly, proportions are always positive or zero, and so long as the amounts a and b which we bet for and against an event are positive, the probability of the event, $\frac{a}{a+b}$, cannot be negative.

If we think of probabilities in terms of proportions, we are immediately led to several other basic rules. For instance:

> **Probabilities can never exceed 1; a probability of 1 implies that the corresponding event is certain to occur; and a probability of 0 implies that it is certain not to occur.**

Evidently, an event cannot possibly happen more than 100% of the time; an event which happens 100% of the time is certain to occur, and an event which happens 0% of the time is certain *not* to occur. (In actual practice, we also assign a probability of 1 to an event for which we are "practically certain" that it will occur, and a probability of 0 to an event for which we are "practically certain" that it will not occur. Thus, we would assign a probability of 1 to the event that at least one person will make a mistake in calculating his income tax for the year 1978, and we would assign a probability of 0 to the event that there will not be a single automobile accident in the United States during the next Labor Day weekend.) So far as subjective probabilities are concerned, we would be willing to give "better and better" odds when we become more and more certain that an event will occur—say, 100 to 1, 1,000 to 1, or perhaps even 1,000,000 to 1. The corresponding probabilities are $\frac{100}{100+1}$, $\frac{1{,}000}{1{,}000+1}$, and $\frac{1{,}000{,}000}{1{,}000{,}000+1}$ (or approximately 0.99, 0.999, and 0.999999), and it can be seen that *the more certain we are that an event will occur, the closer its probability will be to 1.* A corresponding argument leads to the result that *the more certain we are that an event will not occur, the closer its probability will be to 0.*

The next rule is especially important, and it applies only to events which cannot both occur at the same time, namely, events which are *mutually exclusive*:

If two events are mutually exclusive, the probability that one or the other will occur equals the sum of their probabilities.

For instance, if the probabilities that a student will be studying or playing tennis a given afternoon are, respectively, 0.34 and 0.12, then the probability that she will be doing one or the other is $0.34 + 0.12 = 0.46$. This definitely agrees with the frequency interpretation of probability: If one kind of event occurs 34% of the time, another kind of event occurs 12% of the time, and they cannot both occur at the same time, then one or the other will occur $34 + 12 = 46\%$ of the time. So far as subjective probabilities are concerned, this last rule does not follow from the definition which we gave in Chapter 3, but it is generally imposed as the **consistency criterion**. In other words, if a person's subjective probabilities "behave" in accordance with the rule, he is said to be *consistent*; otherwise, he is said to be *inconsistent* and his probability judgments must be taken with a grain of salt (see Exercises 4 and 5 on page 104).

Following the practice of denoting events with capital letters as in Chapter 3, and writing the probability of event A as $P(A)$, the probability of event B as $P(B)$, . . . , we can express the last rule symbolically as

$$P(A \cup B) = P(A) + P(B)$$

keeping in mind, of course, that it applies only when A and B are mutually exclusive events. This formula is sometimes referred to as the **Special Addition Rule** for probabilities; several other addition rules, namely, more general formulas, will be discussed on pages 98 and 103.

Another basic rule of probability concerns the probability of the *complement* A' of an event A (which we defined on page 62). Leaving the proof of the rule to the reader in Exercise 19 on page 107, let us merely state it as follows:

The sum of the probabilities of an event A and its complement A' is always equal to 1; symbolically,

$$P(A) + P(A') = 1$$

for any event A, and hence $P(A') = 1 - P(A)$.

This certainly agrees with the frequency concept of probability, for if the mail is delivered on time 37% of the time, then it is not delivered on time 63% of the time, and if 74% of the students will pass a given test, then 26% will not pass the test. So far as subjective probabilities are concerned, if the odds *for* the occurrence of an event are a to b, then the odds *against* its occurrence are b to a, the corresponding probabilities are $\frac{a}{a+b}$ and $\frac{b}{b+a}$ in accordance with the rule on page 84, and

$$\frac{a}{a+b} + \frac{b}{b+a} = \frac{a+b}{a+b} = 1.$$

Further addition rules

The addition rule of the preceding section was limited to two mutually exclusive events, but it can easily be generalized so that it applies also when there are more than two mutually exclusive events. In that case we have:

If k events are mutually exclusive, the probability that one of them will occur equals the sum of their respective probabilities; symbolically, if the events are denoted $A_1, A_2, \ldots,$ and A_k, we have

$$P(A_1 \cup A_2 \cup \ldots \cup A_k) = P(A_1) + P(A_2) + \ldots + P(A_k)$$

As we indicated in the footnote on page 61, the symbol $\cup$ is commonly read as "or."

Example 1. If the probabilities that someone dining at a certain restaurant will have cherry pie, chocolate cake, or cheese cake for dessert are, respectively, 0.14, 0.25, and 0.09, then the probability that he or she will have one of these desserts is $0.14 + 0.25 + 0.09 = 0.48$.

Example 2. If the probabilities that Ms. F. will find a new tennis dress at her club, at a sporting goods store, at a dress shop, or at a department store are, respectively, 0.08, 0.15, 0.12, and 0.17, then the probability that she will find a new tennis dress at one of these places is $0.08 + 0.15 + 0.12 + 0.17 = 0.52$.

The job of assigning probabilities to *all possible events* connected with a given situation can be very tedious, to say the least.

Example 3. Even if a sample space has as few as *five* points corresponding, say, to a student's getting an A, B, C, D, or F in a philosophy course, there are already $2^5 = 32$ possibilities (see Exercise 14 on page 78). There is the event that the student will get an A; the event that the student will get a B or a C; the event that the student will get a B, C, or D; the event that the student will get an A, B, C, or D; and so forth. Things get worse very rapidly when the number of points increases beyond five—for 12 points, for example, there are 4,096 subsets or events, and for 20 points there are 1,048,576.

Fortunately, it is seldom necessary to assign probabilities to all possible events, and the following rule makes it quite easy to find the probability of any event (relating to a given problem) on the basis of the probabilities which are assigned to the individual points (outcomes) of the sample space:

The probability of any event A is given by the sum of the probabilities of all the individual outcomes that are included in A.

This rule is illustrated in Figure 4.1, where the dots represent the individual (mutually exclusive) outcomes. The fact that the probability of A is given by the sum of the probabilities of all the points in A follows immediately from the addition rule on page 98.

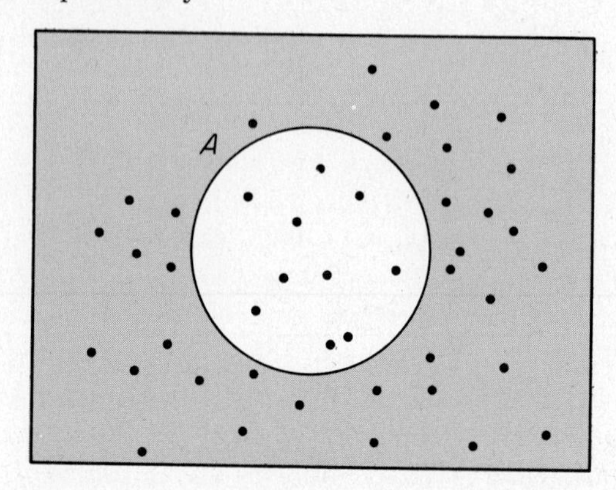

Figure 4.1 Sample space.

Example 3 (Continued). Suppose, for instance, that the probabilities of the student getting an A, B, C, D, or F in the philosophy course are, respectively, 0.12, 0.26, 0.43, 0.12, and 0.07. Then, the probability that the student will get a B or a C is $0.26 + 0.43 = 0.69$; the probability that the student will get a C, D, or F is $0.43 + 0.12 + 0.07 = 0.62$; the probability that the student will get an A, B, C, or D is $0.12 + 0.26 + 0.43 + 0.12 = 0.93$; and so forth.

Example 4. To consider another illustration, let us refer to Example 5 of Chapter 3, which concerned two archaeological expeditions looking for the ruins of two ancient cities. There were the six possible outcomes, (0, 0), (1, 0), (0, 1), (2, 0), (1, 1), and (0, 2), with the coordinates indicating the number of cities discovered by the respective expeditions, and we shall now suppose that these outcomes have the probabilities shown in Figure 4.2 (which is otherwise identical with Figure 3.3 on page 59). If we are interested in the probability that *the first expedition will not discover either city*, we have only to add the probabilities associated with the points (0, 0), (0, 1), and (0, 2), and we get $0.06 + 0.17 + 0.22 = 0.45$. Similarly, if we are interested in the probability that *both cities will be discovered*, we simply add the probabilities associated with the points (2, 0), (1, 1), and (0, 2), and we get $0.22 + 0.16 + 0.22 = 0.60$, and if we are interested in the probability that *the second expedition will discover at least one of the cities*, we have only to add the probabilities associated with the points (0, 1), (1, 1), and (0, 2), and we get $0.17 + 0.16 + 0.22 = 0.55$.

The situation is even simpler when the outcomes are all **equiprobable**, which is often the case in games of chance. In that case it follows from the rule on page 99 that

If there are n possible outcomes which are all equiprobable and s of these outcomes are included in event A, then the probability of A is $\frac{s}{n}$.

The ratio of the number of "successes" to the total number of outcomes is sometimes used as a *definition* of probability, but aside from the logical difficulty of defining "probability" in terms of "equiprobable events," it has the shortcoming that it applies only when the individual outcomes all have the same probability. Nevertheless, this *is* the case in most games of chance, where we assume, for example, that each card in a deck has the same chance of being drawn, each side of a coin has the same probability of facing up, and likewise for each side of a die.

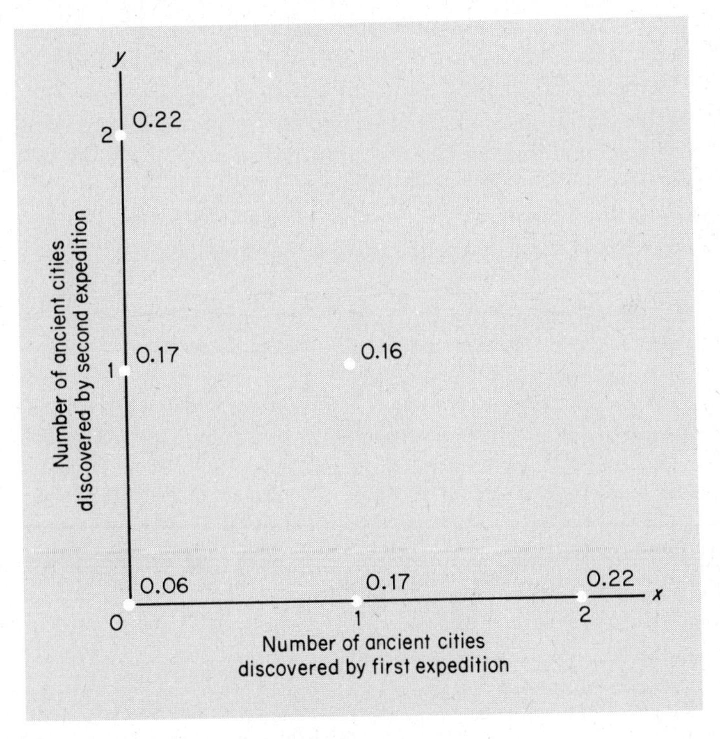

Figure 4.2 Sample space with probabilities.

Example 5. Thus, the probability of drawing a *club* from an ordinary deck of 52 playing cards is $\frac{13}{52}$ (there are 13 clubs among the 52 cards), the probability of getting *heads* with a balanced coin is $\frac{1}{2}$, and the probability of rolling *a 5 or a 6* with a balanced die is $\frac{2}{6}$.

Note that the special rule for equiprobable outcomes applies also when selections are made by means of lots or other gambling devices to decide, say, which cities are to be included in a survey, which items coming off an assembly line are to be inspected, which plots of farm land are to be used for a special variety of wheat, or which families are to be interviewed in a poll.

Since the various addition rules which we have studied apply only to mutually exclusive events, they cannot be used, for example, to determine the probability that at least one of two roommates will pass a certain exam, or the probability that a person involved in an automobile accident will break an arm, a rib, or a leg. In both cases, the alternatives are *not mutually exclusive.* To develop an addition rule which applies also to non-mutually exclusive events, let us consider the following example:

Example 6. Let us take a brief look at a report from a certain school district which claims that 18% of all "problem children" have bad eyesight, and that 14% are hard of hearing. It does not follow from this that 18 + 14 = 32% have bad eyesight or are hard of hearing—these two possibilities are not mutually exclusive—and to find the correct answer let us refer to the Venn diagram of Figure 4.3, where we indicated also that 5% of the "problem children" in this school district have bad eyesight *and* are hard of hearing. Thus, we can

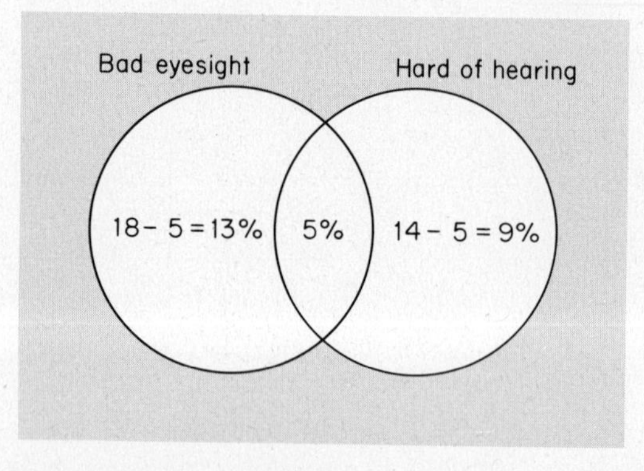

Figure 4.3 Venn diagram.

argue that in addition to the 5% who have bad eyesight and are hard of hearing, 18 − 5 = 13% have bad eyesight but are *not* hard of hearing, 14 − 5 = 9% are hard of hearing but do *not* have bad eyesight, so that 13 + 5 + 9 = 27% will fit either description.

This result could also have been obtained by *subtracting* from the original total of 18 + 14 = 32% the 5% which we *inadvertently included twice,* once among the "problem children" who have bad eyesight, and once among those who are hard of hearing. In fact, if we translate percentages into proportions and, hence, probabilities, and if we let E and H denote the respective events that one of these "problem children" has bad eyesight or is hard of hearing, we can write

$$\begin{aligned} P(E \cup H) &= P(E) + P(H) - P(E \cap H) \\ &= 0.18 + 0.14 - 0.05 \\ &= 0.27 \end{aligned}$$

Since the argument which we have presented in this example holds for any two events A and B, we can now state the following addition rule, called the **General Addition Rule** for probabilities, which applies regardless of whether A and B are mutually exclusive:

$$P(A \cup B) = P(A) + P(B) - P(A \cap B)$$

Note that when A and B *are* mutually exclusive, then $P(A \cap B) = 0$ since A and B have no outcomes in common, and the General Addition Rule reduces to the Special Addition Rule on page 97.

Example 7. To illustrate the use of this formula, let us suppose that the probabilities are, respectively, 0.28, 0.15, and 0.09 that a person traveling through Arizona will visit the Grand Canyon, the Petrified Forest, or both. Substituting these values into the formula, we find that the probability is

$$0.28 + 0.15 - 0.09 = 0.34$$

that a person traveling through Arizona will visit *at least one* of these attractions.

Example 8. Also, if the probabilities are, respectively, 0.20, 0.15, and 0.03 that a certain student will get a failing grade in history, French, or both, then the probability is

$$0.20 + 0.15 - 0.03 = 0.32$$

that he or she will fail *at least one* of these subjects.

Exercises

1. Explain why there must be a mistake in each of the following statements:

a. The probability that a mineral sample will contain copper is 0.18 and the probability that it will not contain copper is 0.92.
b. The probability that Joe will get into law school is 0.32 and the probability that he will not get into law school is -0.68.
c. The probability that the home team will win an upcoming football game is 0.67, the probability that it will tie the game is 0.08, and the probability that it will win or tie the game is 0.79.
d. The probabilities that a visitor to the Hawaiian Islands will visit Oahu or Mauai are, respectively, 0.82 and 0.14, and, hence, the probability that he or she will visit either island is $0.82 + 0.14 = 0.96$.
e. The probabilities that a prospective new-car buyer will buy a subcompact car, a compact car, an intermediate-size car, a large car, or change his mind and not buy a new car at all are, respectively, 0.21, 0.26, 0.18, 0.24, and 0.16.

2. Given the *mutually exclusive* events A and B for which $P(A) = 0.37$ and $P(B) = 0.44$, find

a. $P(A')$;
b. $P(B')$;
c. $P(A \cup B)$;
d. $P(A' \cap B')$.

(*Hint:* If necessary, draw a Venn diagram and fill in the probabilities associated with the various regions.)

3. If C and D are the events that a person will have a hamburger or a frankfurter at a drive-in restaurant, and $P(C) = 0.59$ and $P(D) = 0.27$, find

a. the probability that a person will not order a hamburger at this restaurant;
b. the probability that a person will order either a hamburger or a frankfurter at this restaurant;
c. the probability that a person will order neither a hamburger nor a frankfurter at this restaurant.

What assumption does one have to make to determine these probabilities without any further information?

4. An elected city official feels that the odds are 7 to 5 against her getting a raise in salary of \$1,000, and 11 to 1 against her getting a raise in salary of \$2,000. Furthermore, she feels that she has an *even chance* (the odds are 1 to 1) that she will get a raise of either \$1,000 or \$2,000. Discuss the *consistency* of these subjective probabilities.

5. There are two Porsches in a race, and an expert feels that the odds against their winning are, respectively, 3 to 1 and 4 to 1. To be *consistent*, how should he feel about the chances (that is, the odds) that neither car will win?

6. The probabilities that a consumer testing service will rate a new antipollution device for cars poor, fair, adequate, very good, or excellent are, respectively, 0.12, 0.18, 0.32, 0.21, and 0.17. Find the probabilities that they will rate the device

- **a.** very good or excellent;
- **b.** at least adequate;
- **c.** at best adequate;
- **d.** neither fair nor adequate.

7. The probabilities that a review board will rate a given movie X, R, PG, or G are, respectively, 0.01, 0.28, 0.45, and 0.26. What are the probabilities that they will rate the movie

- **a.** X or R;
- **b.** R or PG;
- **c.** R, PG, or G?

8. The probabilities that a doctor's answering service will receive 0, 1, 2, 3, 4, 5, 6, or 7 *or more* calls for him during the lunch hour are, respectively, 0.002, 0.013, 0.039, 0.081, 0.125, 0.155, 0.160, and 0.425. What are the probabilities that the doctor will receive

- **a.** at most 3 calls;
- **b.** at least 5 calls;
- **c.** anywhere from 2 to 6 calls, inclusive?

9. In a certain town, the probabilities that a driver will receive zero, one, two, three, four, or five *or more* traffic citations within one year are, respectively, 0.11, 0.29, 0.26, 0.19, 0.09, and 0.06. What is the probability that a driver will receive

- **a.** one or two traffic citations;
- **b.** at most one traffic citation;
- **c.** at least two traffic citations;
- **d.** anywhere from one through four traffic citations, inclusive?

10. With reference to Figure 4.2 on page 101, find

- **a.** the probability that the first expedition will discover one and only one of the two ancient cities;
- **b.** the probability that, together, the two expeditions will discover only one of the two ancient cities;
- **c.** the probability that neither expedition will discover both of the ancient cities;
- **d.** the probability that at least one of the ancient cities will be discovered by the two expeditions.

11. An insurance salesman has three customers in Laguna Beach, whom he may or may not visit during a two-week visit to that city. If we are interested only in how many of his customers he visits each week and he does not visit any one of these customers more than once, we can use coordinates (as in Example 5 of Chapter 3) to let (2, 1), for example, represent the event that he visits two of his customers in the first week and one in the second week.

a. Draw a sample space which shows the 10 possible outcomes of this "experiment."

b. If each point of this sample space is assigned the probability $\frac{1}{10}$, find (**1**) the probability that he will visit all three of his customers, (**2**) the probability that he will not visit any of his customers in at least one of the two weeks, and (**3**) the probability that he will visit at least one of his three customers during the first week.

12. If we assume that each card in a deck of 52 playing cards has an equal probability of being dealt first, what is the probability that it will be

a. a black ace;

b. a jack, queen, or king;

c. a club or a spade;

d. a 2, 4, 6, or 8 of any suit?

13. If H stands for *head* and T for *tail*, the 16 possible outcomes for four flips of a coin are HHHH, HHHT, HHTH, HHTT, HTHH, HTHT, HTTH, HTTT, THHH, THHT, THTH, THTT, TTHH, TTHT, TTTH, and TTTT. Assuming that these 16 possibilities are all equally likely, what are the respective probabilities of getting 0, 1, 2, 3, or 4 heads in four flips of a balanced coin?

14. When we roll a pair of dice, one red and one green, there are 36 possible outcomes, which can be denoted by means of coordinates so that (3, 5), for example, denotes the outcome where we roll a 3 with the red die and a 5 with the green die. Assuming that these 36 possibilities are all equally likely, what is the probability of rolling

a. a 7;

b. an 11;

c. 7 or 11;

d. 2, 3, or 12?

15. Given two events F and G for which $P(F) = 0.63$, $P(G) = 0.55$, and $P(F \cap G) = 0.38$, find

a. $P(F')$;

b. $P(G')$;

c. $P(F \cup G)$;

d. $P(F' \cap G)$;

e. $P(F \cup G')$;

f. $P(F' \cap G')$.

(*Hint:* Draw a Venn diagram and fill in the probabilities associated with the various regions.)

16. A movie critic feels that the probabilities are, respectively, 0.18, 0.33, and 0.15 that a certain movie will get an award for good acting, for good directing, or both. What is the corresponding probability that the movie will get at least one of the two awards?

17. An artist, who has entered an oil painting and a watercolor in a show, feels that the probabilities are, respectively, 0.17, 0.11, and 0.08 that she will sell the oil painting, the watercolor, or both.

a. What is the probability that she will sell either of these two works of art?

b. What is the probability that she will sell neither of these two works of art?

c. What is the probability that she will sell the watercolor but not the oil painting?

18. The probability that a person stopping at a given gas station will ask to have his battery checked is 0.13, the probability that he will have the tires checked is 0.19, and the probability that he will have his battery and his tires checked is 0.04. What are the probabilities that a person stopping at this gas station will

a. have his battery but not his tires checked;
b. have his battery and/or his tires checked;
c. have neither his battery nor his tires checked;
d. have his battery or tires, but not both, checked?

19. Prove the formula $P(A') = 1 - P(A)$ for any event A by making use of the fact that A and A' are *by definition* mutually exclusive and $A \cup A'$ is certain to occur.

Conditional probabilities

Very often it is meaningless (or at least very confusing) to speak of the probability of an event without specifying the sample space with which we are concerned. For instance, if we ask for the probability that a doctor makes more than $30,000 a year, we may well get many different answers *and they can all be correct.* One figure might apply to all doctors licensed in the United States, another might apply only to specialists, a third might apply only to doctors engaged in research, and so on. Since the choice of the sample space (namely, the set of all possibilities under consideration) is rarely self-evident, it helps to use the symbol $P(A|S)$, which denotes the **conditional probability** of event A relative to the sample space S. In other words, the symbol $P(A|S)$ indicates not only that we are concerned with event A, but also that we are referring to the specific sample space S. This makes it preferable to the abbreviated symbol $P(A)$ unless the tacit choice of S is clearly understood; the more explicit notation $P(A|S)$ is also of special value *when we are dealing with more than one sample space in one and the same problem.* (For instance, we may be interested in all doctors who are heart specialists and *also* in all doctors licensed to practice in California in one and the same problem.)

Example 9. To elaborate on the idea of a *conditional probability*, let us consider the following experiment: In a study of the effectiveness of a new tranquilizer, 40 patients with severe anxiety symptoms were given a 5-mg dosage of the new tranquilizer, while 40 others were given a placebo (that is, a preparation not containing the medication). Tests performed an hour later showed an improvement in 32 of the 40 patients who received the new tranquilizer and in 16 of the 40 patients who received the placebo. (In the latter case, the improvement can be attributed to the psychological effect of the idea of having taken a medication.) Schematically, all this information can be presented as follows:

	Received Tranquilizer	*Received Placebo*
Improvement	32	16
No Improvement	8	24
	40	40

To be examined further, these patients' names are drawn by lot, so that each of them has a probability of $\frac{1}{80}$ of being chosen first. Then, if we let T denote the event that the first patient thus selected actually received the tranquilizer and I the event that the patient showed improvement, it follows from the table (and the rule for equiprobable outcomes on page 101) that

$$P(T) = \frac{32 + 8}{80} = \frac{1}{2} \quad \text{and} \quad P(I) = \frac{32 + 16}{80} = \frac{3}{5}$$

Also, the probability that the patient thus selected will have received the tranquilizer and will also have shown improvement is

$$P(T \cap I) = \tfrac{32}{80} = \tfrac{2}{5}$$

Suppose now that it is decided to limit all further tests to patients who actually received the tranquilizer. This means that the number of patients (the number of possibilities, or the *size* of the sample space) is reduced to 40, and if we assume that each of these patients still has an equal chance of being chosen first, we now have

$$P(I|T) = \tfrac{32}{40} = \tfrac{4}{5}$$

This is the conditional probability that the patient who is selected first will have shown improvement *given that he or she actually received the tranquilizer*. Note that $P(I|T)$ is greater than $P(I)$, and this suggests that the tranquilizer *is* effective, namely, that its effect is not merely psychological [see also part (a) of Exercise 7 on page 278].

The result which we obtained for $P(I|T)$ is actually the *ratio* of the number of patients who received the tranquilizer and showed improvement to the number of patients who received the tranquilizer. If we replace both of these numbers by the corresponding proportions (that is, if we divide both figures by 80), it can be seen that the conditional probability $P(I|T)$ can also be written as

$$P(I|T) = \frac{\frac{32}{80}}{\frac{40}{80}} = \frac{P(T \cap I)}{P(T)}$$

Generalizing from this example, let us make the following definition of conditional probability which applies to any two events A and B:

If $P(B)$ is not equal to zero, then the conditional probability of A relative to B is given by

$$P(A|B) = \frac{P(A \cap B)}{P(B)}$$

Although the outcomes were all equiprobable in Example 9, it should be observed that the above definition applies regardless of what probabilities we assign to the individual outcomes that are included in A and B. The only restriction is that $P(B)$ must not equal zero, for division by zero is never permissible.

Example 6 (Continued). If we refer to the information on page 102, we find that the probability that one of the "problem children" will have bad eyesight *given that he or she is hard of hearing* is

$$P(E|H) = \frac{P(E \cap H)}{P(H)} = \frac{0.05}{0.14} = \frac{5}{14}$$

or approximately 0.36.

Example 7 (Continued). If we refer to the information on page 103, we find that the probability that a person traveling through Arizona will visit the Petrified Forest *given that he or she will visit the Grand Canyon* is $\frac{0.09}{0.28}$, or approximately 0.32.

To introduce another concept which is important in the study of probability, let us refer again to Example 8, which dealt with the possibility that a certain student might get failing grades in history and French:

Example 8 (Continued). If we let H denote the event that the student will get a failing grade in history and F that he or she will get a failing grade in French, the information on page 103 can be written as $P(H) = 0.20$, $P(F) = 0.15$, and $P(H \cap F) = 0.03$. Then, the probability that this student will get a failing grade in history *given that he or she will get a failing grade in French* is

$$P(H|F) = \frac{P(H \cap F)}{P(F)} = \frac{0.03}{0.15} = 0.20$$

and what is *special* (and interesting) about this result is that

$$P(H|F) = 0.20 = P(H)$$

namely, that *the probability of event H is the same regardless of whether event F has occurred (occurs, or will occur).*

In general, if $P(A|B) = P(A)$, we say that event A is **independent** of event B, and since it can be shown that *event B is independent of event A whenever event A is independent of event B*, namely, that $P(B|A) = P(B)$ whenever $P(A|B) = P(A)$, it is customary to say simply that ***A* and *B* are independent** whenever one is independent of the other.*

Example 8 (Continued). For the student taking history and French, we have

$$P(F|H) = \frac{P(F \cap H)}{P(H)} = \frac{0.03}{0.20} = 0.15$$

which verifies that F is also independent of H; indeed, in this example, the two events F and H *are independent.*

*Intuitively, we might say that two events are independent if the occurrence of either is in no way affected by the occurrence or nonoccurrence of the other (see also Exercise 15 on page 116).

If two events A and B are *not* independent, we say that they are **dependent**.

So far we have used the formula

$$P(A|B) = \frac{P(A \cap B)}{P(B)}$$

only to calculate conditional probabilities, and this was, of course, the purpose for which it was introduced. However, if we multiply the expressions on both sides of the equation by $P(B)$, we get

$$P(A \cap B) = P(B) \cdot P(A|B)$$

and we now have a formula, called the **General Multiplication Rule**, for determining the probability that two events will *both occur*. In words,

> **The probability that two events will both occur is the product of the probability that one of the events will occur and the conditional probability that the other event will occur given that the first event has occurred (occurs, or will occur).**

The above multiplication rule can also be written as

$$P(A \cap B) = P(A) \cdot P(B|A)$$

because it does not matter which event we refer to as A and which we refer to as B.

Example 10. To illustrate the use of these formulas, suppose that the editor of a student newspaper wants to interview two of the ten faculty members who serve on the student publications advisory committee, and ask them about a new rule which would give the college more control over student publications. If three of the faculty members are for the new rule while the other seven are against it, and the two to be interviewed are selected at random, *what is the probability that both will be against the new rule*? If we assume equal probabilities for each choice (which is, in fact, what we mean by "the two to be interviewed are selected at random"), the probability that the first faculty member interviewed will be against the new rule is $\frac{7}{10}$, and the probability that the second faculty member interviewed will be against the new rule *given that the first one was against the new rule* is $\frac{6}{9}$. Clearly, there are only six faculty members who are against the new rule among the nine faculty members who remain after one opposing the new rule has been selected. Hence, the probability that the editor of the student newspaper will randomly select two faculty members opposing the new rule is

$$\frac{7}{10} \cdot \frac{6}{9} = \frac{7}{15}$$

Using the same kind of argument, we also find that the probability of the student editor getting two faculty members who favor the new rule is

$$\frac{3}{10} \cdot \frac{2}{9} = \frac{1}{15}$$

and it follows, by subtraction, that the probability for one faculty member favoring the new rule and the other one opposing it is $1 - \frac{7}{15} - \frac{1}{15} = \frac{7}{15}$ (see also Exercise 11 on page 115).

If two events A and B are *independent*, we can substitute $P(A)$ for $P(A|B)$ in the first of the multiplication rules on page 111, or $P(B)$ for $P(B|A)$ in the second, and we obtain the **Special Multiplication Rule**,

$$P(A \cap B) = P(A) \cdot P(B)$$

This rule tells us that *for independent events we simply multiply their respective probabilities.*

Example 11. For instance, the probability of getting two heads in two flips of a balanced coin is $\frac{1}{2} \cdot \frac{1}{2} = \frac{1}{4}$, which agrees with the observation that among the four possibilities, *heads* and *heads*, *heads* and *tails*, *tails* and *heads*, *tails* and *tails*—only the first is a "success." Also, if two cards are drawn from an ordinary deck of 52 playing cards, and the first card is replaced before the second card is drawn, the probability of getting two aces in a row is

$$\frac{4}{52} \cdot \frac{4}{52} = \frac{1}{169}$$

Note, however, that *if the first card is not replaced*, the probability for getting two aces would be

$$\frac{4}{52} \cdot \frac{3}{51} = \frac{1}{221}$$

for there are only three aces among the 51 cards which remain after one ace has been removed from the deck.

Example 12. Similarly, if the probability is 0.12 that a person will make a mistake in his income tax return, then the probability that two totally unrelated persons (who do not use the same accountant) will both make a mistake is $(0.12)(0.12) = 0.0144$.

The special multiplication rule can easily be generalized so that it applies to the occurrence of more than two independent events—again, *we simply multiply all their respective probabilities.*

Example 13. For instance, the probability of getting three *heads* in a row with a balanced coin is

$$\tfrac{1}{2} \cdot \tfrac{1}{2} \cdot \tfrac{1}{2} = \tfrac{1}{8}$$

and the probability for first rolling four *threes* and then another number in five rolls of a balanced die is

$$\frac{1}{6} \cdot \frac{1}{6} \cdot \frac{1}{6} \cdot \frac{1}{6} \cdot \frac{5}{6} = \frac{5}{7{,}776}$$

For dependent events, the formula becomes somewhat more complicated, as is illustrated in Exercise 20 on page 117.

Exercises

1. If R is the event that the fire alarm was sounded right away and W is the event that the fire was put out without too much damage, state *in words* what probability is expressed by

a. $P(W|R)$;
b. $P(W'|R')$;
c. $P(R|W)$;
d. $P(R|W')$.

2. If A is the event that an astronaut is a member of the armed forces and T is the event that he was once a test pilot, express in symbolic form the probability that

a. an astronaut who is a member of the armed forces was once a test pilot;
b. an astronaut who was once a test pilot is not a member of the armed forces;
c. an astronaut who is not a member of the armed forces was once a test pilot;
d. an astronaut who was never a test pilot is not a member of the armed forces.

3. If G is the event that an ancient coin is in good condition, V is the event that it is very rare, and C is the event that it is a counterfeit, state *in words* what probability is expressed by

a. $P(G|V)$;
b. $P(G|C')$;
c. $P(C|V')$;
d. $P(G \cap C|V)$;
e. $P(R|G')$;
f. $P(G|V \cap C)$.

4. If R is the event that a person votes regularly, A is the event that he or she is active in a political organization, and T is the event that he or she is thinking of running for office, express in symbolic form the probability that

a. a person who is active in a political organization is thinking of running for office;
b. a person thinking of running for office is a regular voter;
c. a person who votes regularly but is not active in a political organization is thinking of running for office;
d. a person thinking of running for office does not vote regularly but is active in a political organization.

5. As part of a promotional scheme in California and Oregon, a company making cake mixes will award a grand prize of \$50,000 (and several other prizes) to persons sending in their names on entry blanks, with the option of also including a box top of one of the company's products.

	With Box Top	*Without Box Top*
California	31,000	12,000
Oregon	9,000	8,000

If the winner of the grand award is chosen by lot, C represents the event that it will be won by an entry from California, and B represents the event that it will be won by an entry which included a box top, find each of the following probabilities *directly* from the above table of entries:

a. $P(C)$;
b. $P(C')$;
c. $P(B)$;
d. $P(B')$;
e. $P(C \cap B)$;
f. $P(C \cap B')$;
g. $P(C \cup B)$;
h. $P(C' \cap B')$;
i. $P(C|B)$;
j. $P(B|C)$;
k. $P(C'|B')$;
l. $P(B'|C')$.

6. Use the result of Exercise 5 to verify that

a. $P(C|B) = \dfrac{P(C \cap B)}{P(B)}$;

b. $P(C'|B') = \dfrac{P(C' \cap B')}{P(B')}$;

c. $P(B|C) = \dfrac{P(C \cap B)}{P(C)}$;

d. $P(B'|C') = \dfrac{P(C' \cap B')}{P(C')}$.

7. The probability that a communication system will have high selectivity is 0.42, the probability that it will have high fidelity is 0.63, and the probability that it will have both is 0.14.

a. What is the probability that it will have high fidelity given that it has high selectivity?

b. What is the probability that it will have high selectivity given that it has high fidelity?

8. In Exercise 16 on page 106 we said that a movie critic felt that the probabilities are, respectively, 0.18, 0.33, and 0.15 that a certain movie will get an award for good acting, for good directing, or both.

a. What is the probability that the movie will get the award for good acting given that it is going to get the award for good directing?

b. What is the probability that the movie will get the award for good directing given that it is *not* going to get the award for good acting?

9. The probability that a certain psychological experiment will be well planned is 0.80, and the probability that it will be well planned and well executed is 0.72.

a. What is the probability that the experiment will be well executed given that it is well planned?

b. If the probability that the experiment will be well executed is 0.75, what is the probability that it will have been well planned given that it is well executed?

10. Police records show that in a certain city the probability is 0.40 that a burglar will be caught, and 0.65 that, if caught, a burglar will be convicted. What is the probability that in this town a burglar will be caught and convicted?

11. In connection with Example 10, we showed on page 112 that the probability is $\frac{7}{15}$ that one faculty member will favor the new rule while the other one will oppose it. Verify that result by *adding* the probabilities of the mutually exclusive alternatives of choosing first the faculty member who favors the new rule or choosing first the faculty member who opposes it.

12. Among the 20 trucks owned by a contractor, 12 violate pollution standards while the others do not. If 2 of these trucks are stopped at a checkpoint, what are the probabilities that

a. both will violate pollution standards;

b. neither will violate pollution standards;

c. one will violate pollution standards while the other does not?

Assume that each of the 20 trucks has the same chance of being stopped.

13. If a zoologist has six male guinea pigs and nine female guinea pigs, and randomly selects two of them for an experiment, what are the probabilities that

a. both will be males;
b. both will be females;
c. there will be one of each sex?

14. Which of the following pairs of events would you suppose are independent and which are dependent:

a. forgetting to study and failing an examination;
b. being an accountant and having green eyes;
c. being born in August and having flat feet;
d. being wealthy and collecting works of art;
e. getting 7's in successive rolls of a pair of dice;
f. being over 20 and smoking a pipe;
g. being hungry and being able to afford a meal;
h. living in San Francisco and reading comic books;
i. any two mutually exclusive events?

15. With reference to the continuation of Example 8 on page 110, verify that

a. $P(H|F') = P(H)$, namely, that event H is also independent of event F';
b. $P(F|H') = P(F)$, namely, that event F is also independent of event H'.

16. What is the probability of drawing two diamonds in a row from an ordinary deck of 52 playing cards, of which 13 are diamonds, if

a. the drawings are without replacement;
b. the drawings are with replacement?

17. If, for each tournament she enters, a tennis star's probability of taking first place is 0.64, what is the probability that she will win two tournaments in a row? Assume independence.

18. If the probability that Mr. Jones will be alive ten years from now is 0.66 and the probability that Mr. Brown will be alive ten years from now is 0.45, what is the probability that they will both be alive ten years from now? What assumption do we have to make to be able to determine this probability? Would it be reasonable to make this assumption if Mr. Jones and Mr. Brown worked together as a team setting dynamite charges in road construction?

19. As we indicated on page 113, the probability that any number of independent events will occur is given by the product of their respective probabilities. Use this rule to find

a. the probability of getting six *heads* in a row with a balanced coin;

b. the probability of drawing (with replacement) four *spades* in a row from an ordinary deck of 52 playing cards;

c. the probability that a student will answer five true–false questions correctly, if for each question the probability is 0.80 that he or she will know or guess the right answer;

d. the probability that three totally unrelated persons who are 30 years old will all be alive at age 65, given that for any of them the probability is 0.58;

e. the probability that a person shooting at a target will hit it twice in a row and then miss it twice in a row, given that the probability of his or her hitting the target on any one try is 0.90.

20. GENERAL MULTIPLICATION RULE FOR MORE THAN TWO EVENTS The problem of determining the probability that any number of events will occur becomes more complicated when the events are *not independent.* For three events, for example, which we shall arbitrarily refer to as the first, second, and third, the probability that they will all occur is obtained by multiplying the probability that the first event will occur by the probability that the second event will occur *given that the first has occurred,* and then multiplying by the probability that the third event will occur *given that the first and second have occurred.* For instance, the probability of drawing (without replacement) three aces in a row from an ordinary deck of 52 playing cards is $\frac{4}{52} \cdot \frac{3}{51} \cdot \frac{2}{50} = \frac{1}{5{,}525}$; clearly, there are only three aces among the 51 cards which remain after the first ace has been drawn, and only two aces among the 50 cards which remain after the first two aces have been drawn.

a. If the parents of 6 of 18 students are divorced and 3 of the students are randomly selected for a sociological study, what is the probability that the parents of all 3 of them are divorced?

b. Referring to Example 10 on page 111, where three of the faculty members on the committee were for the new rule and the other seven were against it, what is the probability that the editor of the student newspaper will randomly select and interview four of the committee members who are all against the new rule?

c. If six bullets, of which three are blanks, are randomly inserted into a gun, what is the probability that the first three bullets fired will all be blanks?

d. The only supermarket in a small town offers two brands of frozen orange juice, brand A and brand B. Among its customers who buy brand A one week, 80% will buy brand A and 20% will buy brand B the next week, and among its customers who buy brand B one week, 40% will buy brand B and 60% will buy brand A the next week. To simplify matters, it will be assumed that each customer buys frozen orange juice once a week.

(1) What is the probability that a customer who buys brand A one week will buy brand B the next week, brand B the week after that, and brand A the week after that?

(2) What is the probability that a customer who buys brand B one week will buy brand B the next two weeks, and brand A the week after that?

(3) What is the probability that a customer who buys brand A in the first week of a month will also buy brand A in the third week of that month? (*Hint:* Add the probabilities associated with the two mutually exclusive possibilities corresponding to his buying brand A or brand B in the second week.)

Bayes' rule

Although the symbols $P(A|B)$ and $P(B|A)$ may seem very much alike, the same cannot be said for the probabilities which they represent.

Example 9 (Continued). In the illustration on page 108, we showed that $P(I|T)$, the probability that a chosen patient will have shown improvement *given that he or she actually received the tranquilizer*, was $\frac{4}{5}$. On the other hand, $P(T|I)$ is the probability that the chosen patient will have received the tranquilizer *given that he or she showed improvement*, and it can be seen from the table on page 108 that its value is $\frac{32}{32 + 16} = \frac{2}{3}$. Note that whereas $P(I|T)$ is a probability about the *effect* of the tranquilizer, $P(T|I)$ is a probability about the *cause* of the improvement.

Example 14. If R is the event that a person committed a certain robbery and G is the event that he is judged guilty, then $P(G|R)$ is the probability that the person will be judged guilty *given that he actually committed the robbery*, and $P(R|G)$ is the probability that the person actually committed the robbery *given that he is judged guilty*.

Since there are many situations in which we must consider such pairs of probabilities, let us derive a formula which expresses $P(A|B)$ in terms of $P(B|A)$, and vice versa. All we really have to do is combine the formula for

$P(A|B)$ on page 109 with the second form of the General Multiplication Rule on page 111, namely, $P(A|B) = \frac{P(A \cap B)}{P(B)}$ and $P(A \cap B) = P(A) \cdot P(B|A)$. Substituting $P(A) \cdot P(B|A)$ for $P(A \cap B)$ into the formula for $P(A|B)$, we thus get

$$P(A|B) = \frac{P(A) \cdot P(B|A)}{P(B)}$$

which is a very simple (special) form of a rule called the **Rule of Bayes.**

Example 15. To illustrate the use of this formula, suppose that the loan officer of a bank is advised by a credit investigator that a loan applicant is a bad risk. Knowing that credit checks are not infallible, the loan officer of the bank will have to judge whether the loan applicant actually *is* a bad risk—in other words, he will have to determine the probability $P(B|R)$, where B denotes the event that the loan applicant *is* a bad risk, while R denotes the event that the loan applicant *is rated* a bad risk. In order to calculate this probability the loan officer of the bank will have to know the values of the probabilities $P(R|B)$, $P(B)$, and $P(R)$, and, for the sake of argument, let us suppose that the values of these probabilities are

$$P(R|B) = 0.96, \qquad P(B) = 0.05, \qquad P(R) = 0.08$$

This means that 96% of the loan applicants who are bad risks will also be rated bad risks, 5% of the loan applicants whose credit is checked by the investigator actually are bad risks; and 8% of the loan applicants whose credit is thus checked are rated bad risks. Now, if we substitute all these probabilities into the formula, we get

$$P(B|R) = \frac{P(B) \cdot P(R|B)}{P(R)} = \frac{(0.05)(0.96)}{0.08} = 0.60$$

for the probability that a loan applicant who is rated a bad risk actually *is* a bad risk. Since this value is fairly low, the loan officer of the bank would probably be well advised to conduct further inquiries before he refuses to grant the loan.

Even though the rule we used here was easy to derive and there is no question about its *validity*, extensive criticism has always surrounded its application. This is due largely to the fact that the rule involves an "inverse" sort of reasoning, namely, reasoning *from effect to cause.* In our numerical example we used the formula to calculate the probability that a loan applicant who is rated a bad risk actually is a bad risk, and earlier, in Example 14, we suggested that we might be interested in the probability that a person who is judged guilty of a robbery actually did commit the crime. All we can really say is that *the formula must be used with care* and that *the significance of the probabilities which we calculate as well as the probabilities whose values we substitute must be clearly understood.*

In most practical applications there are more than two possible "causes" and the formula for Bayes' Rule must be written in a slightly different (that is, more general) form. If $A_1, A_2, \ldots,$ and A_k denote the k mutually exclusive "causes" which could have led to the event B, then the probability that it was A_i (for $i = 1, 2, \ldots,$ or k) is given by

$$P(A_i|B) = \frac{P(A_i) \cdot P(B|A_i)}{P(A_1) \cdot P(B|A_1) + P(A_2) \cdot P(B|A_2) + \ldots + P(A_k) \cdot P(B|A_k)}$$

Referring to Figure 4.4, we might say that $P(A_i|B)$ is the probability that event B is reached along the ith branch of the "tree" *given that the event must be reached along one of the k branches of the tree.* Note, furthermore, that the value of $P(A_i|B)$ is given by the *ratio* of the probability associated with the ith branch to the *sum* of the probabilities associated with all the k branches of the "tree."

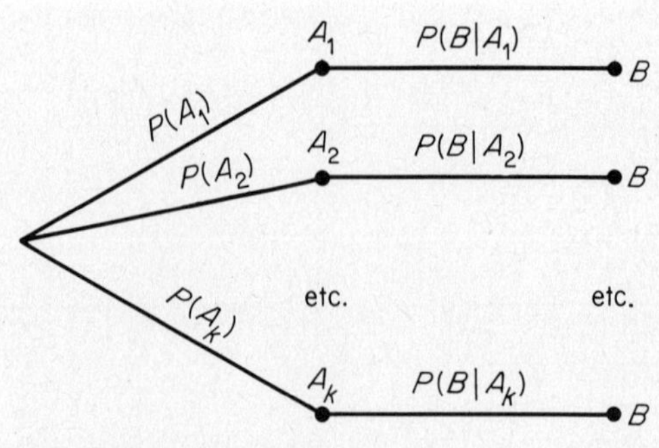

Figure 4.4 Bayes' rule.

Example 16. To illustrate this more general form of Bayes' rule, suppose that in a cannery, assembly lines I, II, and III account, respectively, for 37%, 42%, and 21% of the total output. If 0.6% of the cans from assembly line I are improperly sealed, while the corresponding percentages for assembly lines II and III are, respectively, 0.4% and 1.2%, what we would like to know is *the probability that an improperly sealed can (discovered at the final inspection of outgoing products) came from assembly line III*. Letting B denote the event that a can is improperly sealed, and A_1, A_2, and A_3 denote the respective events that a can comes from assembly lines I, II, or III, we find that the given information can be written as $P(A_1) = 0.37$, $P(A_2) = 0.42$, $P(A_3) = 0.21$, $P(B|A_1) = 0.006$, $P(B|A_2) = 0.004$, and $P(B|A_3) = 0.012$. Thus, substitution into the formula yields

$$P(A_3|B) = \frac{(0.21)(0.012)}{(0.37)(0.006) + (0.42)(0.004) + (0.21)(0.012)}$$
$$= 0.39 \text{ (approximately)}$$

for the probability that the improperly sealed can came from assembly line III. This figure is *high* compared to $P(A_3) = 0.21$, but this is accounted for by the fact that the percentage of improperly sealed cans is much greater for assembly line III than for the other two assembly lines.

If we picture this problem as in Figure 4.5, it can be seen that the probabilities associated with the three branches are $(0.37)(0.006) = 0.00222$, $(0.42)(0.004) = 0.00168$, and $(0.21)(0.012) = 0.00252$. The

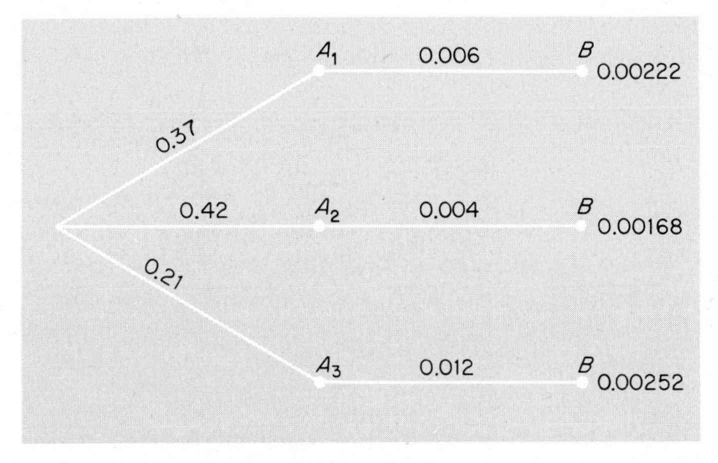

Figure 4.5 Tree diagram for Example 16.

sum of these three probabilities, namely, the quantity which goes into the denominator in the formula for Bayes' rule, is 0.00642, and we get

$$P(A_1|B) = \frac{0.00222}{0.00642} = 0.35 \quad \text{and} \quad P(A_2|B) = \frac{0.00168}{0.00642} = 0.26$$

in addition to $P(A_3|B) = 0.39$. Thus, the odds are almost 2 to 1 that an improperly sealed can *does not* come from assembly line I, almost 3 to 1 that it *does not* come from assembly line II, and slightly better than 3 to 2 that it *does not* come from assembly line III.

Exercises

1. For a student enrolling as a freshman at a certain university, the probability is 0.40 that he or she will graduate and 0.16 that he or she will get a scholarship. Furthermore, if the probability that a student who gets a scholarship will graduate is 0.85, what is the probability that a student who graduates actually got a scholarship?

2. In a very large freshman English class, 52% of the students turn in all assignments and 78% pass the course. Furthermore, if the probability that a student who turns in all assignments will pass the course is 0.87, what is the probability that a student who passes the course actually turned in all assignments?

3. In a certain community, 10% of all adults over 50 have diabetes. If a doctor in this community *correctly* diagnoses 95% of all persons with diabetes as having the disease and *incorrectly* diagnoses 2% of all persons without diabetes as having the disease, what is the probability that a person diagnosed by this doctor as having diabetes actually has the disease? (*Hint:* Draw a tree diagram like that of Figure 4.5.)

4. Three attendants at a gasoline service station are supposed to check the oil of every customer's car. Tom, who services 30% of the cars, fails to check the oil 2% of the time; Robert, who services 40% of the cars, fails to check the oil 5% of the time; and Frank, who services 30% of the cars, forgets to check the oil 1% of the time. If a customer complains later that his oil was not checked, what is the probability that her car was serviced by Robert?

5. It is a toss-up whether Mr. Cooper will stay in Rome, Naples, or Milan during a forthcoming trip to Italy. If the probabilities of his having a good time staying in Rome, Naples, and Milan are, respectively, 0.60,

0.40, and 0.80, and later we hear indirectly that he is having a good time, what are the odds that he is *not* staying in Milan?

6. A resort hotel gets cars for its guests from three rental agencies, 20% from agency D, 20% from agency E, and 60% from agency F. If 10% of the cars from rental agency D have bad tires, while the corresponding percentages for agencies E and F are, respectively, 4 and 12, what is the probability that a car with bad tires delivered to a guest at this resort hotel came from rental agency F?

5

Chance variation: probability functions

Introduction

We might well have called this chapter "How to Live with Uncertainties," and this is probably the most important lesson to be learned from modern statistics. Uncertainties face us wherever we turn—we cannot be certain that an experiment will succeed, we cannot be certain that the weather will stay nice, we cannot be certain that the milk we buy in a sealed carton is fresh, we cannot be certain that a rocket will get off the ground, we cannot be certain that a marriage will last, we cannot be certain that an investment will pay off, and so on. Actually, life without uncertainties would probably be quite dull and most uncertainties cannot be eliminated, but we shall see that **chance fluctuations** (namely, changes, variations, or differences "caused" by chance) are quite often *predictable*, that they can be *measured*, and that sometimes they can even be *controlled* to some extent.

If the values which a quantity takes on are somehow dependent on chance, we refer to it as a **random variable**, and this definition makes the term applicable to a great variety of phenomena: It applies to the price of a share of A.T. & T.; it applies to the moisture content of the air, the wind velocity at Kennedy Airport, and the annual production of corn; it applies to the number of eggs a chicken will lay each week or the number of mistakes a secretary will make in typing a report; and it also applies to the size of an audience at a football game or the number of leaves that are left on a tree after a mild frost. Needless to say, perhaps, the fluctuations of such random variables are *not* predictable in the same way in which an "infallible forecaster" predicts events—all we can do is determine the probabilities that random variables will take on specific values, the probabilities that their values will fall on given intervals, or the probabilities that their values will be less than (or greater than) given numbers. For instance, we *cannot* say for sure whether we will get two *heads* and two *tails* in four flips of a balanced coin, but we *can* determine the probabilities of getting no *heads* and four *tails*, one *head* and three *tails*, two *heads* and two *tails*, and so forth. Similarly, we *cannot* say for sure whether a new tire we buy for our car will last 15,000 miles, but we *can* quote probabilities that without unduly rough treatment it will last at most 15,000 miles, anywhere from 12,000 to 18,000 miles, at least 10,000 miles, and so on. Finally, we *cannot* predict for sure whether a newly married couple will eventually have two children, but we can determine the probabilities that they will have 0, 1, 2, 3, 4, or more.

The problem of determining the probabilities which are associated with the different values of a random variable will be the subject matter of the next four sections. Then we shall turn to the problem of constructing a "yardstick" which will enable us to *measure* the size of chance fluctuations, and in the last two sections we shall see how this "yardstick" can actually be used to *predict* and to some extent *control* chance variations to which we are exposed.

Probability functions

The tables in the two examples which follow serve to illustrate what we mean by a **probability function**, namely, *a correspondence which assigns probabilities to the values of a random variable.*

Example 1. For the number of points rolled with a die, we assume that each face has the same probability of $\frac{1}{6}$, and we, therefore, get the correspondence shown in the following table:

Number of Points Rolled with a Die	*Probability*
1	$\frac{1}{6}$
2	$\frac{1}{6}$
3	$\frac{1}{6}$
4	$\frac{1}{6}$
5	$\frac{1}{6}$
6	$\frac{1}{6}$

Example 2. For the number of *heads* we might get in four flips of a coin, we consider as equally likely the sixteen possibilities HHHH, HHHT, HHTH, HHTT, HTHH, HTHT, HTTH, HTTT, THHH,

THHT, THTH, THTT, TTHH, TTHT, TTTH, and TTTT, where H stands for *head* and T for *tail.* Then we count the number of heads in each case (see also Exercise 13 on page 106), and get

Number of Heads	*Probability*
0	$\frac{1}{16}$
1	$\frac{4}{16}$
2	$\frac{6}{16}$
3	$\frac{4}{16}$
4	$\frac{1}{16}$

Although these examples may seem trivial and of interest only to someone addicted to gambling, *the tables which we have constructed can provide answers to questions relating to entirely different matters.*

Example 2 (Continued). Consider, for example, a boy who has pet gerbils (little hamster-like animals of Mongolian origin) and sells their offspring to a pet shop, whose owner gives him \$8.00 for a matched pair, \$3.60 for a female, and \$3.00 for a male. Suppose now that he has a litter of four baby gerbils, and wants to know immediately how much money he can expect to get; in other words, he is too impatient to wait the several weeks it takes before one can identify their sex. If the probabilities that any one of the baby gerbils is male or female are $\frac{1}{2}$ and $\frac{1}{2}$, we can copy the above table with "Number of Males" substituted for "Number of Heads," calculate the corresponding amounts the boy will receive, and we get

Number of Males	*Probability*	*Amount the Boy Will Receive*
0	$\frac{1}{16}$	\$14.40
1	$\frac{4}{16}$	\$15.20
2	$\frac{6}{16}$	\$16.00
3	$\frac{4}{16}$	\$14.00
4	$\frac{1}{16}$	\$12.00

Note that in the right-hand column, $4(3.60) = \$14.40$ is the amount he will receive for no males and four females; $8.00 + 2(3.60) = \$15.20$ is the amount he will receive for one male and three females, namely, for one matched pair and two females; $2(8.00) = \$16.00$ is the amount he will receive for two males and two females, namely, for two matched pairs, and so on. Using the formula for a *mathematical expectation* on page 88, we can thus say that the boy can expect to get

$$14.40(\tfrac{1}{16}) + 15.20(\tfrac{4}{16}) + 16.00(\tfrac{6}{16}) + 14.00(\tfrac{4}{16}) + 12.00(\tfrac{1}{16}) = \$14.95$$

for his litter of four baby gerbils.

Whenever possible, we try to express probability functions by means of mathematical formulas which enable us to calculate the probabilities associated with the various values of a random variable.

Example 1 (Continued). For the number of points which we roll with a die, we can, thus, write*

$$f(x) = \tfrac{1}{6} \qquad \text{for } x = 1, 2, 3, 4, 5, \text{ and } 6$$

where $f(1)$ represents the probability of rolling a 1, $f(2)$ represents the probability of rolling a 2, . . . , in the usual functional notation.

In most instances, the formulas are more complicated than in this example, and in Exercise 3 on page 136 the reader will be asked to show that for Example 2 the formula is given by $f(x) = \frac{1}{16} \cdot \binom{4}{x}$ for $x = 0, 1, 2, 3,$ and 4, where $\binom{4}{x}$ is a binomial coefficient as defined on page 75.

To conclude this preliminary discussion of probability functions, let us point out the following general rules which the values of any probability function must obey:

> **Since the values of probability functions are probabilities, they must always be numbers between 0 and 1, inclusive; also, since a random variable has to take on one of its values (for instance, we have to get 0, 1, 2, 3, or 4 heads in four flips of a coin), the sum of all the values of a probability function must always be equal to 1.**

Let us also point out that although some statisticians make a rather fine distinction between probability functions and **probability distributions**, we shall use these terms interchangeably in this book; indeed, a probability

*Most of the time we shall write the probability that a random variable takes on the value x as $f(x)$, but we could just as well write it as $g(x)$, $h(x)$, $b(x)$, and so on.

function tells us how the total probability of 1 is *distributed* among the possible values a random variable can take on.

The binomial distribution

There are many problems in which we are interested in the probability that an event will occur x times out of n. For instance, we may be interested in getting 36 responses to 200 mail questionnaires sent out as part of a sociological survey, the probability that 82 of 400 tagged wild geese will be recaptured, the probability that 27 of 60 drivers stopped at a road block are not wearing their seatbelts, and so on. To borrow from the language of games of chance, in each of these examples we are interested in the probability of getting x **successes** in n **trials**, or in other words, the probability of x **successes** and $n - x$ **failures**. In the problems which we shall study in this section it will be assumed that:

> **There is a fixed number of trials; the probability of a success is the same for each trial; and the trials are, furthermore, all independent.**

This means that our methods will *not* apply, for example, if we alternately ask men and women whether they regularly watch professional football on television, or if we are interested in the chances that a "senior citizen" will have voted for the Democratic candidate six times in the last eight presidential elections. In the first case, men and women are apt to feel differently about watching professional football on television, so that the probability of a positive response is *not the same for each trial*; in the second case, the trials are *not independent* since a person who has voted for the Democratic candidate in one presidential election is more apt to do so again in the next election.

To handle problems which *do* meet the conditions listed in the preceding paragraph, we use a special kind of probability function called the **binomial probability function**, or simply the **binomial distribution**. It includes Example 2 on page 125, and its formula is fairly easily obtained. If p and $1 - p$ are the probabilities for a success and a failure on any given trial, then the probability of getting x successes and $n - x$ failures *in some specific order* is $p^x(1 - p)^{n-x}$; clearly, in this product of p's and $(1 - p)$'s there is *one* factor p for each success, *one* factor $1 - p$ for each failure, and the x factors p and $n - x$ factors $1 - p$ are all multiplied together by virtue of the Special Multiplication Rule, which we generalized for more than two independent events on page 113. Since this probability applies to any point of the sample space which represents x successes and $n - x$ failures (namely, to any specific arrangement of x successes and $n - x$ failures), we have only to see how many points of this kind there are, and then multiply their number by $p^x(1 - p)^{n-x}$. Clearly, there are $\binom{n}{x}$, the number of combinations of x elements selected from a set of n elements (namely, the number of ways in which we can select the x trials on which there is to be a success), and we have thus arrived at the result that the probability for x successes in n trials is given by

$$f(x) = \binom{n}{x} p^x(1 - p)^{n-x} \qquad \textit{for } x = 0, 1, 2, \ldots, \textit{ or } n$$

This is the formula for the *binomial distribution*, and to remind the reader, $\binom{n}{x}$ is a binomial coefficient whose value can be looked up in Table VIII, or it can be calculated directly according to the formula on page 75. The reason why we refer to this probability function as the *binomial distribution* is that the values we get for $x = 0$, $x = 1$, $x = 2, \ldots,$ and $x = n$ are the successive terms of the *binomial expansion* of $[(1 - p) + p]^n$, with which the reader may be familiar from elementary algebra (see also Exercise 4 on page 136). Incidentally, since

$$[(1 - p) + p]^n = 1^n = 1$$

this demonstrates that the sum of *all* the values of a binomial distribution is always equal to 1, as it should be.

Example 3. To illustrate the use of the formula for the binomial distribution, let us calculate the probability that two of five students with high grades will get into medical school, given that the probability for each is 0.70. Substituting $x = 2$, $n = 5$, $p = 0.70$, and $\binom{5}{2} = \frac{5 \cdot 4}{2!} = 10$ into the formula, we get

$$\begin{aligned} f(2) &= \binom{5}{2}(0.70)^2(1 - 0.70)^{5-2} \\ &= 10(0.70)^2(0.30)^3 \\ &= 0.1323 \end{aligned}$$

or approximately 0.13. Similarly, to find the probability that six of eight reviewers will like a new movie, when the probability that any one of them will like it is 0.80, we substitute $x = 6$, $n = 8$, $p = 0.80$, and $\binom{8}{6} = \frac{8 \cdot 7 \cdot 6 \cdot 5 \cdot 4 \cdot 3}{6!} = 28$ into the formula, and get

$$\begin{aligned} f(6) &= \binom{8}{6}(0.80)^6(1 - 0.80)^2 \\ &= 28(0.80)^6(0.20)^2 \\ &= 0.294 \end{aligned}$$

rounded to three decimals.

Note that in both of these problems the binomial coefficients, $\binom{5}{2}$ and $\binom{8}{6}$, could have been obtained directly from Table VIII.

Example 4. To give an example in which we calculate *all* the values of a binomial distribution, suppose that an athlete claims that he can run the mile in less than 4 minutes 60% of the time. Assuming that this figure is correct, let us determine the probabilities that in five races the athlete will finish 0, 1, 2, 3, 4, or 5 times in less than 4 minutes. Substituting $n = 5$, $p = 0.60$, and, respectively, $x = 0, 1, 2, 3, 4$, and 5, we get

$$\begin{aligned} f(0) &= \binom{5}{0}(0.60)^0(1 - 0.60)^{5-0} = 0.010 \\ f(1) &= \binom{5}{1}(0.60)^1(1 - 0.60)^{5-1} = 0.077 \\ f(2) &= \binom{5}{2}(0.60)^2(1 - 0.60)^{5-2} = 0.230 \\ f(3) &= \binom{5}{3}(0.60)^3(1 - 0.60)^{5-3} = 0.346 \\ f(4) &= \binom{5}{4}(0.60)^4(1 - 0.60)^{5-4} = 0.259 \\ f(5) &= \binom{5}{5}(0.60)^5(1 - 0.60)^{5-5} = 0.078 \end{aligned}$$

where all the answers are rounded to three decimals. Thus, it can be seen that the probability is $1 - 0.010 = 0.990$ and the odds are 99 to1 that he will finish in less than 4 minutes *at least once*; also, the probability is 0.010 + 0.077 + 0.230 + 0.346 = 0.663 and the odds are about 2 to 1 that he will finish in less than 4 minutes *at most three times*.

In actual practice, problems pertaining to binomial distributions are seldom solved by direct substitution into the formula. Sometimes we use the approximation discussed in Exercise 23 on page 139 or that given in Chapter 6, but more often we refer to special tables such as Table XI or the more detailed tables listed in the Bibliography at the end of the book. Table XI is limited to the binomial probabilities for $n = 2$ to $n = 15$, and $p = 0.05, 0.1, 0.2, 0.3, 0.4, 0.5, 0.6, 0.7, 0.8, 0.9$, and 0.95. All these probabilities are given to three decimals, and where values are omitted, they are 0.0005 or less.

Example 5. Using Table XI, it is easy to verify the results of Examples 3 and 4; also, we can immediately read off the result that the probability of getting four *heads* and eight *tails* in 12 flips of a balanced coin is 0.121, while the corresponding probability for six *heads* and six *tails* is 0.226.

Example 6. Similarly, if the probability is 0.40 that a car stolen in a given city will be recovered, then the probability that *at most three* of ten cars stolen in this city will be recovered is

$$0.006 + 0.040 + 0.121 + 0.215 = 0.382$$

and the probability that *at least seven* will be recovered is

$$0.042 + 0.011 + 0.002 = 0.055$$

The hypergeometric distribution

To illustrate another kind of probability function, let us consider the following problem of *quality control.*

Example 7. A factory ships certain tape recorders in lots of 16. When they arrive at their destination, an inspector randomly selects 3 from each lot, and the whole lot is accepted if all three are in good working condition; otherwise, the whole lot is inspected. It is easy to see that this kind of sampling inspection involves certain risks; in fact, a lot could be accepted even though 13 of the 16 tape recorders do not work. More realistically, it may be of interest to know the probability that a lot will be accepted even though, say, 4 of the 16 tape recorders are defective. Since we would thus be interested in the probability of getting 3 successes (nondefective tape recorders) in 3 trials (among the tape recorders inspected), we might be tempted to argue that the probability of a nondefective tape recorder is $\frac{12}{16} = 0.75$, and, hence that the desired probability is

$$\begin{aligned} f(3) &= \binom{3}{3}(0.75)^3(1 - 0.75)^{3-3} \\ &= 1 \cdot (0.75)^3 \\ &= 0.422 \end{aligned}$$

rounded to three decimals. This result would be correct if we sampled **with replacement**, namely, if each tape recorder were replaced before the next one is selected; otherwise, the basic assumption of the binomial distribution that *the trials must be independent* would be violated. Although the probability that the first tape recorder chosen for inspection is not defective is $\frac{12}{16}$, the probability that the second tape recorder is not defective *given that the first one was not defective* is $\frac{11}{15}$, and the probability that the third tape recorder is not defective *given that the first two were not defective* is $\frac{10}{14}$. Consequently, if we follow the multiplication rule explained in Exercise 20 on page 117, we find that the *correct* value of the probability (of getting three nondefective tape recorders) is

$$\frac{12}{16} \cdot \frac{11}{15} \cdot \frac{10}{14} = \frac{11}{28}$$

or approximately 0.393. Thus, the error which we made by using the binomial distribution is quite considerable—it is $0.422 - 0.393 = 0.029$.

The correct probability function for this kind of problem (namely, for *sampling without replacement*) is that of the **hypergeometric distribution**. It applies whenever n elements are randomly selected from a set containing a elements of one kind (successes) and b elements of another kind (failures), and we are interested in the probability of getting x successes and $n - x$ failures. The formula for this probability function is

$$f(x) = \frac{\binom{a}{x} \cdot \binom{b}{n-x}}{\binom{a+b}{n}} \quad \text{for } x = 0, 1, 2, \ldots, \text{ or } n$$

and it should be noted that when x is greater than a or $n - x$ is greater than b, the respective binomial coefficients $\binom{a}{x}$ and $\binom{b}{n-x}$ are equal to 0. Clearly, there is *no way* in which we can select more elements from a set than there are in the set.

Example 7 (Continued). If we apply this formula to our example with the tape recorders, where we had $x = 3$, $n = 3$, $a = 12$, and $b = 4$, we get

$$f(3) = \frac{\binom{12}{3} \cdot \binom{4}{0}}{\binom{16}{3}} = \frac{220 \cdot 1}{560} = \frac{11}{28}$$

The binomial coefficients were obtained from Table VIII, and the answer is, of course, the same as before.

To prove the formula for the hypergeometric distribution, we have only to observe that the x successes can be selected from among a possibilities in $\binom{a}{x}$ ways, the $n - x$ failures can be selected from among b possibilities in $\binom{b}{n-x}$ ways, so that the x successes *and* $n - x$ failures can be selected in $\binom{a}{x} \cdot \binom{b}{n-x}$ ways in accordance with the rule on page 71. The total number of ways in which we can select n elements from a set of $a + b$ elements is $\binom{a+b}{n}$, and according to the special rule for equiprobable events on page 101, the desired probability is given by the ratio of $\binom{a}{x} \cdot \binom{b}{n-x}$ to $\binom{a+b}{n}$.

Example 8. To consider another example, suppose that 100 of the 240 immates of a federal prison for women have radical political views. If 5 of them are randomly chosen to appear before a legislative committee, what is the probability that *at least one* of them will have radical political views? The probability that at least one of them will have radical political views is 1 *minus* the probability that *none* of them has radical political views; thus, we get

$$1 - f(0) = 1 - \frac{\binom{100}{0} \cdot \binom{140}{5}}{\binom{240}{5}} = 1 - \frac{140 \cdot 139 \cdot 138 \cdot 137 \cdot 136}{240 \cdot 239 \cdot 238 \cdot 237 \cdot 236}$$

$$= 1 - 0.066$$

$$= 0.934$$

rounded to three decimals.

Example 9. Finally, suppose that a customs official suspects that 5 of 16 ships due to arrive in port carry contraband. He does not know which ones carry the contraband, and since his office is understaffed, he can search only 8 of these ships. What he would like to know is the probability that *if his suspicion is correct*, his staff will catch *at least* 3 of the 5 ships that carry contraband. The probability he wants to know is given by $f(3) + f(4) + f(5)$, where each term in this sum is to be calculated by means of the formula for the hypergeometric distribution, with $a = 5$, $b = 11$, and $n = 8$. Substituting these values together with $x = 3$, $x = 4$, and $x = 5$, respectively, into the formula, we get

$$f(3) = \frac{\binom{5}{3} \cdot \binom{11}{5}}{\binom{16}{8}} = \frac{10 \cdot 462}{12{,}870} = 0.359$$

$$f(4) = \frac{\binom{5}{4} \cdot \binom{11}{4}}{\binom{16}{8}} = \frac{5 \cdot 330}{12{,}870} = 0.128$$

$$f(5) = \frac{\binom{5}{5} \cdot \binom{11}{3}}{\binom{16}{8}} = \frac{1 \cdot 165}{12{,}870} = 0.013$$

and the probability that at least 3 of the 5 ships with contraband will be caught is $0.359 + 0.128 + 0.013 = 0.500$. This suggests that the customs official should have his staff search more than 8 of the ships, and it will be left to the reader to show in Exercise 19 on page 138 that the probability of catching *at least 3* of the 5 ships with contraband is increased to 0.758 if his staff searches 10 of the ships.

In the beginning of this section we introduced the hypergeometric distribution in connection with a problem in which we *erroneously* used the binomial distribution. Actually, when n is small compared to $a + b$, the binomial distribution often provides a very good *approximation* to the hypergeometric distribution. It is generally agreed that this approximation can be used so long as n constitutes less than 5% of $a + b$; this is helpful because the binomial distribution has been tabulated much more extensively than the hypergeometric distribution, and it is generally easier to use. As the reader will be asked to verify in Exercise 20 on page 138, the answer would have been the same if we had used this approximation in Example 8.

Exercises

1. Check whether the following can be probability functions, and explain your answers:

a. $f(x) = \frac{1}{3}$ for $x = 0$, 1, and 2;

b. $f(x) = \frac{x}{12}$ for $x = 1$, 2, 3, and 4;

c. $f(x) = \frac{x - 2}{5}$ for $x = 1$, 2, 3, 4, and 5;

d. $f(x) = \frac{x^2}{30}$ for $x = 0$, 1, 2, 3, and 4.

2. In each case check whether the given values can be looked upon as the values of the probability function of a random variable which can take on only the values 1, 2, 3, and 4, and explain your answers:

a. $f(1) = 0.21$, $f(2) = 0.23$, $f(3) = 0.27$, and $f(4) = 0.29$;
b. $f(1) = 0.13$, $f(2) = 0.43$, $f(3) = 0.29$, and $f(4) = 0.17$;
c. $f(1) = 0.32$, $f(2) = 0.25$, $f(3) = -0.17$, and $f(4) = 0.60$;
d. $f(1) = \frac{1}{2}$, $f(2) = \frac{1}{4}$, $f(3) = \frac{1}{8}$, and $f(4) = \frac{1}{16}$.

3. Verify that $f(x) = \frac{1}{16} \cdot \binom{4}{x}$, for $x = 0, 1, 2, 3$, and 4, gives the probabilities for Example 2 on page 126. Look up the necessary binomial coefficients in Table VIII.

4. Using the *binomial expansions* $(a + b)^3 = a^3 + 3a^2b + 3ab^2 + b^3$ and $(a + b)^4 = a^4 + 4a^3b + 6a^2b^2 + 4ab^3 + b^4$ (which can be checked by performing the necessary term-by-term multiplications), show that

- **a.** the successive terms in the expansion of $(\frac{9}{10} + \frac{1}{10})^3$ equal the probabilities given in Table XI for getting 0, 1, 2, and 3 successes in three trials when $p = \frac{1}{10}$;
- **b.** the successive terms in the expansion of $(\frac{1}{5} + \frac{4}{5})^4$ equal the probabilities given in Table XI for getting 0, 1, 2, 3, and 4 successes in four trials when $p = \frac{4}{5}$.

5. A multiple-choice test consists of 8 questions and 3 answers to each question (of which only one is correct). If a student answers each question by rolling a balanced die and checking the first answer if he gets a 1 or a 2, the second answer if he gets a 3 or a 4, and the third answer if he gets a 5 or a 6, find (by means of the formula for the binomial distribution) the probability of getting

- **a.** exactly 3 correct answers;
- **b.** no correct answers;
- **c.** at least 6 correct answers.

6. Find the probability that 2 of the 4 members of an expedition will get frostbite, when the probability that any one of them will get frostbite is 0.25.

7. It is expected that 30% of the mice used in an experiment will become very aggressive within 1 minute after having been administered an experimental drug. Find the probability that exactly 3 of 6 mice which have been administered the drug become very aggressive within 1 minute

- **a.** by using the formula for the binomial distribution;
- **b.** by referring to Table XI.

8. Assuming that it is true that 2 in 10 automobile accidents are due to driver fatigue, find the probability that among 8 automobile accidents 3 will be due to driver fatigue

- **a.** by using the formula for the binomial distribution;
- **b.** by referring to Table XI.

9. If the probability is 0.10 that a hockey playoff game will go into overtime, find the probability that *at most 2* (0, 1, or 2) of 5 playoff games will go into overtime

- **a.** by using the formula for the binomial distribution;
- **b.** by referring to Table XI.

10. If it is true, as someone claims, that 80% of all industrial accidents can be prevented by paying strict attention to safety regulations, find the probability that *at least 6* of 7 industrial accidents can thus be prevented.

- **a.** Use the formula for the binomial distribution;
- **b.** Refer to Table XI.

11. Suppose that a civil service examination is designed so that 70% of all persons with an I.Q. of 90 can pass it. Use Table XI to find the probabilities that among 12 persons with an I.Q. of 90 who take the test

- **a.** at most 8 will pass;
- **b.** at least 10 will pass.

12. A social scientist claims that only 40% of all high school seniors who are capable of doing college work actually go to college. Use Table XI to find the probabilities that among 15 high school seniors capable of doing college work

- **a.** at most 4 will go to college;
- **b.** exactly 7 will go to college;
- **c.** at least 9 will go to college;
- **d.** anywhere from 5 through 10, inclusive, will go to college.

13. A quality control inspector wants to check whether (in accordance with specifications) 95% of the electronic components shipped by his company are in good working condition. To this end, he randomly selects 14 from each large lot ready to be shipped and passes the lot only if they are *all* in good working condition; otherwise, each of the components in the lot is checked.

- **a.** What is the probability that he will commit the error of holding a lot for further inspection even though 95% of the components are in good working condition?
- **b.** What is the probability that he will commit the error of letting a lot pass through even though only 80% of the components are in good working condition?

14. A study shows that 60% of all patients have to wait in their doctor's waiting room for at least 45 minutes. Use Table XI to find the probabilities that 0, 1, 2, 3, . . . , or 10 patients out of 10 have to wait in their doctor's waiting room for at least 45 minutes, and draw a *histogram* of this probability distribution.

15. Among the 16 cities which a political organization is considering for its next three annual conventions, 7 are in the western part of the United States. If, to avoid arguments, the selection is left to chance, what is the probability that none of these conventions will be held in the western part of the United States?

16. Among the 12 applicants for a research job, 8 have Ph.D.'s. If 2 of the applicants are randomly chosen for further consideration, find the probabilities that

a. none of them have Ph.D.'s;
b. only one has a Ph.D.;
c. both have Ph.D.'s.

17. When she buys a dozen eggs, Ms. Murphy always inspects 3 of the eggs carefully for cracks, and if at least one of them has a crack she looks for another carton. If she randomly selects the eggs which she inspects, what are the probabilities that Ms. Murphy will buy a carton with

a. 2 cracked eggs;
b. 3 cracked eggs;
c. 4 cracked eggs?

18. A shipment of 120 burglar alarms contains 5 that are defective. If 3 of these burglar alarms are randomly selected and shipped to a customer, find the probability that he will get exactly one bad unit.

19. Verify the statement made in Example 9 on page 135 that the probability of catching *at least 3* of the 5 ships with contraband would be increased to 0.758 if they searched 10 of the 16 ships.

20. Show that if we had used the binomial approximation with $p = \frac{100}{240} = 0.42$ in Example 8, the probability of *zero successes* would also have been 0.066 rounded to three decimals, and the answer would have been the same.

21. Among 600 plants exposed to excessive radiation, 90 show abnormal growth. If a scientist collects the seed of three of these plants chosen at random, find the probability that he will get the seed from one plant with abnormal growth and two plants with normal growth

a. by using the formula for the hypergeometric distribution;
b. by approximating the probability with the use of the formula for the binomial distribution having $n = 3$ and $p = \frac{90}{600} = 0.15$.

22. Among the 200 employees of a company, 160 are union members while the others are not. If 4 of the employees are to be chosen by lot to serve on a committee which administers the pension fund, find the probability that two of them will be union members while the others are not

a. by using the formula for the hypergeometric distribution;
b. by approximating the probability with the use of the formula for the binomial distribution having $n = 4$ and $p = \frac{160}{200} = 0.80$.

23. THE POISSON APPROXIMATION OF THE BINOMIAL DISTRIBUTION When n is large and p is small, binomial probabilities are often approximated by means of the values of the **Poisson distribution**, given by*

$$f(x) = \frac{(np)^x \cdot e^{-np}}{x!} \qquad \text{for } x = 0, 1, 2, 3, \ldots$$

Here e is the number 2.71828 . . . used in connection with natural logarithms; there is no need to be concerned about this, however, since the required values of e^{-np} are given in Table IX at the end of the book. To illustrate this technique, suppose that a very large shipment of books contains 2% with defective bindings. To find the probability that among 400 books taken at random from this shipment exactly 5 will have imperfect bindings, we substitute $x = 5$, $n = 400$, $p = 0.02$, and, hence, $np = 400(0.02) = 8$ into the formula for the Poisson distribution, getting

$$f(5) = \frac{8^5 \cdot e^{-8}}{5!} = \frac{32{,}768(0.00034)}{120} = 0.093$$

a. If 5% of the drivers on the road are not properly licensed, use the Poisson approximation to find the probability that among 100 drivers stopped at a road block, 3 will not be properly licensed.
b. If 1.6% of the fuses delivered to an arsenal are defective, use the Poisson approximation to find the probability that among 200 fuses (randomly selected), exactly 4 will be defective.
c. If 2.4% of the calls arriving at a switchboard are wrong numbers, use the Poisson approximation to find the probability that among 250 calls received by this switchboard, exactly 5 will be wrong numbers.

The mean of a probability distribution

On page 89 we said that a child in the age group from 6 to 16 goes to the dentist 1.9 times a year, and we pointed out that this figure is the sum of the products obtained by multiplying 0, 1, 2, 3, 4, . . . , by the respective probabilities that a child in this age group will visit a dentist that many times a year.

*Note that the sample space, consisting of all the nonnegative integers, is *infinite*; however, this is merely a matter of mathematical convenience, and in actual practice the probabilities become negligible (very close to 0) after a fairly small set of values of x.

Example 2 (Continued). To give a similar example in which we already know all the probabilities, let us find the expected number of heads for four flips of a balanced coin. Multiplying 0, 1, 2, 3, and 4 by the corresponding probabilities shown in the table on page 126, we find that the sum of these products is

$$0(\tfrac{1}{16}) + 1(\tfrac{4}{16}) + 2(\tfrac{6}{16}) + 3(\tfrac{4}{16}) + 4(\tfrac{1}{16}) = 2$$

and this result should not come as a surprise—as we pointed out on page 87, a mathematical expectation represents an *average*, or as we referred to it in Chapter 2, a *mean*.

Thus, let us state formally that

The mean of a probability distribution is the mathematical expectation (also called the expected value) of a random variable having the particular distribution.

Symbolically, the mean of a probability distribution is given by the formula

$$\mu = \sum x \cdot f(x)$$

where the summation extends over all values of x which the random variable can take on, and the quantities $f(x)$ are the corresponding probabilities. The symbol μ which we use to denote the mean of a probability distribution is the Greek letter *mu*, and it should be noted that the formula for μ is very similar to the one for the weighted mean on page 33. The weights which we assign to the x's are simply their probabilities, and the denominator in the formula on page 33 can be omitted since the sum of the probabilities (of a probability function) must always be equal to 1.

Example 4 (Continued). To give another example, let us calculate the mean of the probability distribution on page 130, which pertained to the number of times, in five races, an athlete can run the mile in less than 4 minutes. Substituting the probabilities which we calculated on page 130 into the formula for μ, we get

$$\begin{aligned}\mu &= 0(0.010) + 1(0.077) + 2(0.230) + 3(0.346) \\ &\qquad + 4(0.259) + 5(0.078) \\ &= 3.001\end{aligned}$$

or approximately 3. Thus, if the athlete's claim is correct, namely, if he can run the mile in less than 4 minutes 60% of the time, we can expect him to run the mile in less than 4 minutes *three times out of five.*

When a random variable can take on many different values, the calculation of μ becomes very laborious unless we use simplifications. For instance, if we wanted to determine how many of the 360 persons at a movie can be *expected* to buy popcorn (when the probability is 0.30 that any one of them will buy popcorn), we would have to calculate the 361 probabilities corresponding to 0, 1, 2, 3, . . . , and 360 of them buying popcorn, multiply each of these probabilities by the corresponding value of the random variable (the number of persons who will buy popcorn), and then add the 361 products. However, if we think for a moment, we might argue that *in the long run* 30% of the persons going to this movie will buy popcorn, 30% of 360 is 0.30(360) = 108, and, hence, we might conclude that 108 of the 360 persons can be expected to buy popcorn. Similarly, if a balanced coin is flipped 800 times, we can argue that *heads* should come up about 50% of the time, and, hence, that we can *expect* to get 800(0.50) = 400 heads in 800 flips of a balanced coin. Without this argument, we would have had to determine the 801 probabilities of getting 0, 1, 2, 3, . . . , and 800 heads, and it is fortunate, indeed, that there is the special formula

$$\mu = n \cdot p$$

for the mean of a binomial distribution. In words, this formula expresses the fact that

The mean of a binomial distribution is given by the product of the number of trials and the probability of success on each individual trial.

Proofs of this result may be found in most textbooks on the theory of statistics, for instance, in those listed in the Bibliography at the end of the book.

Using the formula $\mu = n \cdot p$, we can now verify the results obtained earlier in this section.

Example 2 (Continued). For the number of heads in four flips of a balanced coin, we have $n = 4$, $p = \frac{1}{2}$, and, hence, $\mu = 4 \cdot \frac{1}{2} = 2$.

Example 4 (Continued). For the number of times the athlete can run the mile in less than 4 minutes in 5 races, we have $n = 5$, $p = 0.60$ (according to his claim), and, hence, $\mu = 5(0.60) = 3$. This also agrees with the result which we obtained before (except for the rounding error of 0.001).

It is important to remember, of course, that the formula $\mu = n \cdot p$ applies only to binomial distributions. Fortunately, there are other special formulas for other special distributions; for the *hypergeometric distribution*, for example, the formula is

$$\mu = \frac{n \cdot a}{a + b}$$

Example 9 (Continued). If 5 of the 16 ships carry contraband and 8 of them are searched, we have $a = 5$, $b = 11$, $n = 8$, and substitution into the formula yields

$$\mu = \frac{8 \cdot 5}{5 + 11} = 2.5$$

This is the number of ships with contraband that we can expect to get caught when 8 are searched.

Measuring chance variation

In the introduction to this chapter we claimed that it is possible to *predict*, *measure*, and sometimes even *control* chance variations. This is tremendously important, for we are often faced with situations in which we must decide whether differences (discrepancies) between *what we expect* and *what we get* can reasonably be attributed to chance.

Example 10. For instance, when a balanced coin is flipped 100 times, we would expect to get 50 *heads* and 50 *tails*, but what can we say about the coin (or the manner in which it was flipped) if we get 42 *heads* and 58 *tails*, or perhaps 26 *heads* and 74 *tails*?

Example 11. Similarly, if the starter of a car is said to be performing satisfactorily if it fails only one-half of 1% of the time, what can we say about a starter which fails 5 times in 400 tries (which is more than one-half of 1%)? Is the difference small enough to be attributed to chance, or is it big enough to say that the starter is of inferior quality, perhaps due to poor workmanship?

To answer questions of this kind we shall first have to develop a way of measuring the fluctuations of random variables, or as we put it on page 125, we need a "yardstick" for measuring variations that are due to chance. Now, if a random variable takes on the value x and the mean of its probability distribution is μ, then the difference $x - \mu$ is called the **deviation from the mean**, and it measures by how much the value of the random variable differs from what we would expect.

Example 10 (Continued). For instance, if we get 42 *heads* in 100 flips of a balanced coin, then $\mu = 50$ and the amount by which we are "off" is

$$x - \mu = 42 - 50 = -8$$

Thus, we got 8 fewer *heads* than expected, and it would be of interest to know whether this discrepancy can be attributed to chance, or whether the coin is not really balanced or was improperly flipped.

This leaves us with the question of *how we can know what to expect*, namely, *by how much we can expect to be "off" in a given situation*; to look for an answer, let us return to the example which dealt with the number of *heads* we get in 4 flips of a coin.

Example 2 (Continued). As we saw on page 141, $\mu = 2$ for the number of *heads* we get in 4 flips of a coin. Thus, if we get 0 heads we are off by $0 - 2 = -2$, if we get 1 head we are off by $1 - 2 = -1$, if we get 2 heads we are off by $2 - 2 = 0$, if we get 3 heads we are off by $3 - 2 = 1$, and if we get 4 heads we are off by $4 - 2 = 2$. Since the corresponding probabilities are $\frac{1}{16}$, $\frac{4}{16}$, $\frac{6}{16}$, $\frac{4}{16}$, and $\frac{1}{16}$ according to the table on page 126, we find that the (average) amount by which we can *expect* to be off is

$$(-2)(\tfrac{1}{16}) + (-1)(\tfrac{4}{16}) + 0(\tfrac{6}{16}) + 1(\tfrac{4}{16}) + 2(\tfrac{1}{16}) = 0$$

The trouble with the answer to this example is that some of the deviations from the mean are *positive* while others are *negative*, so that the amount by which we can *expect* to be off is *zero*, and, hence, *not indicative of the size of the chance fluctuations in the number of heads we might get in 4 flips of a coin.*

There are several ways in which we can get rid of the minus signs, in which we are really not interested. One possibility is to ignore the signs of the amounts by which we are off (namely, take their *absolute values*), but mathematically more convenient is to *square* the amounts by which we are off—after all, the squares of real numbers are always positive or zero.

Example 2 (Continued). Referring again to the number of *heads* which we get in 4 flips of a balanced coin, we can thus say that the *average* (namely, the *mathematical expectation*) of the squares of the amounts by which we are off is

$$(-2)^2(\tfrac{1}{16}) + (-1)^2(\tfrac{4}{16}) + 0^2(\tfrac{6}{16}) + 1^2(\tfrac{4}{16}) + 2^2(\tfrac{1}{16}) = 1$$

The quantity which we have calculated is called the **variance** of the probability distribution, and in general it is defined as follows:

The variance of a probability distribution is the mathematical expectation of the squared deviations from the mean (namely, the squares of the amounts by which we are off); symbolically, it is given by the formula

$$\sigma^2 = \sum (x - \mu)^2 \cdot f(x)$$

where the summation extends over all values of x for which the probability function is defined.

The symbol which we use to denote the variance of a probability distribution is σ^2, where σ is the lowercase Greek letter *sigma*. If we want to compensate for the fact that the variance is the expected value of the *squared* deviations from the mean, we can simply take the square root of the variance and we will get another widely used measure of chance variation, the **standard deviation**:

$$\sigma = \sqrt{\sum (x - \mu)^2 \cdot f(x)}$$

The following example will give the reader some idea of how the variance (or the standard deviation) reflects the average size of chance fluctuations:

Example 12. Figure 5.1 contains the histograms of four probability distributions defined for $x = 1, 2, 3, \ldots$, and 9. They all have the mean $\mu = 5$, as can easily be verified using the formula on page 140, but their respective variances are $\sigma^2 = 5.26$, $\sigma^2 = 3.18$, $\sigma^2 = 1.66$, and $\sigma^2 = 0.88$. Aside from the obvious differences in the "spread" of these four distributions, the reader can verify for himself that the probabilities of getting a value which differs from $\mu = 5$ by *two or more* are, respectively, 0.58, 0.40, 0.24, and 0.08. Similarly, the probabilities of getting a value which differs from $\mu = 5$ by *three or more* are, respectively, 0.34, 0.20, 0.04, and 0.04.

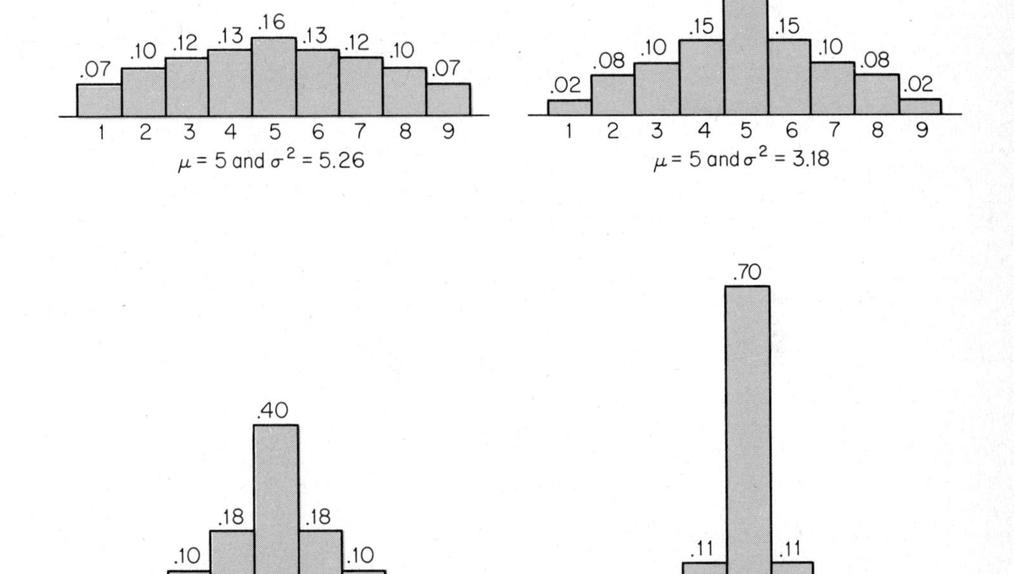

Figure 5.1 Histograms of four probability distributions.

As is illustrated by this example, a small value of σ^2 implies that we are more likely to get a value close to the mean, while a large value of σ^2 implies that we are more likely to get a value far away from the mean. This will be discussed further in the next section.

Example 4 (Continued). To give another illustration of the calculation of the variance of a probability distribution, let us refer again to the example on page 130, which dealt with the number of times, in five races, an athlete can run the mile in less than 4 minutes. Since we have already shown that the mean of this distribution is $\mu = 3$, we can arrange the necessary calculations as in the following table, where x is the number of times the athlete will run the miles in less than 4 minutes:

x	$x - \mu$	$(x - \mu)^2$	*Probability* $f(x)$	$(x - \mu)^2 f(x)$
0	−3	9	0.010	0.090
1	−2	4	0.077	0.308
2	−1	1	0.230	0.230
3	0	0	0.346	0.000
4	1	1	0.259	0.259
5	2	4	0.078	0.312
				$\sigma^2 = 1.199$

The values in the right-hand column were obtained by multiplying each squared deviation from the mean by the corresponding probability, and their sum gives the variance of the distribution. To find the standard deviation of this distribution, we have only to look up the square root of 1.199, and the nearest value in Table X at the end of the book is $\sqrt{1.20} = 1.0954$.

As in the case of the mean, the calculation of the variance or the standard deviation can be greatly simplified when we deal with special kinds of probability distribution. For instance, for the *binomial distribution* we have the formula

$$\sigma^2 = np(1 - p)$$

which we shall not prove, but which can easily be verified for our two examples.

Example 2 (Continued). For the distribution of the number of heads in 4 flips of a coin we have $n = 4$ and $p = \frac{1}{2}$, and the formula yields

$$\sigma^2 = 4 \cdot \tfrac{1}{2} \cdot (1 - \tfrac{1}{2}) = 4 \cdot \tfrac{1}{2} \cdot \tfrac{1}{2} = 1$$

which agrees with the result on page 144.

Example 4 (Continued). For the distribution of the number of times the athlete will run the mile in less than 4 minutes we have $n = 5$ and $p = 0.60$, and the formula yields

$$\sigma^2 = 5(0.60)(1 - 0.60) = 5(0.60)(0.40) = 1.20$$

which agrees with the result obtained above (except for the rounding error of 0.001).

Chebyshev's theorem

The variance and the standard deviation of a probability distribution were introduced as measures of the *expected chance fluctuations* of a random variable having that distribution, and as we already pointed out on page 145, a small value of σ or σ^2 implies that we are likely to get a value which is very close to the mean, while a large value of σ or σ^2 implies that we are more likely to get a value far away from the mean. Formally, this important idea is expressed by the following rule, called **Chebyshev's theorem:**

The probability that a random variable will take on a value within k standard deviations of the mean is always at least

$$1 - \frac{1}{k^2}$$

where k can be any positive number.

Thus, the probability of getting a value within *two* standard deviations of the mean (namely, a value between $\mu - 2\sigma$ and $\mu + 2\sigma$) is always *at least* $1 - \frac{1}{2^2} = 0.75$. Similarly, the probability of getting a value within *five* standard deviations of the mean (namely, a value between $\mu - 5\sigma$ and $\mu + 5\sigma$) is always *at least* $1 - \frac{1}{5^2} = 0.96$, and the probability of getting a value within *ten* standard deviations of the mean is always *at least* $1 - \frac{1}{10^2} = 0.99$.

Example 13. To consider a concrete application, suppose that the number of telephone calls which a doctor receives between 9 A.M. and 10 A.M. is a random variable whose distribution has the mean $\mu = 17$ and $\sigma = 4.1$. Thus, if we use Chebyshev's theorem with $k = 2$, we can say that the probability is *at least 0.75* (or that the odds are *at least 3 to 1*) that the doctor will receive between $17 - 2(4.1) = 8.8$ and $17 + 2(4.1) = 25.2$ calls, namely, that he will receive anywhere from 9 to 25 calls. Similarly, for $k = 5$ we can say that the probability is *at least* 0.96 (or that the odds are *at least 24 to 1*) that the doctor will receive less than $17 + 5(4.1) = 37.5$, namely, at most 37 calls. [We did not have to mention the lower limit in this case since $17 - 5(4.1)$ is *negative*.]

Example 14. To give another example, suppose that we have won only 139 times betting on *heads* in 400 flips of a coin, and that we are beginning to wonder whether we should raise any questions about the conduct of the game. If the game is honest, we should have a fifty-fifty chance of winning on each flip of the coin, and the number of heads in 400 flips of a balanced coin is a random variable having the binomial distribution with $n = 400$, $p = \frac{1}{2}$, and, hence,

$$\mu = 400 \cdot \tfrac{1}{2} = 200$$

and

$$\sigma = \sqrt{400 \cdot \tfrac{1}{2} \cdot (1 - \tfrac{1}{2})} = 10$$

Since the 139 heads we got differs from $\mu = 200$ by 61, which is *more than six standard deviations*, we could argue that the probability is *at least* $1 - \frac{1}{6^2} = \frac{35}{36} = 0.972$ that we should get a value *within six standard deviations of the mean*, and, hence, that something may well be wrong with the coin or the way in which it is flipped.

When we are dealing with random variables having binomial distributions and we are interested in the *proportion* of "successes," Chebyshev's theorem leads to the **Law of Large Numbers**, which laymen often refer to as the **Law of Averages**.

Example 13 (Continued). Using the results obtained above, we can say that for 400 flips of a coin the probability is *at least* 0.972 (Chebyshev's theorem with $k = 6$) that the number of heads will be between $200 - 6(10) = 140$ and $200 + 6(10) = 260$, and, hence, that the *proportion* of heads will be between $\frac{140}{400} = 0.35$ and $\frac{260}{400} = 0.65$. To continue this argument, the reader will be asked to show in Exercise 14 on page 151 that the probability is *at least* 0.972 that for 10,000 flips of a coin the *proportion* of heads will be between 0.47 and 0.53, or that for 1,000,000 flips of a coin the *proportion* of heads will be between 0.497 and 0.503.

In general, the Law of Large Numbers states that

When the number of trials grows larger and larger, the proportion of "successes" will tend to come closer and closer to the probability of a "success" for each trial.

This is true, incidentally, regardless of whether or not the probability of a success on each individual trial happens to be 0.50. A very important aspect of the Law of Large Numbers is that it justifies the *frequency interpretation of probability* which we studied in Chapter 3—after all, it was there that we talked about what happens to a proportion "in the long run."

In actual practice, Chebyshev's theorem is very rarely used; the very fact that it applies to *any* probability distribution has the unfortunate side effect that the probability "at least $1 - \frac{1}{k^2}$" is often *unnecessarily small*. For instance, in Example 13 we stated that the probability of getting a value within six standard deviations of the mean is *at least* 0.972, whereas the *actual* probability that this will happen for a random variable having a binomial distribution with $n = 400$ and $p = \frac{1}{2}$ is about 0.999999998. "*At least* 0.972" is, of course, correct, but it does not tell us quite enough.

Exercises

1. The probabilities that there are, respectively, 0, 1, 2, or 3 bank robberies in a western city in any given month are 0.4, 0.3, 0.2, and 0.1.

a. Find the mean of this probability distribution.
b. Find the variance of this probability distribution.

2. The following is the probability distribution for the number of trout, x, a person will catch in one hour while fishing in the early morning at a certain lake:

x	0	1	2	3	4	5	6
$f(x)$	0.13	0.27	0.27	0.19	0.09	0.04	0.01

a. Find the mean of this distribution.
b. Find the standard deviation of this distribution.

3. The following table gives the probabilities that a computer will malfunction 0, 1, 2, 3, 4, 5, or 6 times on any given day:

Number of Malfunctions x	0	1	2	3	4	5	6
Probability f(x)	0.17	0.29	0.27	0.16	0.07	0.03	0.01

a. Find the mean of this distribution.
b. Find the variance of this distribution.

4. A real estate salesman figures that 80% of all the women to whom he shows a certain model home will complain about the size of the kitchen. Referring to Table XI, find the probabilities that 0, 1, 2, . . . , or 12 women out of 12 to whom he shows the model home will complain about the size of the kitchen, and use these probabilities to calculate the mean. Verify the result by means of the formula $\mu = n \cdot p$.

5. Find the mean and the standard deviation of each of the following binomial random variables:

a. the number of *heads* obtained in 900 flips of a balanced coin;
b. the number of *threes* obtained in 720 rolls of a balanced die;
c. the number of seeds that will germinate in a package of 36, if the probability that any one of the seeds will germinate is 0.20;
d. the number of students (among 400 interviewed) who prefer to live in a coed dormitory, if the probability that any one student will prefer to live in a coed dormitory is 0.70.

6. In Example 9 on page 134 we showed that if 5 of the 16 ships carry contraband and 8 are searched, the probabilities that 3, 4, or 5 of the ships with contraband will be caught are, respectively, 0.359, 0.128, and 0.013. Given that the corresponding probabilities for 0, 1, or 2 of the ships with contraband getting caught are, respectively, 0.013, 0.128, and 0.359, calculate the mean of this hypergeometric distribution, and compare this result with that obtained on page 142.

7. A collection of 15 silver dollars includes 4 that are counterfeits.

a. If 3 of these coins are randomly selected to be sold at auction, find the probabilities that 0, 1, 2, or all 3 of them are counterfeits.
b. Use the probabilities obtained in part (a) to calculate the mean of this hypergeometric distribution, and use the special formula on page 142 to check the result.

8. As can readily be verified by means of the formula for the binomial distribution (or by listing all 32 possibilities), the probabilities of getting 0, 1, 2, 3, 4, or 5 heads in 5 flips of a balanced coin are, respectively, $\frac{1}{32}$, $\frac{5}{32}$, $\frac{10}{32}$, $\frac{10}{32}$, $\frac{5}{32}$, and $\frac{1}{32}$.

a. Use these probabilities to show that the mean of this probability distribution is $\mu = 2.5$.

b. Use the given probabilities and the result of part (a) to show that the standard deviation of this probability distribution is $\sigma = \sqrt{5}/2$ (or approximately 1.12).

c. According to Chebyshev's theorem, the probability of getting a value within $k = 2$ standard deviations of the mean is *at least* $1 - \dfrac{1}{2^2} = 0.75$. What is the exact value of this probability for the given distribution?

9. A student answers the 120 questions on a true–false test by flipping a coin (*heads* is "true" and *tails* is "false").

a. Use the special formulas for the mean and the standard deviation of a binomial distribution to calculate μ and σ for the distribution of the number of correct answers the student will get.

b. If we use Chebyshev's theorem with $k = 3$, what can we assert about the number of correct answers the student will get?

c. If we use Chebyshev's theorem, what can we assert with a probability of *at least* $\frac{15}{16}$ about the number of correct answers the student will get?

10. With reference to part (c) of Exercise 5, what does Chebyshev's theorem with $k = 2$ assert about the number of seeds that will germinate?

11. With reference to part (d) of Exercise 5, what does Chebyshev's theorem with $k = 8$ assert about the number of students who prefer to live in coed dormitories?

12. With reference to part (b) of Exercise 5, what does Chebyshev's theorem tell us about the probability of getting anywhere between 70 and 170 *threes* in 720 rolls of a balanced die?

13. Suppose that Mr. Jones flips a coin 100 times and that he gets 30 heads, which is 20 less than the 50 he can expect. Then he flips the coin another 100 times and gets 44 heads, so that altogether he has 74 heads, which is 26 less than the 100 he can expect. Discuss his complaint that the "Law of Averages," namely, the Law of Large Numbers, is letting him down.

14. Duplicating the work on page 148, show that

a. we can assert with a probability of at least $\frac{35}{36} = 0.972$ that in 10,000 flips of a balanced coin the *proportion* of heads will be between 0.47 and 0.53;

b. we can assert with a probability of at least $\frac{35}{36} = 0.972$ that in 1,000,000 flips of a balanced coin the *proportion* of heads will be between 0.497 and 0.503.

6

The normal distribution

Introduction

Continuous sample spaces arise whenever we deal with quantities that are measured on a *continuous scale*—for instance, when we measure the speed of a car, the amount of alcohol in a person's blood, when we measure the weight of a package of frozen food, when we measure the amount of tar in a cigarette, and so forth. It is true, of course, that in actual practice we always round our answers to the nearest whole unit or to a few decimals, but this does not take away from the fact that in each of these examples there *is* a a continuum of possibilities, and that we *could* ask for probabilities associated with individual numbers or points. However, if we did this, we would find that (for all practical purposes) *the answers would always be zero* —surely, we should be willing to give *any odds* that a car will not be traveling exactly 20π miles per hour (where π is the irrational number 3.1415926 . . . which arises in connection with the area of a circle), and we should be willing to give *any odds* that the weight of a package of frozen food will not be exactly $\sqrt{64.1} = 8.0062475605 \ldots$ ounces. *Anyhow, in the continuous case, we are not really interested in probabilities associated with individual points of the sample space, but in probabilities associated with intervals or regions.* For instance, we might ask for the probability that (at a certain check point) a car will be traveling anywhere from 55 to 65 miles per hour, or that a package of frozen food weighs anywhere from 7.95 to 8.05 ounces.

Continuous distributions

When we first discussed histograms in Chapter 2, we pointed out that the frequencies, percentages, or proportions associated with the various classes are given by the *areas* of the rectangles, and with reference to Figure 5.1 on page 145 we might add that this is true also for the probabilities associated with the values of a random variable. In the continuous case, we also represent probabilities by means of areas as is illustrated in Figure 6.1, but instead of the areas of rectangles we now use areas under continuous curves. The first diagram of Figure 6.1 represents the probability

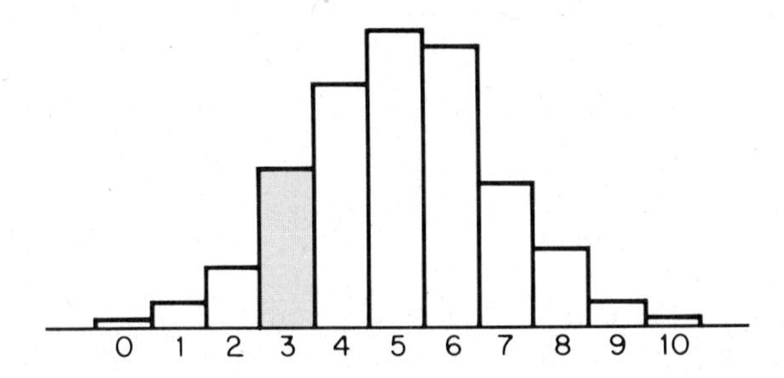

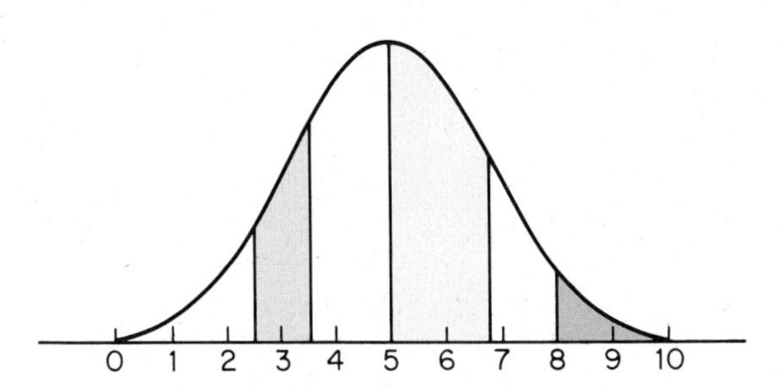

Figure 6.1 Histogram of probability function and graph of continuous distribution.

distribution of a random variable which takes on the values 0, 1, 2, . . . , 9, and 10, and the probability of getting a 3, for example, is given by the area of the shaded rectangle. The second diagram refers to a random variable which can take on any value on the *continuous* interval from 0 to 10, and the probability of getting a value between 2.5 and 3.5 is given by the medium green area under the curve. Similarly, the light green area under the curve gives the probability of getting a value between 5 and 6.8, while the dark green area gives the probability of getting a value greater than 8.

Curves like the one shown in the second diagram of Figure 6.1 are the graphs of functions which we refer to as **probability densities**, or more informally as **continuous distributions**. (The first of these terms is borrowed from the language of physics, where the terms "weight" and "density" are used in very much the same way in which we use the terms "probability" and "probability density" in statistics.) What characterizes a probability density is the fact that

> **The area under the curve between any two values a and b (for instance, those of Figure 6.2) gives the probability that a random variable having this "continuous distribution" will take on a value on the interval from a to b.**

It follows from this that the values of a probability density function *cannot be negative*, and that *the total area under the curve* (*representing the certainty that a corresponding random variable must take on one of its values*) *is always equal to 1*.

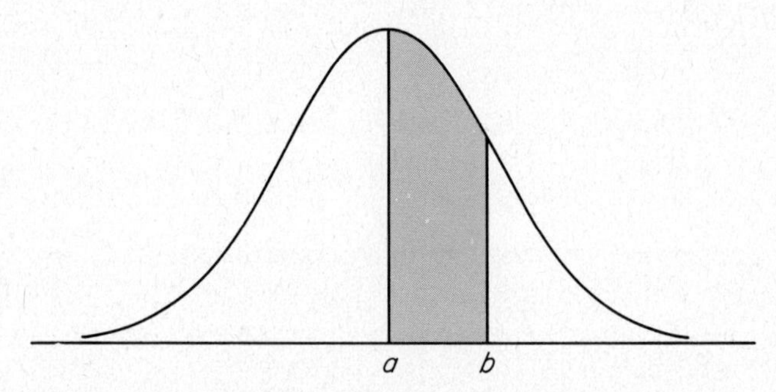

Figure 6.2 A probability density.

It is impossible to define the mean and the standard deviation (or variance) of a continuous distribution without the use of calculus, but we can always picture continuous distributions as approximated by means of histograms of probability functions for which the mean and the standard deviation can be calculated in accordance with the formulas on pages 140 and 144 (see Figure 6.3). Then, if we choose histograms with narrower and narrower classes, the means and the standard deviations of the corresponding probability functions will approach the mean the standard deviation of the continuous distribution. Actually, the mean and the standard deviation of a continuous distribution measure the same features as the mean and the standard deviation of a probability function—the *expected value* of a random variable having the given distribution, and the *average* (that is, the mathematical expectation) *of its squared deviations from the mean.* More

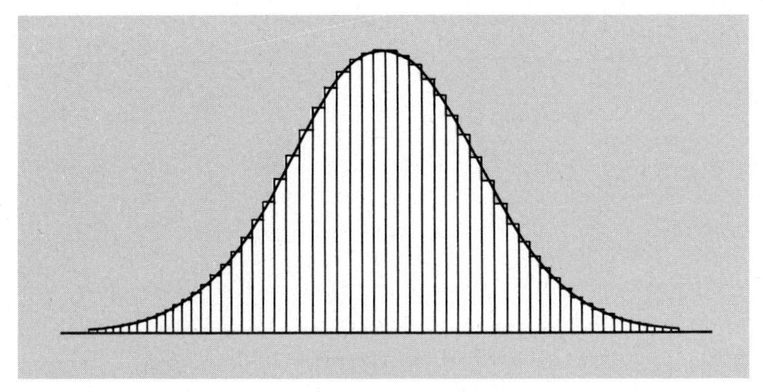

Figure 6.3 Curve of continuous distribution approximated by means of histogram of probability function.

intuitively, the mean μ of a continuous distribution is indicative of its "center" or "middle," while the standard deviation σ of a continuous distribution measures its "dispersion" or "spread."

The standard normal distribution

Among the many special kinds of continuous distributions that are used in statistics, **normal distributions** are by far the most important. Their study dates back to the eighteenth century and investigations into the nature of experimental errors. It was observed that variations among repeated measurements of the same physical quantity displayed a surprising degree of regularity, and it was found that their pattern (distribution) could be closely approximated by a certain kind of continuous distribution. This distribution was referred to as the "normal curve of errors" and it was attributed to the laws of chance.

The graph of a normal distribution is a bell-shaped curve that extends indefinitely in both directions, coming closer and closer to the horizontal axis without ever reaching it. Actually, it is seldom necessary to extend the "tails" of the curve very far, since the area under the curve becomes negligible when we go more than four or five standard deviations away from the mean. A very important feature of any normal distribution is that it is *symmetrical about its mean*; with reference to Figure 6.4 this means that if we folded the diagram along the dashed line, the two halves of the curve would coincide. Another important feature is that *there is one and only one normal distribution which has a given mean μ and a given standard deviation σ*. For instance, the normal distribution which has

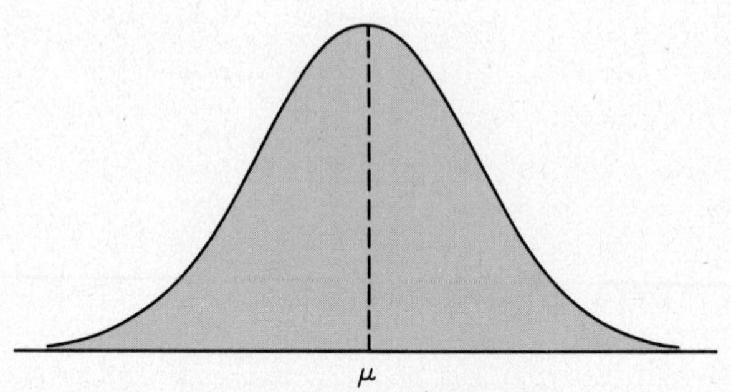

Figure 6.4 Graph of a normal distribution.

$\mu = 10$ and $\sigma = 5$ is shown in Figure 6.5 together with the normal distribution which has $\mu = 20$ and $\sigma = 10$. Thus,

> **Normal curves can have different shapes and they can be moved to the left or to the right, but there is one and only one normal distribution for any given pair of values of μ and σ.**

In actual practice, probabilities related to normal distributions (that is, areas under the corresponding curves) are obtained from special tables, such as Table I at the end of this book. As is apparent from Figure 6.5,

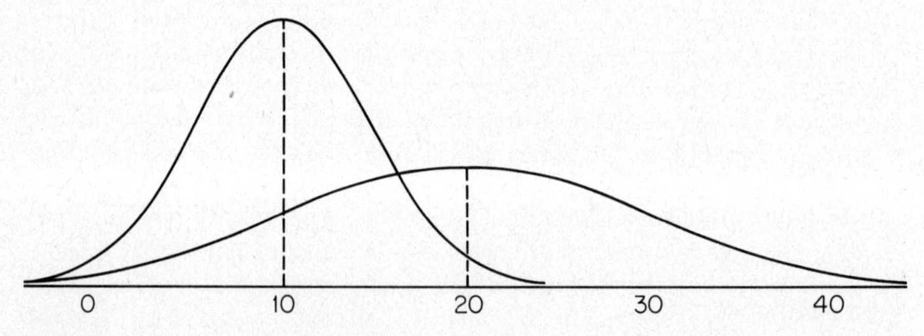

Figure 6.5 Two normal distributions.

areas under normal distributions (and, hence, probabilities related to these distributions) will differ depending on their means and standard deviations, but this difficulty is resolved by always working with the **standard normal distribution**, namely, the normal distribution which has $\mu = 0$ and $\sigma = 1$. Areas under any other normal curve can then be obtained by performing the change of scale illustrated in Figure 6.6. All we are really doing here is converting the units of measurement into **standard units** by means of the formula

$$z = \frac{x - \mu}{\sigma}$$

Note that a value of z simply tells us *how many standard deviations* the corresponding x value is above or below the mean.

Example 1. For instance, to find the area between 12 and 15 under the *first* normal curve of Figure 6.5 (the one with $\mu = 10$ and $\sigma = 5$), we simply look for the area under the standard normal distribution between

$$z = \frac{12 - 10}{5} = 0.40 \quad \text{and} \quad z = \frac{15 - 10}{5} = 1.00$$

Similarly, to find the area between 12 and 15 under the *second* curve of Figure 6.5 (the one with $\mu = 20$ and $\sigma = 10$), we look for the area under the standard normal distribution between

$$z = \frac{12 - 20}{10} = -0.80 \quad \text{and} \quad z = \frac{15 - 20}{10} = -0.50$$

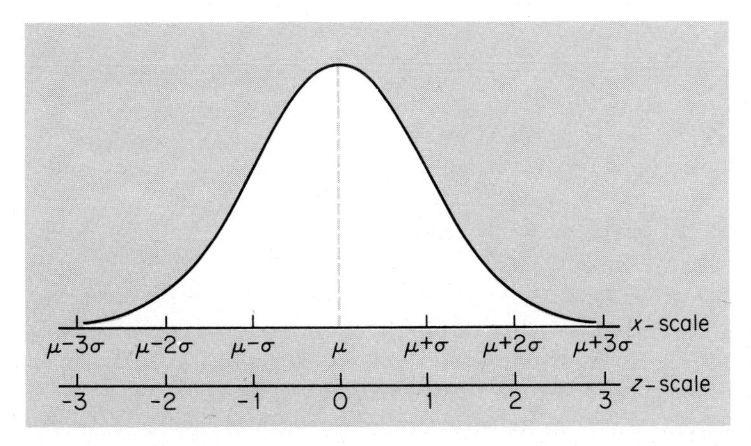

Figure 6.6 Change of scale.

This will actually be done in the continuation of this example on page 160. Note that in the first case the two z-values are *positive* because 12 and 15 both exceed the mean $\mu = 10$, while in the second case they are *negative* because 12 and 15 are both less than $\mu = 20$. The two areas, themselves, will be positive, of course, because they are probabilities.

The table to which we shall have to refer in problems like this is Table I, *whose entries are the areas under the standard normal distribution between the mean* ($z = 0$) *and* $z = 0.00, 0.01, 0.02, 0.03, \ldots, 3.08,$ *and* 3.09. In other words, the entries in Table I are areas like the one tinted in Figure 6.7.

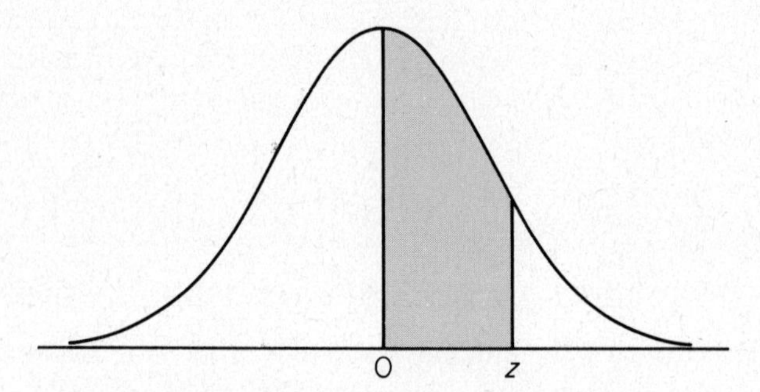

Figure 6.7 Tabulated areas under the graph of the standard normal distribution.

Also observe that Table I has no entries corresponding to *negative* values of z, for these are not needed by virtue of the symmetry of the standard normal distribution about its mean, namely, about $z = 0$.

Example 2. For instance, to find the area between $z = -1.20$ and $z = 0$, we simply look up the area between $z = 0$ and $z = 1.20$ (see Figure 6.8), and to determine the area between $z = -0.85$ and $z = -0.15$, we simply find the area between $z = 0.15$ and $z = 0.85$. As can easily be verified, the answer to the first problem is 0.3849, namely, the entry of Table I which corresponds to $z = 1.20$. So far as the second problem is concerned, the area cannot be looked up directly, but we can look up the area under the curve between $z = 0$ and $z = 0.85$, the area between $z = 0$ and $z = 0.15$, and then take the *difference* between the two. We thus get $0.3023 - 0.0596 = 0.2427$, as the reader will be asked to verify in Exercise 12 on page 165.

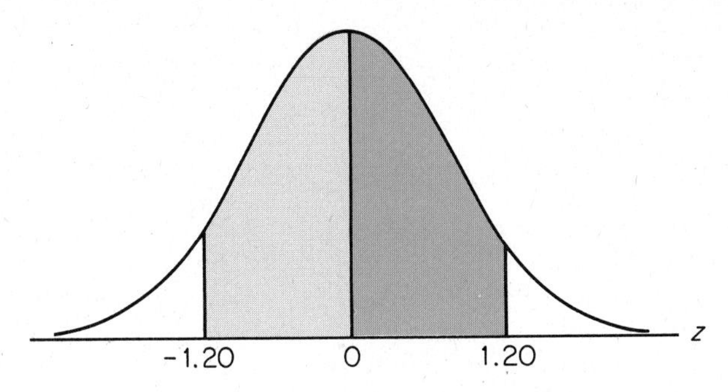

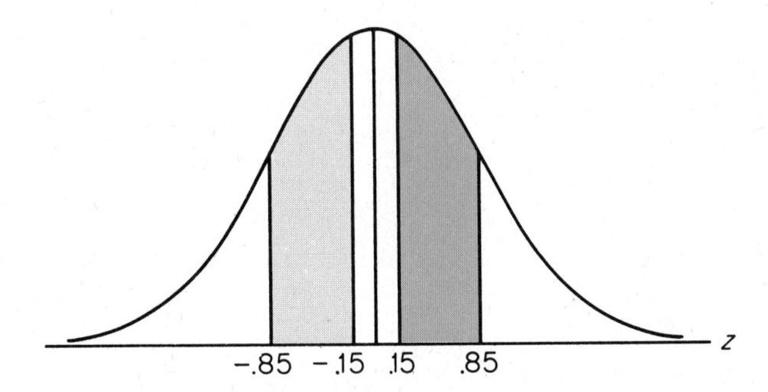

Figure 6.8 Diagram for Example 2.

Questions concerning areas under normal distributions arise in various ways, and the ability to find any desired area quickly can be a big help. Although the table gives only areas between the mean $z = 0$ and selected positive values of z, we often have to find areas to the left or to the right of given positive or negative values of z, or areas between two given values of z (as in the second example of the preceding paragraph). Finding any of these areas is easy, provided we remember exactly what areas are represented by the entries of Table I, and make use of the fact that the standard normal distribution is symmetrical about $z = 0$, so that the area to the left of $z = 0$ and the area to the right of $z = 0$ are both equal to 0.5000.

Example 3. For instance, we find that the probability of getting a z less than 0.87 (namely, the area under the curve to the left of $z = 0.87$, which is tinted in the first diagram of Figure 6.9) is $0.5000 + 0.3078 = 0.8078$, and that the probability of getting a z greater than -0.45

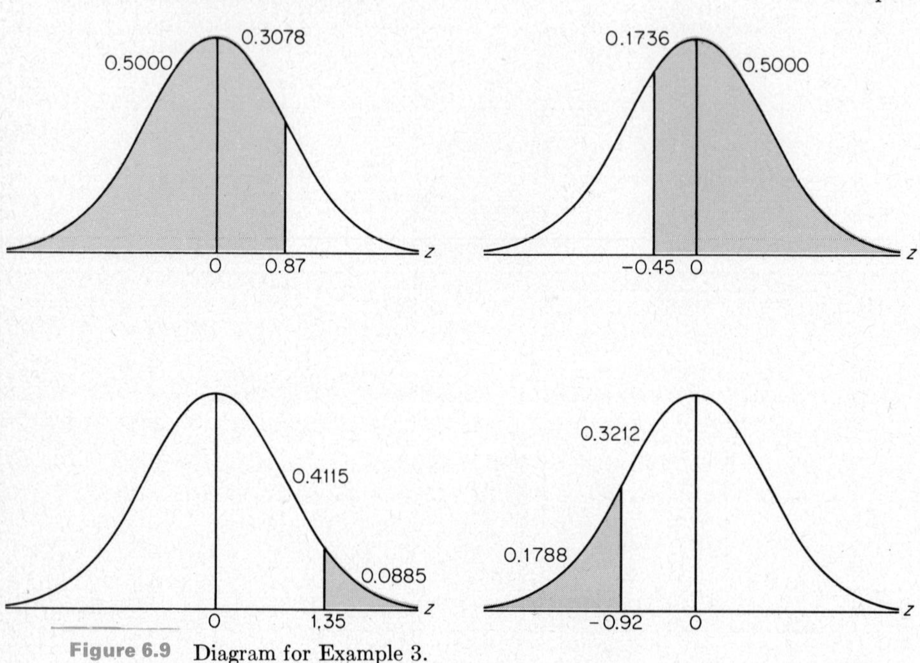

Figure 6.9 Diagram for Example 3.

(namely, the area under the curve to the right of $z = -0.45$, which is tinted in the second diagram of Figure 6.9) is $0.5000 + 0.1736 = 0.6736$. Similarly, we find *by subtraction* that the probability of getting a z greater than 1.35 (namely, the area under the curve to the right of $z = 1.35$, which is tinted in the third diagram of Figure 6.9) is $0.5000 - 0.4115 = 0.0885$, and that the probability of getting a z less than -0.92 (namely, the area under the curve to the left of $z = -0.92$, which is tinted in the fourth diagram of Figure 6.9) is $0.5000 - 0.3212 = 0.1788$.

Example 1 (Continued). Returning now to Example 1 on page 157, we find that for a normal distribution with $\mu = 10$ and $\sigma = 5$, the probability of getting a value between 12 and 15 is given by the area of the tinted region of the *first* diagram of Figure 6.10. Looking up the area between $z = 0$ and $z = 1.00$ as well as the area between $z = 0$ and $z = 0.40$, we find *by subtraction* that the area between $z = 0.40$ and $z = 1.00$ is $0.3413 - 0.1554 = 0.1859$. Similarly, for a normal distribution with $\mu = 20$ and $\sigma = 10$, the probability of getting a value between 12 and 15 is given by the area of the tinted region of the second diagram of Figure 6.10. Looking up the area between

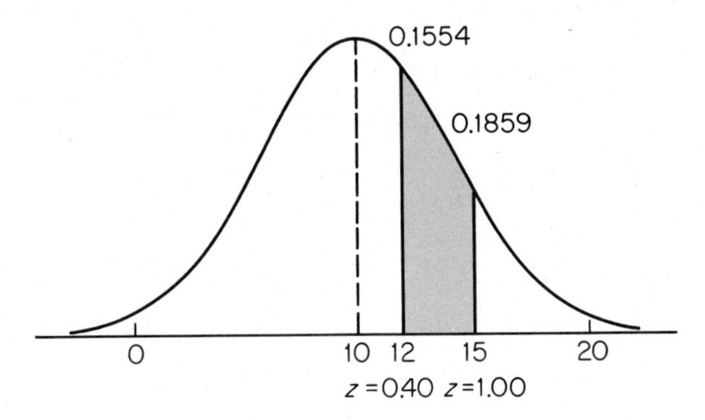

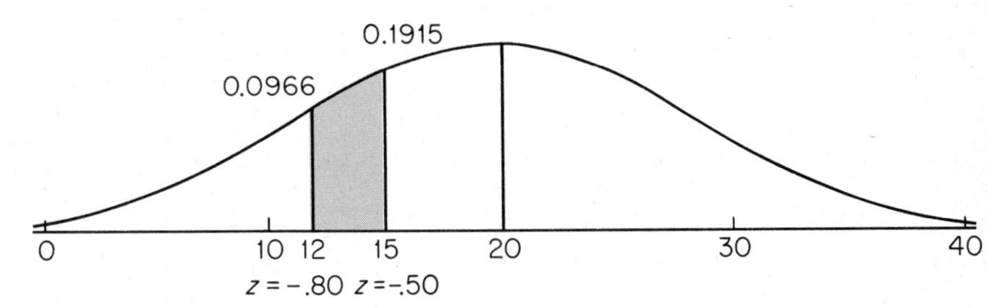

Figure 6.10 Diagram for continuation of Example 1.

$z = 0$ and $z = 0.50$ as well as the area between $z = 0$ and $z = 0.80$ (instead of that between $z = 0$ and $z = -0.50$ and that between $z = 0$ and $z = -0.80$), we find *by subtraction* that the area between $z = -0.80$ and $z = -0.50$ is $0.2881 - 0.1915 = 0.0966$.

When one z-value is positive and the other is negative, the normal curve area between the two is always given by the *sum* of the corresponding entries in Table I.

Example 4. For instance, the probability of getting a z between -0.48 and 0.76 (namely, the area under the curve tinted in Figure 6.11) is $0.1844 + 0.2764 = 0.4608$, where 0.1844 and 0.2764 are, respectively, the entries corresponding to 0.48 and 0.76 in Table I.

There are also problems in which we are given areas under normal curves and are asked to find the corresponding values of z.

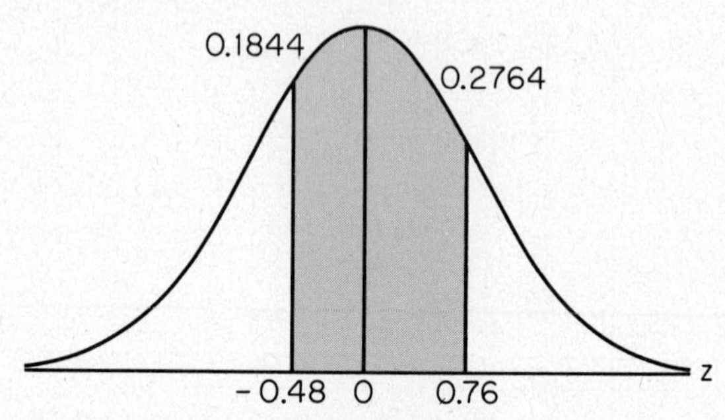

Figure 6.11 Diagram for Example 4.

Example 5. For instance, if we want to find a z which is such that the total area to its right equals 0.1500 (see Figure 6.12), it is apparent that it must correspond to an entry of 0.3500 in Table I; hence, the result is approximately $z = 1.04$ (which corresponds to an entry of 0.3508).

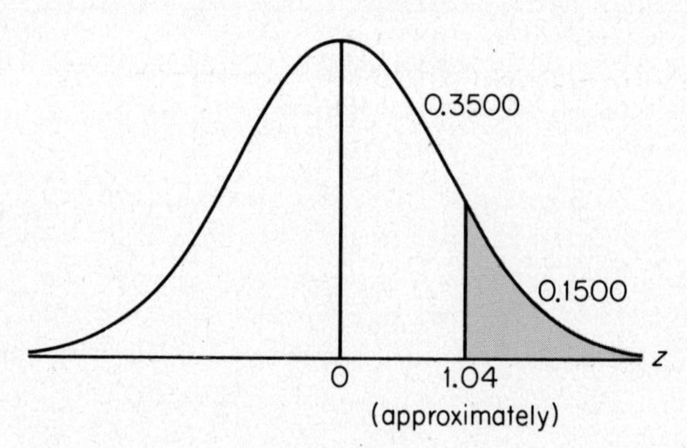

Figure 6.12 Diagram for Example 5.

Exercises

1. Suppose that a random variable can only take on values on the continuous interval from 2 to 10, and that the graph of its probability density, called a **uniform density**, is given by the horizontal line of Figure 6.13.

   a. What probability is represented by the white region of the diagram and what is its value?

   b. What is the probability that the random variable will take on a

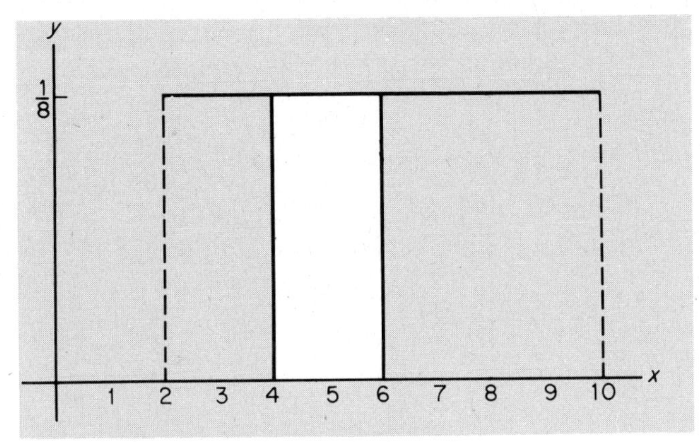

Figure 6.13 Diagram for Exercise 1.

value less than 7? Would the answer be the same if we asked for the probability that the random variable will take on a value less than or equal to 7?

c. What is the probability that the random variable will take on a value between 1.7 and 6.8?

d. What is the probability that the random variable will take on a value greater than 7.6?

2. Suppose that a random variable can only take on values in the continuous interval from 0 to 1, and that the graph of its probability density (called a **triangular density**) is given by the line of Figure 6.14.

a. Verify that the total area under the curve (line) is equal to 1.

b. What probability is represented by the white region of the diagram and what is its value?

c. What is the probability that the random variable will take on a value between 0.50 and 0.75?

d. What is the probability that the random variable will take on a value greater than 0.80?

[*Hint:* In parts (c) and (d) subtract the areas of appropriate triangles or use the formula for the area of a trapezoid; incidentally, the area of a triangle is one half times the product of its base and its height.]

3. Find the area under the graph of the standard normal distribution which lies

a. between $z = 0$ and $z = 0.82$;

b. between $z = -1.65$ and $z = 0$;

c. to the left of $z = 0.61$;

d. to the right of $z = 0.10$;

e. to the right of $z = -0.55$;

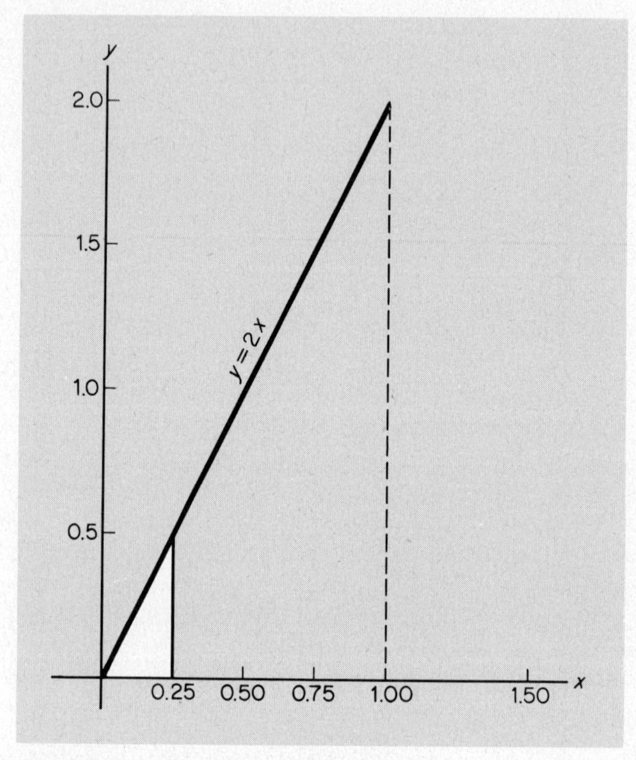

Figure 6.14 Diagram for Exercise 2.

f. to the left of $z = -1.49$;
g. between $z = 0.63$ and $z = 2.55$;
h. between $z = -1.79$ and $z = -0.12$;
i. between $z = -1.18$ and $z = 0.55$;
j. between $z = -0.90$ and $z = 0.90$.

4. Find the area under the graph of the standard normal distribution which lies

a. between $z = 0$ and $z = 2.25$;
b. between $z = -0.17$ and $z = 0$;
c. to the left of $z = 1.11$;
d. to the left of $z = -0.83$;
e. to the right of $z = 1.59$;
f. to the right of $z = -1.33$;
g. between $z = 0.15$ and $z = 1.15$;
h. between $z = -2.00$ and $z = -1.00$;
i. between $z = -0.66$ and $z = 1.53$;
j. between $z = -1.53$ and $z = 0.66$.

5. Find z if

a. the normal curve area between 0 and z is 0.4726;
b. the normal curve area to the left of z is 0.7704;
c. the normal curve area to the right of z is 0.9838;
d. the normal curve area to the right of z is 0.1314;
e. the normal curve area to the left of z is 0.0023;
f. the normal curve area between $-z$ and z is 0.8502.

6. Find the probability that a random variable having the standard normal distribution takes on a value between $-z$ and z if

a. $z = 0.675$;
b. $z = 1.00$;
c. $z = 2.00$;
d. $z = 3.00$.

[*Hint:* In part (a) *interpolate*; namely, take the mean of the results which would be obtained for $z = 0.67$ and 0.68.]

7. Find z (rounded to two decimals) so that

a. the normal curve area to the right of z is 0.05;
b. the normal curve area between $-z$ and z is 0.95;
c. the normal curve area to the right of z is 0.02;
d. the normal curve area between $-z$ and z is 0.98;
e. the normal curve area to the right of z is 0.01;
f. the normal curve area between $-z$ and z is 0.99.

8. A random variable has a normal distribution with the mean $\mu = 120.4$ and the standard deviation $\sigma = 3.6$. What is the probability that this random variable will take on a value

a. less than 125.8;
b. greater than 117.7;
c. between 124.9 and 128.5;
d. between 114.1 and 122.2?

9. A random variable has a normal distribution with the mean $\mu = 75.2$ and the standard deviation $\sigma = 4.4$. What is the probability that this random variable will take on a value

a. less than 84.0;
b. less than 67.5;
c. between 68.6 and 81.8;
d. between 69.7 and 85.1?

10. A normal distribution has the mean $\mu = 82.4$. Find its standard deviation if 20% of the area under the curve lies to the right of 99.2.

11. A random variable has a normal distribution with the standard deviation $\sigma = 10$. Find its mean if the probability that the random variable will take on a value less than 82.5 is 0.8264.

12. Verify the result on page 158, namely, that the area under the standard normal curve between $z = 0.15$ and $z = 0.85$ equals 0.2427.

Some applications

Let us now consider some examples dealing with random variables having at least approximately normal distributions.

Example 6. Suppose that the amount of cosmic radiation to which a person is exposed while flying by jet across the United States can be looked upon as a random variable having a normal distribution with a mean of 4.35 mrem and a standard deviation of 0.59 mrem.* *With this information we can determine all sorts of probabilities, or percentages, about the amount of cosmic radiation to which a person is exposed while flying by jet across the United States.* For instance, let us ask for the probability that a person will be exposed to more than 5.00 mrem of cosmic radiation on such a flight. The answer to this question is given by the area of the tinted region of the first diagram of Figure 6.15, namely, that to the right of

$$z = \frac{5.00 - 4.35}{0.59} = 1.10$$

Since the corresponding entry in Table I is 0.3643, we find that the desired probability is $0.5000 - 0.3643 = 0.1357$, or approximately 0.14.

Continuing with the same example, let us also determine what percentage of all passengers on such flights are exposed to anywhere from 3.00 to 4.00 mrem of cosmic radiation. This time the answer is given by the area of the tinted region of the second diagram of Figure 6.15, namely, that between

$$z = \frac{3.00 - 4.35}{0.59} = -2.29 \quad \text{and} \quad z = \frac{4.00 - 4.35}{0.59} = -0.59$$

Since the corresponding entries in Table I are 0.4890 for $z = 2.29$ and 0.2224 for $z = 0.59$, we find that $0.4890 - 0.2224 = 0.2666$, or approximately 26.7% of all passengers on such flights are exposed to anywhere from 3.00 to 4.00 mrem of cosmic radiation.

Example 7. To consider another example, suppose that the actual amount of instant coffee which a filling machine puts into a food processor's 4-ounce jars varies somewhat from jar to jar, and that it can be looked upon as a random variable having a normal distribution with a

*This unit of radiation stands for "milli (1/1,000) roentgen equivalent man."

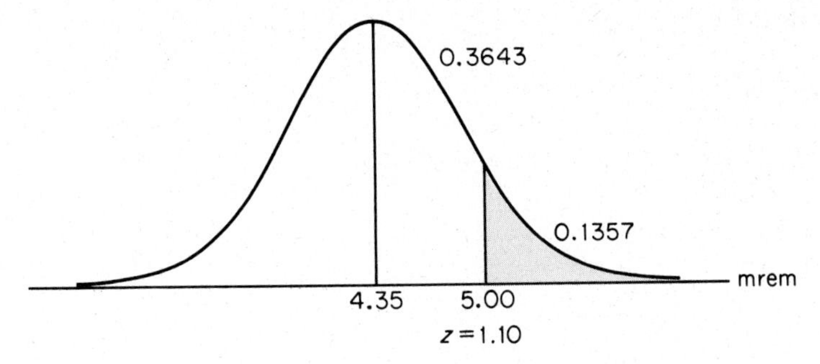

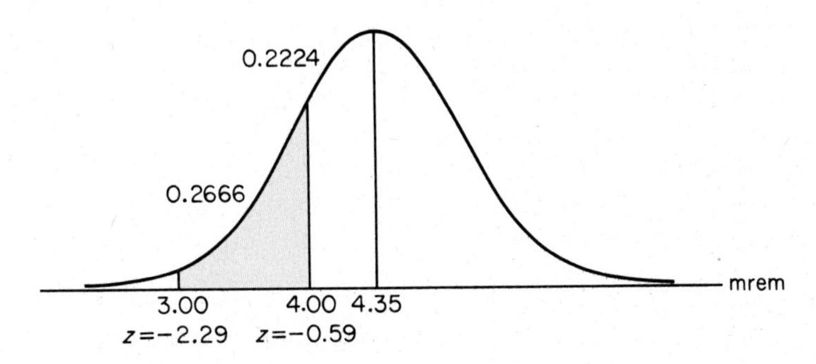

Figure 6.15 Diagram for Example 6.

standard deviation of 0.04 ounce. If only 2% of the jars are to contain less than 4.00 ounces, let us determine *the average amount which the filling machine will have to put into these jars.* This example differs from the preceding one insofar as we are given a normal curve area (the one tinted in Figure 6.16), we are given $\sigma = 0.04$, and we are

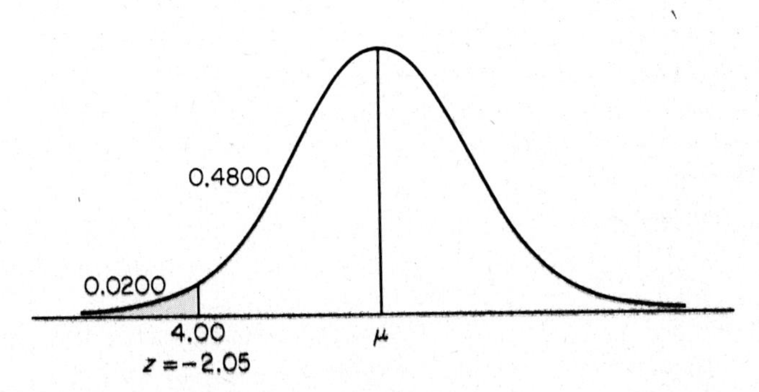

Figure 6.16 Diagram for Example 7.

asked to find μ. Since the value of z for which the entry in Table I comes closest to $0.5000 - 0.0200 = 0.4800$ is 2.05, we have

$$-2.05 = \frac{4.00 - \mu}{0.04}$$

and, solving for μ, we get

$$4.00 - \mu = -2.05(0.04) = -0.082$$

and

$$\mu = 4.00 + 0.082 = 4.082 \text{ ounces}$$

This would seem to be very unsatisfactory so far as the food processor is concerned, and in Exercise 4 on page 173 the reader will be asked to show that if the variability of the filling machine is reduced so that $\sigma = 0.025$ ounce, this will lower the required average amount of coffee per jar to $\mu = 4.05$ ounces, yet keep just about 98% of the jars above 4 ounces.

Although, strictly speaking, the normal distribution applies to *continuous measurements*, it is often used to approximate distributions of random variables which can take on only a finite number of values. This is perfectly all right in many situations, but we have to be careful to make the **continuity correction** illustrated in the following example.

Example 8. Suppose that a sociological study of aggressive behavior showed that male white mice, returned to the group in which they live after 5 weeks of isolation, averaged 18.6 fights in the first 5 minutes with a standard deviation of 3.3. What we would like to know is the probability that, under the stated conditions, a male white mouse will get into at least 15 fights in the first 5 minutes. Assuming that it is reasonable to approximate the distribution of such data with a normal distribution, we shall have to look for the area of the tinted region of Figure 6.17, namely, that to the right of 14.5. The reason we look for the area to the right of 14.5 and *not* for that to the right of 15 is that *the number of fights in which one of these mice gets involved must be a whole number*. Thus, if we want to approximate the distribution of the number of fights with a normal curve, we have to "spread" the values of this random variable over a *continuous* scale, and we can do this by representing each whole number k by the interval from $k - \frac{1}{2}$ to $k + \frac{1}{2}$. In particular, 5 is thus represented by the interval from 4.5 to 5.5, 10 is represented by the interval from 9.5 to 10.5,

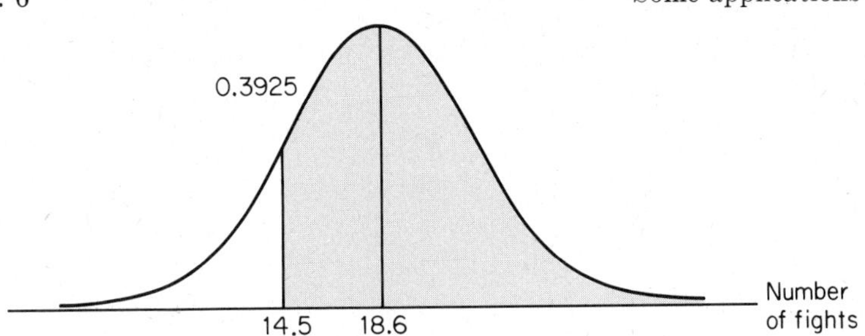

Figure 6.17 Diagram for Example 8.

15 is represented by the interval from 14.5 to 15.5, . . . , and the probability of getting *at least* 15 is given by the area under the curve to the right of 14.5. We thus get

$$z = \frac{14.5 - 18.6}{3.3} = -1.24$$

and the area of the tinted region of Figure 6.17 turns out to be $0.5000 + 0.3925 = 0.8925$. This means that the probability that one of the mice will get involved in at least 15 fights is approximately 0.89.

The *continuity correction* which we have introduced in this example will be discussed further on page 176.

In Chapter 5 we showed by means of Chebyshev's theorem that the probability is *at least* 0.75 that a random variable will take on a value within two standard deviations of the mean. It is important to note that for a random variable having a normal distribution, we can now make the much stronger statement that the probability of getting a value within two standard deviations of the mean is 0.9544. This is precisely what the reader was asked to show in part (c) of Exercise 6 on page 165. In fact, it follows from parts (b), (c), and (d) of that exercise that

If a random variable has a normal distribution, approximately 68% of the time its values will fall within one standard deviation of the mean, approximately 95% of the time its values will fall within two standard deviations of the mean, and approximately 99.7% of the time its values will fall within three standard deviations of the mean.

These percentages are referred to in many applications.

Probability graph paper*

As in the examples of the preceding section, we often assume that observed data follow the overall pattern of a normal distribution, and this raises the question: How can we tell whether this is actually the case? Among the various methods that can be used to test whether the assumption is reasonable, or justifiable, the one we shall give here is *not* the best—it is largely subjective, but it has the advantage that it is very easy to perform.

Example 9. To illustrate this method, let us refer again to the distribution of absences, which we used as an example throughout Chapter 2. If we divide each of the cumulative frequencies in the table on page 16 by 80 (the total number of data grouped) and then multiply by 100 to express the figures as percentages, we obtain the following *cumulative percentage distribution*:

Number of Absences	*Cumulative Percentages*
Less than 4.5	0
Less than 9.5	10
Less than 14.5	45
Less than 19.5	78.75
Less than 24.5	93.75
Less than 29.5	98.75
Less than 34.5	100

Note that we substituted the class boundaries for the class limits in the column on the left.

Before we continue with this example and plot this cumulative distribution on the special graph paper of Figure 6.18, let us briefly investigate its scales. As can be seen from Figure 6.18, the cumulative percentage scale is already marked off in the rather unusual pattern which makes the paper suitable for our particular purpose. The other scale consists of equal subdivisions which are not labeled, and in our example we used them to represent the successive class boundaries of the distribution. The graph paper of Figure 6.18 is generally referred to as **arithmetic probability paper** and it can be bought in most college bookstores; actually, it would be much more appropriate to refer to it as **normal probability paper**.†

*This section may be omitted without loss of continuity.

†The paper is referred to as *arithmetic* to distinguish it from **logarithmic probability paper,** in which the equal subdivisions are replaced by subdivisions like those on the logarithmic scale of a slide rule.

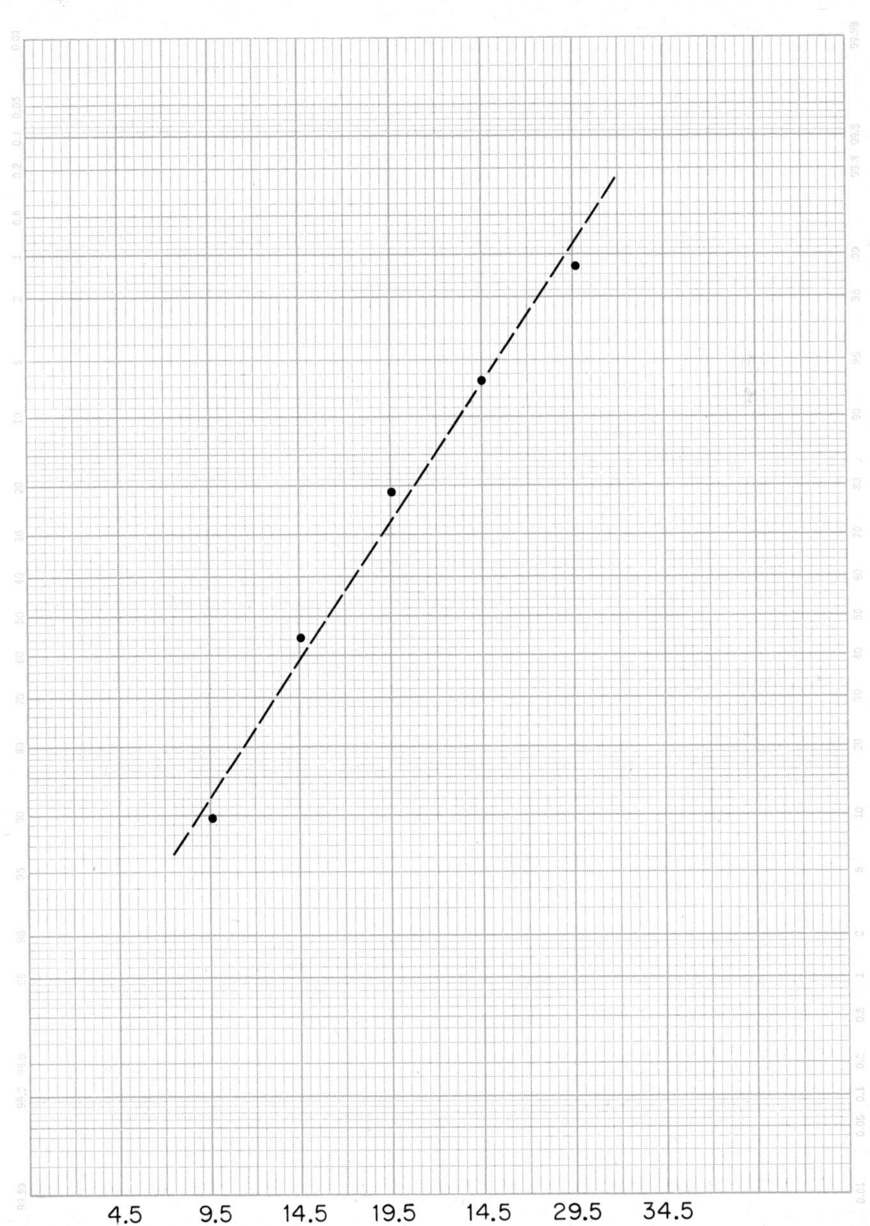

Figure 6.18 Arithmetic probability paper.

Example 9 (Continued). Now, if we use the right-hand scale and plot the cumulative "less than" percentages which correspond to 9.5, 14.5, 19.5, 24.5, and 29.5 on this special kind of graph paper, we obtain the five points shown in Figure 6.18, and it should be observed that they fall fairly close to a straight line.

This, in fact, is the criterion:

> **If the cumulative "less than" percentages of a distribution are plotted on arithmetic probability paper and the resulting points lie very close to a straight line, then this is construed as positive evidence that the distribution follows the general pattern of a normal curve.**

So far as our example is concerned, it would seem reasonable to conclude that the distribution of absences "more or less" follows the general pattern of a normal curve.

Note that in Figure 6.18 we did not plot points corresponding to 0% at 4.5 and 100% at 34.5. As we pointed out on page 155, the normal curve never quite reaches the horizontal axis no matter how far we go away from the mean; thus, we never quite reach 0% or 100% of the area under the curve no matter how far we go in either direction.

Let us repeat that the method which we have described here is only a crude (and highly subjective) way of checking whether a distribution follows the pattern of a normal curve. Also, in a graph like that of Figure 6.18 *only pronounced departures from a straight line are real evidence that the data do not follow the pattern of a normal distribution.* A more objective way of testing whether observed data follow the pattern of a normal curve may be found in more advanced texts in statistics.

Exercises

1. With reference to Example 6 on page 166, find the probability that

a. a person flying by jet across the United States will be exposed to at least 3.50 mrem of cosmic radiation;

b. a person flying by jet across the United States will be exposed to anywhere from 4.00 to 4.50 mrem of cosmic radiation.

2. Suppose that a study has shown that during periods of transcendental meditation, the reduction of a person's oxygen consumption may be looked upon as a random variable having a normal distribution with a mean of 38.6 cc per minute and a standard deviation of 4.3 cc per minute. Find the probability that

a. during a period of transcendental meditation a person's oxygen consumption is reduced by at least 50.0 cc per minute;
b. during a period of transcendental meditation a person's oxygen consumption is reduced by anywhere from 30.0 to 40.0 cc per minute;
c. during a period of transcendental meditation a person's oxygen consumption is reduced by at most 35.0 cc per minute.

3. Suppose that in a photographic process, the developing time of prints may be looked upon as a random variable having a normal distribution with a mean of 8.28 seconds and a standard deviation of 0.03 second. Find the probability that it will take
a. anywhere from 8.25 to 8.30 seconds to develop one of the prints;
b. at least 8.33 seconds to develop one of the prints;
c. at most 8.20 seconds to develop one of the prints.

4. With reference to Example 7 on page 166, show that if the filling machine is modified so that $\mu = 4.05$ ounces and $\sigma = 0.025$ ounce (for the amount of instant coffee it puts into the 4-ounce jars), then 98% of the jars will contain at least 4 ounces.

5. In a very large class in world history, the final examination grades have a mean of 68.3 and a standard deviation of 14.5. Assuming that it is reasonable to treat the distribution of these grades (which, incidentally, are all *whole numbers*) as if it were a normal distribution, find
a. what percentage of the students should get a grade of 70;
b. what percentage of the students should get a grade of 50 or less;
c. the lowest A, if the highest 10% of the grades are to be regarded as A's;
d. the lowest passing grade, if the lowest 20% of the grades are to be regarded as failing grades.

6. Suppose that the yearly number of major earthquakes, the world over, is a random variable whose distribution can be closely approximated with a normal distribution having the mean $\mu = 20.8$ and the standard deviation $\sigma = 4.5$. Find the probability that
a. there will be exactly 20 major earthquakes in any given year;
b. there will be at least 12 major earthquakes in any given year.

7. The head of the complaint department of a department store knows from experience that the number of complaints he receives per 8-hour day is a random variable with the mean $\mu = 31.4$ and the standard deviation $\sigma = 5.5$. Assuming that the distribution of the number of complaints has roughly the shape of a normal distribution, find the probability that he will receive
a. exactly 30 complaints on any one day;
b. anywhere from 25 to 35 complaints, inclusive, on any one day.

8. Convert the grade distribution of Exercise 10 on page 24 into a cumulative "less than" percentage distribution, and use arithmetic probability paper to judge whether it is reasonable to treat the data as having roughly the shape of a normal distribution.

9. Convert the service time distribution of Exercise 8 on page 35 into a cumulative "less than" percentage distribution, and use arithmetic probability paper to judge whether the distribution has roughly the shape of a normal curve.

10. Plot the cumulative "less than" percentage distribution of whichever data you grouped among those of Exercises 11 through 14 on pages 25 and 26 and judge whether the distribution originally obtained has roughly the shape of a normal curve.

Approximating the binomial distribution

The normal distribution is often introduced as a continuous distribution which provides a very close approximation to the binomial distribution when n (the number of trials) is large, and p (the probability of a success on an individual trial) is close to 0.50. This is illustrated in Figure 6.19, which contains the histograms of binomial distributions having $p = 0.50$ and $n = 2, 5, 10$, and 25—evidently, when n becomes large, the distributions rapidly approach the bell-shaped pattern of a normal curve. In reality, normal distributions with

$$\mu = np \qquad \text{and} \qquad \sigma = \sqrt{np(1 - p)}$$

are often used to approximate corresponding binomial distributions even when n is "not too large" and p differs from 0.50, but is *not too close* to either 0 or 1. It is a good rule of thumb to use this approximation only when np and $n(1 - p)$ are *both* greater than 5.

Example 10. To illustrate the normal curve approximation of a binomial distribution, let us use it to determine the probability of getting 6 *heads* and 10 *tails* in 16 flips of a balanced coin. The normal curve approximation of this probability is given by the area of the tinted region of Figure 6.20, namely, that between 5.5 and 6.5, since the interval from 5.5 to 6.5 represents 6 *heads* in accordance with the

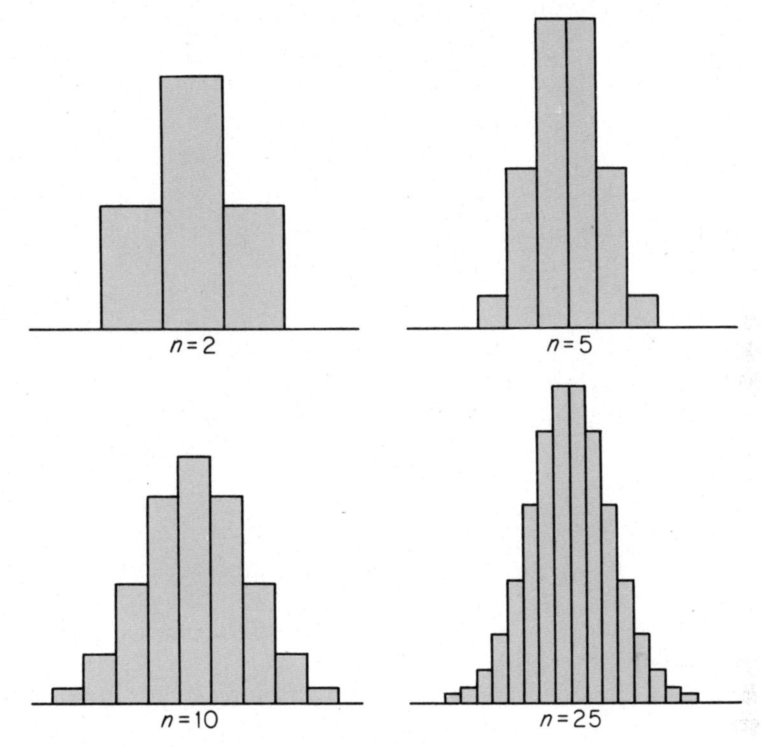

Figure 6.19 Binomial distributions with $p = 0.50$.

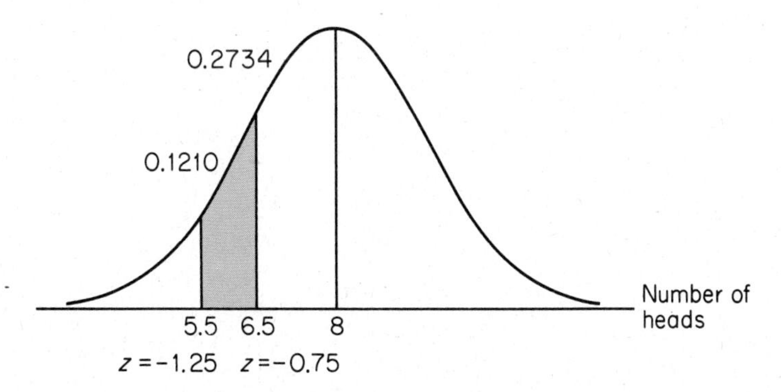

Figure 6.20 Diagram for Example 10.

continuity correction which we explained on page 168. For $n = 16$ and $p = \frac{1}{2}$ we get

$$\mu = 16 \cdot \tfrac{1}{2} = 8$$

and

$$\sigma = \sqrt{16 \cdot \tfrac{1}{2} \cdot \tfrac{1}{2}} = 2$$

in accordance with the formulas on pages 141 and 146. Thus, we shall have to determine the normal curve area between

$$z = \frac{5.5 - 8}{2} = -1.25 \qquad \text{and} \qquad z = \frac{6.5 - 8}{2} = -0.75$$

and since the corresponding tabular values are 0.3944 and 0.2734, we get $0.3944 - 0.2734 = 0.1210$. To see whether this approximation is any good, let us calculate the *exact* probability of getting 6 *heads* and 10 *tails* in 16 flips of a balanced coin by substituting $x = 6$, $n = 16$, and $p = \frac{1}{2}$ into the formula for the binomial distribution on page 129. This gives

$$\binom{16}{6}(\tfrac{1}{2})^6(1 - \tfrac{1}{2})^{10} = 8{,}008(\tfrac{1}{2})^6(\tfrac{1}{2})^{10} = \frac{8{,}008}{65{,}536}$$

or 0.1222 rounded to four decimals. Since the difference between 0.1210 and 0.1222 is very small, only 0.0012, we have thus shown that *in this case* the normal curve approximation is very close.

The normal curve approximation of binomial distributions is especially useful in problems in which we would otherwise have to calculate the probabilities of many different values of a random variable.

Example 11. Suppose, for instance, that we want to determine the probability that at least 70 of 100 mosquitos will be killed by a new insect spray, when the probability that any one of them will be killed by the spray is 0.75. In other words, we want to know the probability of getting *at least* 70 successes in 100 trials when the probability of a success on any one trial is 0.75. If we tried to solve this problem by using the formula for the binomial distribution, we would have to find the *sum* of the 31 probabilities which correspond to $x = 70, 71, 72, \ldots, 99$, and 100, and it would hardly seem necessary to point out that this would require an enormous amount of

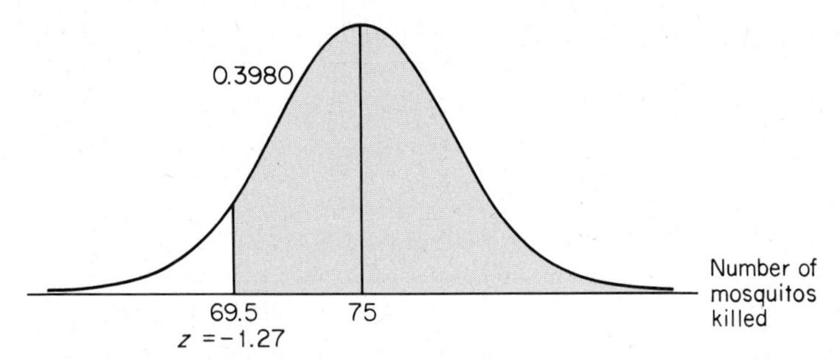

Figure 6.21 Diagram for Example 11.

work. However, since $np = 100(0.75) = 75$ and $n(1 - p) = 100(0.25) = 25$ both exceed 5 (the minimum suggested on page 174), we can use the normal curve approximation and find the answer by determining the area of the tinted region of Figure 6.21, namely, that to the right of 69.5. Note that we are again using the *continuity correction* according to which 70 is represented by the interval from 69.5 to 70.5. Since

$$\mu = 100(0.75) = 75$$

and

$$\sigma = \sqrt{100(0.75)(0.25)} = 4.33$$

we find that *in standard units* 69.5 becomes

$$z = \frac{69.5 - 75}{4.33} = -1.27$$

and it follows (according to Table I) that the desired probability is $0.5000 + 0.3980 = 0.8980$. As a matter of interest, let us point out that the *actual* value of the probability (looked up in an appropriate table and rounded to four decimals) is 0.8962, so that the error is only 0.0018.

Exercises

1. Use the normal curve approximation to find the probability of getting 7 *heads* and 7 *tails* in 14 flips of a balanced coin, and compare the result with the corresponding value given in Table XI.

2. Use the normal curve approximation to find the probability that a student will answer at least 9 of 15 questions correctly in a true–false test, if he decides how to mark each question by flipping a balanced coin. Compare the result with the corresponding value given in Table XI.

3. A multiple-choice test consists of 27 questions, each with four possible answers. If a student answers each question by drawing a card from an ordinary deck of 52 playing cards and checking the first, second, third, or fourth answer depending on the suit (spades, hearts, diamonds, or clubs), find approximately the probability that the student will get

a. exactly eight correct answers;
b. at least seven correct answers;
c. at most nine correct answers.

4. If 70% of all persons flying across the Atlantic feel the effect of the time difference for at least 24 hours, what is the probability that among 140 persons starting out by jet from New York to Madrid *at least* 100 will feel the effect of the time difference for at least 24 hours?

5. If $\frac{2}{3}$ of all clouds seeded with silver iodide show spectacular growth, what is the probability that among 45 clouds thus seeded *at most* 25 will show spectacular growth?

6. A television station claims that its Tuesday night movie regularly has 32% of the total viewing audience in its area. If this claim is correct, what is the probability that among 300 viewers reached by phone on a Tuesday night, at least 100 will be watching the station's movie?

7. If a study shows that only 18% of drug-treated hypertension patients develop complications during the next five years, what is the probability that among 100 such patients anywhere from 12 to 24, inclusive, will develop complications during the next five years?

8. If the probability that a mountain climber will have to be rescued from a certain trail is 0.03, what is the probability that among 400 totally unrelated mountain climbers on that trail anywhere from 10 to 15, inclusive, will have to be rescued?

9. To illustrate the Law of Large Numbers which we discussed on page 149, find the probability that the *proportion of heads* will be anywhere from 0.49 to 0.51 when a balanced coin is flipped

a. 100 times;
b. 1,000 times;
c. 10,000 times.

7

Chance variation: sampling

Introduction

So far we have often referred to sets of data as **samples** without defining this term in a rigorous way. Intuitively, most persons know what the word "sample" means, and if we look it up we find that a sample is "a part to show what the rest is like" according to one dictionary, and "a portion, part, or piece taken or shown as representative of the whole" according to another. All this will probably make sense to most readers, but difficulties *can* arise when we ask ourselves what is meant here by "the rest" and "the whole." Suppose, for instance, that after listening to two lectures by a professor, you are thoroughly bored and very critical. Well then, *is it your intention to criticize only the two particular lectures which you have heard, all his lectures in the given course, or perhaps all the lectures he has ever made or will make*? In the first case, in which you criticize only the two lectures which you actually heard, you would have a **100% sample**, namely, *complete information.* In the other two cases, however, you would have only *partial information*, and it should be clear that in the second case you would have a *better sample* than in the third, namely, *percentagewise* more information for making any kind of generalization.

The whole question "What is a sample a sample of?" is of critical importance in statistics. For instance, if someone interviews several of the teenagers attending a rock concert, are they to be looked upon as a sample of all the teenagers attending the particular concert, all teenagers who attend rock concerts, all teenagers who enjoy rock music, or perhaps even all teenagers in the United States? Similarly, if a buyer checks the flavor of a few coffee beans, are they to be looked upon as a sample of the sack from which they came, the shipment from which they came, or the entire crop of the country where they were grown. Finally, suppose that a scientist is studying the life style of the Indians who once inhabited a cliff dwelling.

Do his findings apply only to the Indians who once lived in that cliff dwelling, to all the Indians of the tribe to which they belonged, or perhaps to all cliff dwellers, past, present, and future? As may be apparent, each of these questions can be answered by "Take your pick!", but it should be understood that in some instances we would have *good samples* which lend themselves to meaningful generalizations, while in other instances it would be utter folly to make any generalization whatsoever.

In statistics, we refer to the "whole" of which a sample is a part as the **population** from which we are sampling, and we do this even when we are talking about the heights of trees, the weights of pineapples, the heartbeats of birds, the alcohol contents of wines, and so on. As we explained in the beginning of Chapter 1, statistics used to concern itself mostly with the descriptions of *human populations* (we now call this *demography*), and as the scope of the subject grew, the term "population" has assumed this much wider connotation. Whether or not it sounds strange to refer to such things as the heights of all the trees in a forest as a population is beside the point—in statistics, "population" is a technical term with a meaning of its own.

A population is referred to as **finite** if it consists of a *fixed number* (that is, some definite whole number) of elements, items, objects, measurements, or observations. For instance if we are analyzing the final examination grades of all students who took Freshman Writing at Stanford University in the fall of 1974 without any intention of making any generalizations (say, about the grades of students who will take this course in future years), then we consider these grades as a finite population. Similarly, if we are studying the low humidity readings at all official weather stations in the United States on August 1, 1975, without any intention of making any generalizations (say, about the humidity elsewhere), these low humidity readings constitute a finite population. Finally, we would be sampling from a finite population if we randomly chose a number from 1 to 10, if we selected one of the 20 sandwiches listed on a menu, or if we picked a course from a list of courses taught by a community college as part of its adult education program. We have made it a point to refer to these populations as *finite*, because there are also situations where there is (hypothetically, at least) no limit to the number of observations that could be made. This will be discussed further on page 181.

It may have occurred to the reader from the discussion of the preceding paragraph that *whether a set of data constitutes a finite population or a sample depends entirely on what we intend to do with it.* For instance, if we are given complete information about the weights of the lions in the San Diego Zoo, these data constitute a finite population so long as we are not going to generalize, say, to the weights of lions in other zoos, or the weights of lions who roam free in their natural habitat. Once we make any kind of generalization, however, the data must be looked upon as a sample; thus, it is always important to state whether a given set of data is to be regarded as a sample or a population. To emphasize the distinction we even use different symbols for the statistical description of samples and populations. For instance, whereas $\bar{x}$ denotes the mean of a sample, the Greek letter μ (*mu*) is used to denote the mean of a population, and it should be observed that this is the same symbol which we used in Chapter 5 for the mean of a probability distribution.

Example 1. If a population consists of the numbers 18, 21, 20, and 25, which are the ages of four persons, its mean is

$$\mu = \frac{18 + 21 + 20 + 25}{4} = \frac{84}{4} = 21$$

To distinguish between samples and populations, we not only use different symbols for their statistical descriptions, but we also refer to these descriptions by different names: *Descriptions of samples are called* **statistics** (*as we have already indicated on page* 27), *and descriptions of populations are called* **parameters**. Hence, we say that $\bar{x}$ is a statistic whereas μ is a parameter. This terminology has the advantage that it will prevent a good deal of the confusion which might otherwise arise in subsequent chapters, where we shall use statistics to reach decisions about parameters—namely, generalize from descriptions of samples to corresponding descriptions of populations.

To introduce the concept of an **infinite population**, let us consider ten slips of paper, numbered from 1 to 10, which are thoroughly mixed in a hat. Suppose that we repeatedly draw one of the slips of paper, but that each slip is replaced before the next one is drawn. This is called **sampling with replacement**, and *except for the element of time* there is no limit to the number of drawings that could be made. We could make 100 drawings, even 1 million or more, and by replacing each slip before the next one is drawn we create the illusion of an *endless supply*, namely, an *infinite population.*

Generally speaking, we say that we are sampling from an infinite population if there is no limit to the number of observations that *could* be made. For instance, we might say that five flips of a coin constitute a sample from the infinite population of all (hypothetically possible) flips of the coin; we might say that a dozen light bulbs constitute a sample from the (hypothetically infinite) population of *all* the past, present, and future light bulbs that will ever be made by a given firm; and we often look at the results of an experiment as a sample from the (hypothetically infinite) population which consists of the results we would, or could, obtain if the experiment were repeated over and over again.

Exercises

1. Suppose that we are given complete information about the fire insurance claims filed against an insurance company during February, 1975. Under what circumstances would we consider these data to be

a. a sample; **b.** a population?

2. A surgeon knows exactly how much time it took for each appendectomy which he performed during the year 1974. Explain under what circumstances these data would be looked upon as

a. a sample; **b.** a population.

3. A congressman knows exactly how many letters he received from his constituents during each month that he was in office. Give one illustration each of a problem where these data would be looked upon as

a. a sample; **b.** a population.

4. The Game and Fish Department of a state has complete records of how many deer licenses were issued to hunters during each of the years 1965 through 1975. Give one illustration each of a problem where these data would be looked upon as

a. a sample; **b.** a population.

5. In each of the following examples, state whether the sample is taken from a finite population or an infinite population, and describe the population:

a. Five scientists are nominated for the 1979 Nobel prize in chemistry.

b. To estimate the "true" weight of a rock specimen brought back from the moon, it is weighed six times.

c. We select three of the novels on a bookstore's shelves to take along on a trip.

d. A police department keeps records on how many burglaries criminals commit before they finally get caught.

e. A sports fan went to see the Chicago Cubs play eight times during the 1974 season.

Random sampling

The purpose of most statistical studies is to make generalizations from samples to finite or infinite populations, and there are certain rules that must be observed to avoid results which are obviously incorrect, irrelevant, or misleading. In other words, not all samples lend themselves to making valid generalizations, and as we pointed out in Chapter 1, there are many pitfalls. To give a few examples, suppose, for instance, that we want to determine how much the average person spends on his summer vacation. It is very unlikely that we would arrive at anything even remotely close to the correct figure if we sampled only the expenses of passengers on a deluxe ocean cruise, or if we sampled only those who camp out in tents in a national park. Similarly, we can hardly expect to get a reasonable generalization about personal income in the United States from data pertaining only to the incomes of doctors, and we can hardly expect to be able to infer much about the retail prices of foods if we sampled only the price of fresh asparagus out of season. Of course, these examples are extreme, but they serve to illustrate some of the many things we have to watch out for if we want to make meaningful and useful generalizations on the basis of sample data.

The whole problem of when and under what conditions samples permit reasonable generalizations is not easily answered. So far as the work in this book is concerned, we shall partly skirt the issue by limiting our discussion to **random samples**—some other kinds of samples will be mentioned very briefly in the second section of Chapter 11. To illustrate the idea of a random sample, let us begin with samples from *finite populations.*

Example 2. Suppose, for instance, that a finite population is of **size** 6, namely, that it consists of six elements, which we shall label a, b, c, d, e, and f. (These elements of the finite population might be the grades which a student gets in six tests, a person's pulse rate taken at six different times during a day, the weights of six mummies, the root length of six different plants, and so on.) Suppose, furthermore, that we want to select two of the six elements of this population and that, to begin with, we are curious to know in how many different ways this might be done. In other words, we want to find out *how many different samples of* **size** *2 can be selected from a finite population of size 6*. The answer to this question is easily determined; in fact, using the rule on page 75, we immediately get

$$\binom{6}{2} = \frac{6 \cdot 5}{2 \cdot 1} = 15$$

One of these samples consists of the elements a and b, another consists of the elements d and f, a third consists of the elements b and c, and the reader should not find it difficult to list all 15 possibilities.

Now if we select one of the 15 possible samples in such a way that each has the same probability of $\frac{1}{15}$ of being chosen, we say that the sample we get is a **simple random sample** or, more briefly, a **random**

sample. One way in which we could obtain such a sample would be to write the letters a, b, c, d, e, and f on separate slips of paper, mix them thoroughly, and then draw two.

The point we have been trying to make with this example is that *the selection of a random sample from a finite population must somehow be left to chance, with each possible sample having the same probability.* Thus, if we deal with a finite population of size N from which we want to select a sample of size n, we say that

The sample we get is random if it is selected in such a way that each of the $\binom{N}{n}$ possible samples has the same probability of $\frac{1}{\binom{N}{n}}$ of being chosen.

The job of numbering slips of paper, as suggested in Example 2, can be very tedious when N is large, but it can be avoided altogether by using instead a table of **random numbers**, or **random digits**, which consists of pages on which the digits 0, 1, 2, 3, 4, 5, 6, 7, 8, and 9 are recorded in a "random" fashion, much as they would appear if they had been generated by a gambling device. We could construct a table of random numbers by using a spinner such as the one shown in Figure 7.1, but in actual practice this is usually done with electronic computers. Several commercially

Figure 7.1 Spinner which could be used for generating random numbers.

published tables of random numbers are listed in the Bibliography at the end of the book, and it should be observed that before such tables are published they are generally required to "pass" various statistical tests intended (insofar as this is possible) to ensure their randomness. Another point which must not be overlooked is that *tables of random numbers will yield random samples only if they are properly used.* For *instance*, if someone presents us with the numbers 7, 7, 7, 7, 7, 7, 7, 7, 7, 7 and refers to them as a random sample because he copied them from a table of random numbers, this is merely a bad joke if he *intentionally* picked one 7 here, one 7 there, another 7 there, and so on.

Example 3. To illustrate the proper use of a table of random numbers (in this case Table XII at the end of the book), suppose that an instructor wants to choose, at random, 10 of 240 rock specimens for students to identify. Numbering the rock specimens from 1 to 240, or better, numbering them 001, 002, 003, . . . , 239, and 240 (so that each of them is represented by a sequence of three digits), he arbitrarily picks a page in a table of random numbers, he arbitrarily picks three columns (usually adjacent ones as a matter of convenience), and he arbitrarily picks a row from which to start. Then he reads off three-digit numbers from these columns moving down the page (or up if so desired), and if necessary he continues on another page or another set of columns. For instance, if he arbitrarily chose the 6th, 7th, and 8th columns of the table on page 357 starting with the 3rd row, he would get 639, 115, 962, 813, 587, . . . , but recording only the numbers from 001 through 240, he would get 115, 229, 134, 015, 081, 219, 104, 174, 109, and 018. Thus, the random sample consists of the rock specimens whose numbers are 115, 229, 134, 15, 81, 219, 104, 174, 109, and 18. Note that when he made this selection he ignored all numbers greater than 240, and he would also have ignored a number already chosen if it occurred again. Had the instructor wanted to be "even more random," he could have used some gambling device to choose the page, the columns, and the row from which to start.

Now that we have explained what we mean by random sampling from finite populations, we should add that in actual practice this is often much easier said than done. For instance, if we wanted to take a sample in order to estimate the average weight of 100,000 packages of spaghetti packed in large crates, it would hardly be practical to number the packages 00000, 00001, 00002, 00003, . . ., 99998, and 99999, choose five-digit numbers from a table of random numbers, and then weigh the corresponding packages of spaghetti.* Similarly, it would be physically impossible to sample trees in

*Note that we started with 00000 in this example to avoid having to work with six-digit numbers.

the Blue Ridge Mountains by assigning a number to each tree, or to sample mosquitoes infesting a swamp with the use of a table of random digits. In situations like this there is very little choice but to proceed according to the dictionary definition of the word "random," namely, "haphazardly, without definite aim or purpose," or perhaps improve the situation somewhat by using one of the special sampling techniques described briefly in Chapter 11. *Then we keep our fingers crossed that statistical theory intended for random samples (for instance, that of Chapters 8 through 10) can nevertheless be employed.* This is true, particularly, in situations where we have little, if any, control over the selection of the data—as in some medical research where scientists have no choice but to use whatever cases happen to be available.

Because of the difficulty of assuring the randomness of data, we must always be on the lookout for biases such as those mentioned in Chapter 1, and a lack of randomness can sometimes be detected by looking at the data, themselves.

Example 4. If a driver averages 17.8, 17.4, 17.5, 16.8, 16.5, 16.6, 16.2, and 16.0 miles per gallon for eight successive tanks of gas, it would hardly seem reasonable to look upon these figures as a random sample of the performance of his car. Evidently, there is a general *downward trend* (things are gradually getting worse), and it would be very unreasonable to analyze these data by applying theory that is ordinarily reserved for random samples. (As we shall indicate briefly in Chapter 11, there *are* more objective methods which enable us to decide whether any "unusual" features of a set of data are due to chance or whether they are really indicative of a lack of randomness.)

To this point we have discussed only random sampling from *finite populations*, but, as can well be imagined, the concept of a random sample from an *infinite population* is much more difficult to explain.

Example 5. To give a simple illustration, let us consider 10 flips of a balanced coin as a sample from the (hypothetically) infinite population which consists of all possible flips of the coin. We shall consider these 10 flips as a *random sample* provided the probability of getting heads is the same for each flip and if, furthermore, the 10 flips are independent.

Example 6. We would also be sampling from an infinite population if we sampled *with replacement* (see Example 7 on page 132) from a finite population, and our sample would be *random* if, for each draw, each element of the population has the same chance of being selected, and successive draws are independent.

In general, *a sample from an infinite population is said to be random if the selection of each sample value is "controlled" by the same probabilities, and successive selections are independent.* All this is assumed, for example, when we refer to the weights of ten racoons, the I.Q.'s of 40 students, the diameters of 35 trees, the lifetimes of 24 transistors, or the yields of eight test plots of corn as *random samples* from (hypothetically) infinite populations.

Exercises

1. What is the probability of each possible sample if a random sample of size 3 is taken from

a. a finite population of size 8;
b. a finite population of size 12?

2. What is the probability of each possible sample if a random sample of size 5 is taken from

a. a finite population of size 15;
b. a finite population of size 20?

3. In Example 2 on page 183 we showed that there are 15 possible samples of size $n = 2$ from the finite population whose elements are denoted a, b, c, d, e and f.

a. List the 15 possible samples of size 2 from this finite population.
b. If a random sample of size 2 is taken from this finite population, what is the probability that it will include the element denoted by the letter b? Does the same probability apply to the elements denoted a, c, d, e, and f?

4. List all possible choices of three of the following European cities: London, Paris, Rome, Brussels, and Amsterdam. If a person randomly selects three of these cities to visit on a trip, find

a. the probability that any one of these five cities (say, Rome) will be included;
b. the probability that any two of these five cities (say, Paris and London) will be included.

5. List all possible choices of four of the following six corporations: General Motors, Xerox, IBM, Shell Oil, Polaroid, and American Air Lines. If a person randomly selects four of these corporations to invest in its stock, find

a. the probability that any one of these corporations (say, Xerox) will be included;

b. the probability that any two of these corporations (say, General Motors and Shell Oil) will be included;

c. the probability that any three of these corporations (say, IBM, Polaroid, and Xerox) will be included.

6. Making use of the fact that among the $\binom{N}{n}$ possible random samples of size n which can be drawn from a finite population of size N there are $\binom{N-1}{n-1}$ which contain a specific element, show that the probability for any specific element to be contained in a random sample of size n from this finite population is $\frac{n}{N}$.

7. A random sample from a finite population can be obtained by choosing the sample *all at once* (as suggested in the text), but it can also be obtained by choosing the elements *one after the other* (without replacement), making sure that after each selection all the remaining elements of the population have the same probability of being chosen next.

a. Show that if a random sample of size 2 is, thus, chosen from a finite population of size 6, the probability for each possible sample is $\frac{1}{15}$ (as in Example 2 on page 183).

b. Show that if a random sample of size 3 is chosen by this method from a finite population of size 8, the probability of each possible sample is the same as in part (a) of Exercise 1.

8. Suppose that a social scientist wants to select 12 of the 749 school districts in the State of Pennsylvania to be included in a study. Assuming that these school districts are numbered by the Education Department from 1 through 749, which ones (by numbers) would he choose if he used the 16th, 17th, and 18th columns of the table on page 357 starting at the top to select his random sample?

9. Suppose that a bacteriologist wants to doublecheck a subsample of the 550 blood samples sent to a laboratory during August, 1974. Assuming that these blood samples are numbered from 1 through 550, which ones (by numbers) would she choose if she used the 1st, 2nd, and 3rd columns of the table on page 358 starting with the 6th row to select a random sample of size 15?

10. In the office of a county assessor, one-family homes are numbered from 1 through 6,558. Which ones (by numbers) would he choose if he used the 26th, 27th, 28th, and 29th columns of the table on page 357 starting with the 7th row to select a random sample of 20 of these one-family homes to be reassessed?

11. SIMULATION Random numbers can also be used to simulate various probability distributions and games of chance.

a. Letting 0, 2, 4, 6, and 8 represent *heads*, and 1, 3, 5, 7, and 9 *tails*, use the first column of the table on page 358 to simulate 50 flips of a coin. How many heads were there and how many tails? How many times were there *three or more heads in a row* and how many times were there *three or more tails in a row?*

b. Describe in detail how random numbers might be used to simulate 240 rolls of a balanced die. Actually perform this simulation and indicate how many times the "die" came up 1, 2, 3, 4, 5, and 6. Do these results agree more or less with what one would expect to get with a balanced die?

c. To simulate the results (number of heads) we might obtain if we repeatedly flipped a pair of coins, we could proceed as in part (a) and use two random digits, one for each coin. Explain why we could also simulate this "experiment" by using two-digit random numbers and letting 00 through 24 represent *0 heads and 2 tails*, 25 through 74 represent *1 head and 1 tail*, and 75 through 99 represent *2 heads and 0 tails.* Use this alternative method to simulate 200 flips of a pair of coins, and construct a table showing how many times there were, respectively, 0, 1, and 2 heads.

12. SIMULATION (Continued) Suppose that the number of days in court it takes to get auto-theft charges settled is a random variable having the following probability distribution:

Number of Days in Court	*Probability*	*Random Numbers*
1	0.27	00–26
2	0.40	27–66
3	0.21	67–87
4	0.08	88–95
5	0.04	96–99

a. Verify that the above assignment of random numbers will yield the correct probability for each value of the random variable.

b. Use Table XII and the above assignment of random numbers to simulate the days it takes to get 25 auto-theft charges settled. Also calculate the mean of these data and compare it with the mean of the probability distribution determined in accordance with the formula on page 140.

13. **SIMULATION (Continued)** Suppose that the number of hurricanes reaching the U.S. coast per year is a random variable having the following probability distribution:

Number of Hurricanes	*Probability*	*Random Numbers*
0	0.05	00–04
1	0.15	05–19
2	0.22	20–41
3	0.22	42–63
4	0.17	64–80
5	0.10	81–90
6	0.05	91–95
7	0.03	96–98
8	0.01	99

a. Verify that the above assignment of random numbers will yield the correct probability for each value of the random variable.
b. Use Table XII and the above assignment of random numbers to simulate the number of hurricanes that will reach the U.S. coast over a period of 20 years. Also calculate the mean of these data and compare it with the mean of the probability distribution determined in accordance with the formula on page 140.

Chance fluctuations of means

If *several* laboratories were asked to determine the nicotine content of a new kind of cigarette, if *several* pollsters sampled public opinion about the popularity of the president, and if *several* teachers tried a new method (say, computer-assisted instruction) on some of their students, in each case there would probably be differences among the respective results. As we saw in the preceding section, random sampling (by definition) involves an element of chance, and it stands to reason that this will affect whatever quantities we may calculate on the basis of sample data. Thus, whatever differences there may be among the results obtained by the different laboratories, the different pollsters, and the different teachers are what we call **chance variations** or **chance fluctuations**.

Example 7. Suppose that in an air pollution study it is desired to estimate the average daily emission of sulfur oxides by a large industrial plant. Available for this purpose are the following data, in tons, on the plant's emission of sulfur oxides for 40 days:

17	15	20	29	19	18	22	25	27	9
24	20	17	6	24	14	15	23	24	26
19	23	28	19	16	22	24	17	20	13
19	10	23	18	31	13	20	17	24	14

The mean of these figures is $\frac{784}{40} = 19.6$ tons, and (assuming that the sample is random) it provides an *estimate* of the quantity with which we are concerned, namely, the *true* average daily emission of sulfur oxides of the given plant.

Since this estimate might be used to determine whether the plant meets federal regulations or whether further antipollution controls are needed, it is only natural to ask whether the figure which we obtained from the sample is really **reliable**. In other words, we ought to investigate whether we are justified in expecting our result of 19.6 tons to be reasonably close to the *true* value, or whether we should, perhaps, have used more data.

This question of *reliability* is what we shall be concerned with mostly in the remainder of this chapter, and without going into any detailed statistical theory or elaborate mathematics, let us investigate it by repeating the whole sampling procedure several times.

Example 7 (Continued). Suppose, thus, that the management of the plant provides us with ten more random samples, each consisting of the sulfur oxides emission for 40 days, and that these samples yield the following means:

20.4 20.2 20.0 18.9 19.1 22.0 20.6 20.2 21.4 20.2

These figures, which are all in tons, tell us quite a bit about the question with which we are concerned, namely, *the extent to which sample means vary from sample to sample.* Just by looking at the ten means we find that the smallest is 18.9 tons, that the largest is 22.0 tons, and that only one of them falls outside the interval from 19.0 to 22.0 tons. Furthermore, the overall mean of the ten samples is

$$\frac{20.4+20.2+20.0+18.9+19.1+22.0+20.6+20.2+21.4+20.2}{10} = 20.3$$

and if this were the *true mean* (which is not a bad guess as it is based on 400 observations), we would find that *six of the ten sample means differ from the true mean by less than 1 ton* (all those except 18.9, 19.1, 22.0, and 21.4), while *all ten of the sample means differ from the true mean by less than 2 tons.*

If we now return to the original sample on page 191, whose mean was 19.6 tons, we might argue that *6 to 4 would be fair odds* that the interval from $19.6 - 1 = 18.6$ tons to $19.6 + 1 = 20.6$ tons contains the plant's true average daily emission of sulfur oxides. Also, we could argue that we are *practically certain* that the true average is contained in the interval from $19.6 - 2 = 17.6$ tons to $19.6 + 2 = 21.6$ tons.

This example has shown how questions about the reliability of a sample mean *can* be answered, but we might add that it is seldom, if ever, practical to repeat experiments over and over for this purpose. In fact, it would be wasteful, for there exists an alternative approach to the problem which is based partly on statistical theory and partly on an analysis of whatever fluctuations there are *within* a sample, namely, among the sample values, themselves.

Before we introduce this technique, let us observe that there are essentially two factors which affect the chance fluctuations of sample means: *the size of the sample and the variability of the population from which the sample is obtained.* So far as the sample size is concerned, it certainly stands to reason that more and more information should lead to more and more reliable results; in other words, the means of very large samples are more apt to be close to the corresponding true (population) means. So far as the variability of the population is concerned, it also stands to reason that its magnitude should be reflected in the size of the fluctuations among the means of samples.

Example 8. To illustrate this point, consider the following two finite populations, which consist of numbers written on individual slips of paper: *population A* consists of the numbers 1, 2, 3, . . . , 19, and 20, each written on a separate slip of paper, and *population B* consists of the numbers 10 and 11, each written on ten separate slips of paper. Now, if we take a sample of size 3 from population A, the sample mean can vary anywhere from $\bar{x} = 2$ (the mean of 1, 2, and 3) to $\bar{x} = 19$ (the mean of 18, 19, and 20); however, if we take a sample of size 3 from population B, the sample mean can vary only from $\bar{x} = 10$ (the mean of 10, 10, and 10) to $\bar{x} = 11$ (the mean of 11, 11, and 11).

This demonstrated that

> **If there is very little or very much variability in a population, the same will also be true for the fluctuations of sample means.**

To explain how we measure the variability of a population, let us return to the example with which we began this section.

Example 7 (Continued). The random variable with which we are concerned in this example is the plant's daily emission of sulfur oxides, and we can get some idea about the probabilities with which it takes on its values from the *percentage distribution* of Figure 7.2, into which

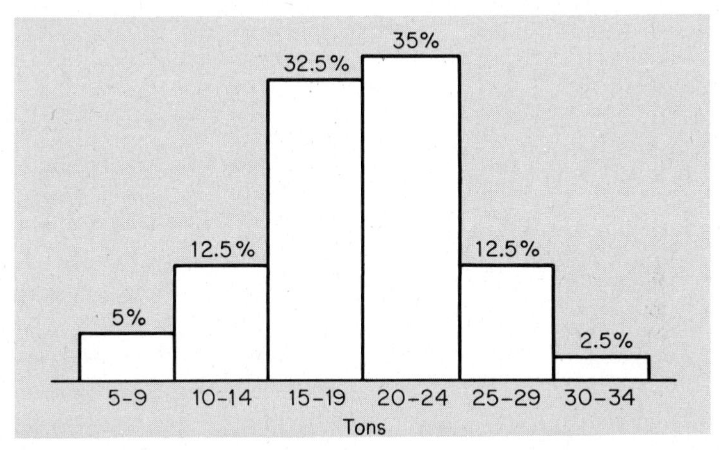

Figure 7.2 Percentage distribution of the data of Example 7.

we have grouped the original data (given on page 191). Looking at this diagram, we might argue that the probability is about 0.35 that a day's emission of sulfur oxides will be anywhere from 20 to 24 tons, that the probability is about 0.05 that it will be anywhere from 5 to 9 tons, that the probability is about 0.125 that it will be anywhere from 25 to 29 tons, and so on. Of course, these figures are based on a sample, and they may be way off, but they enable us to *estimate* the *standard deviation* σ (defined on page 154), which measures the variability of the daily emission data.

Since the estimation of σ is of great importance in most of the work that follows in this chapter and in Chapter 8, we shall discuss this problem separately in the next section. Then, on page 205, we shall return to the *original* problem of studying the chance fluctuations of sample means.

The sample standard deviation

In Chapter 5 we defined σ^2, the variance of a probability distribution, as the expected value (or average) of the *squared deviations from the mean.* Hence, if we actually observed several values of a random variable, it would seem logical to estimate σ^2 in terms of the average of *their* squared deviations from the mean. Symbolically, if $x_1, x_2, \ldots,$ and x_n are n observations of a random variable with the mean μ (say, the 40 emission figures on page 191, the ages of 25 doctors, the I.Q.'s of 120 students, . . .), we could estimate the variance σ^2 of the distribution of the corresponding random variable as

$$\frac{(x_1 - \mu)^2 + (x_2 - \mu)^2 + \ldots + (x_n - \mu)^2}{n}$$

which can also be written as $\frac{\sum (x - \mu)^2}{n}$. This would be fine, if it were not for the fact that *in most practical situations* μ *is unknown and will, therefore, have to be replaced with an estimate.* (For instance, in the problem of the preceding section, μ was the plant's true average daily emission of sulfur oxides.) The most obvious replacement for μ is the sample mean $\bar{x}$ (namely, the average of the data which were actually observed), but if we substitute $\bar{x}$ for μ we shall have to account for the fact that $\sum (x - \bar{x})^2$ is *always less than or equal to* $\sum (x - \mu)^2$, as the reader will be asked to demonstrate in Exercise 14 on page 202. In practice, we take care of this by dividing by $n - 1$ instead of n when we take the average of the squared deviations from the mean, and we thus define the **sample variance** as

$$s^2 = \frac{\sum (x - \bar{x})^2}{n - 1}$$

and, correspondingly, the **sample standard deviation** as*

$$s = \sqrt{\frac{\sum (x - \bar{x})^2}{n - 1}}$$

*In connection with this, we refer to $n - 1$ as the number of **degrees of freedom.** The reason for this is that *the sum of the deviations from the mean is always equal to zero* (which is not very difficult to prove); hence, if we know $n - 1$ of the deviations from the mean (which play such an important role in the definition of the standard deviation), the nth is automatically determined.

In words, the variance of a set of sample data is given by the sum of the squared deviations from their mean divided by $n - 1$.

Example 7 (Continued). If we calculate s^2 for the 40 emission figures on page 191, we get

$$s^2 = \frac{(17 - 19.6)^2 + (15 - 19.6)^2 + \ldots + (14 - 19.6)^2}{40 - 1}$$

$$= 30.4$$

and then, looking up the square root of 30.4 in Table X at the end of the book, we find that the standard deviation of the emission data is approximately

$$s = \sqrt{30.4} = 5.51 \text{ tons}$$

To be honest, we did not really obtain this result by actually calculating the forty deviations from the mean, $17 - 19.6$, $15 - 19.6, \ldots,$ and $14 - 19.6$, dividing the sum of their squares by 39, and then taking the square root. This *could* have been done and the result would have been the same, but there exist shortcut formulas for s^2 and s which can save a great deal of work. All we need for these formulas are $\sum x$ and $\sum x^2$, namely, the sum of the x's and the sum of their squares, which we substitute into the expression

$$s^2 = \frac{n(\sum x^2) - (\sum x)^2}{n(n - 1)}$$

for the sample variance, and into the expression

$$s = \sqrt{\frac{n(\sum x^2) - (\sum x)^2}{n(n - 1)}}$$

for the sample standard deviation.

Example 7 (Continued). As we admitted earlier, we actually used these shortcut formulas in determining s^2 and s for the sulfur oxides emission data. First we found that $\sum x = 784$ and $\sum x^2 = 16{,}552$, and then we obtained by substitution

$$s^2 = \frac{40(16{,}552) - (784)^2}{40 \cdot 39}$$

$$= \frac{47{,}424}{1{,}560}$$

$$= 30.4$$

and, hence, $s = \sqrt{30.4} = 5.51$.

Since beginners often seem to get confused about the difference between $\sum x^2$ and $(\sum x)^2$, let us emphasize the point that *to find* $\sum x^2$ *we first square each individual* x *and then add all the squares*; on the other hand, *to find* $(\sum x)^2$ *we first add all the* x*'s and then square their sum.*

Example 9. To illustrate the calculation of the sample standard deviation with an example in which the arithmetic is somewhat easier to follow, let us refer to the following data on the number of microorganisms found in six cultures:

11 15 5 8 14 7

To be able to use the formula on page 194 and not the shortcut formula, we must first find $\bar{x}$, which in this case equals

$$\bar{x} = \frac{11 + 15 + 5 + 8 + 14 + 7}{6} = 10$$

The work required to find $\sum (x - \bar{x})^2$ is shown in the following table:

x	$x - \bar{x}$	$(x - \bar{x})^2$
11	1	1
15	5	25
5	−5	25
8	−2	4
14	4	16
7	−3	9
	0	80

and we thus get

$$s = \sqrt{\frac{80}{5}} = \sqrt{16} = 4$$

Note that the sum of the entries in the middle column, namely, $\sum (x - \bar{x})$, is equal to zero; as the reader will be asked to demonstrate in Exercise 13 on page 202, *this must always be the case*, and, hence, it provides a check on the calculations.

There would have been no advantage to using the shortcut formula in this example, since the mean was a whole number and the deviations from the mean were easily obtained. The following is an example where the mean is *not* a whole number and the shortcut formula does, indeed, provide simplifications:

Example 10. If an electric company reports, respectively,

11 13 9 7 5 4 2 12 8 9 17 14

power failures caused by storms for twelve weeks, $\sum x = 111$, $\sum x^2 = 1{,}239$, and substitution into the formula yields

$$s^2 = \frac{12(1{,}239) - (111)^2}{12 \cdot 11} = 19.3$$

and, hence,

$$s = \sqrt{19.3} = 4.39$$

Note that the mean is $\frac{111}{12} = 9.25$, so that the calculations with the deviations from the mean would have been much more tedious.

If we have to calculate the standard deviation or the variance of a set of data which is already grouped, we are faced with the same problem as on page 30. However, if we proceed as in the calculation of the mean of a frequency distribution and replace each value with the *class mark* of the class to which it belongs (namely, with the *midpoint* of the corresponding class interval), we can write the formula for s as

$$s = \sqrt{\frac{\sum (x - \bar{x})^2 \cdot f}{n - 1}}$$

where the x's are the class marks, the f's are the corresponding class frequencies, and n is the total number of observations. Thus, for each class mark we multiply the squared deviation from the mean by the corresponding class frequency, then we divide the sum of all these products by $n - 1$, and finally we take the square root to get s rather than s^2.

The formula which we gave in the preceding paragraph serves to *define* s for grouped data, but it is seldom used in practice. To begin with, there is a shortcut formula analogous to the one on page 195, namely,

$$s = \sqrt{\frac{n(\sum x^2 f) - (\sum xf)^2}{n(n-1)}}$$

where the x's are now the class marks and the f's are the corresponding class frequencies. Although this formula may look a bit involved, it makes the calculation of the standard deviation of grouped data fairly easy. Instead of having to work with the squared deviations from the mean, we have only to find $\sum xf$ (the sum of the products obtained by multiplying each class mark by the corresponding class frequency), $\sum x^2 f$ (the sum of the products obtained by multiplying the *square* of each class mark by the corresponding class frequency), and substitute into the formula.

Example 7 (Continued). To illustrate this technique, let us calculate the standard deviation of the 40 air pollution figures on page 191 on the basis of the distribution given in Figure 7.2, using the actual frequencies, 2, 5, 13, 14, 5, and 1, instead of the percentages shown in that diagram. Performing the necessary calculations, we get

Emission (tons)	*Class Mark* x	x^2	f	xf	x^2f
5–9	7	49	2	14	98
10–14	12	144	5	60	720
15–19	17	289	13	221	3,757
20–24	22	484	14	308	6,776
25–29	27	729	5	135	3,645
30–34	32	1,024	1	32	1,024
			Totals	770	16,020

so that

$$s = \sqrt{\frac{40(16{,}020) - (770)^2}{40 \cdot 39}}$$

$$= \sqrt{30.7}$$

$$= 5.54 \text{ tons}$$

which is very close to the value of 5.51 tons obtained on page 195.

As things turned out, the calculations were quite easy in the last example, but this was due mainly to the fact that the class marks were whole numbers and the frequencies were relatively small. Otherwise, it is preferable to use the same kind of *coding* which we suggested in Exercise 11 on page 35; how this is done will be explained in Exercise 11 on page 200.

In conclusion, let us point out that the various formulas which we have given for the standard deviation of a sample can serve also to *define* the standard deviation of a *finite population*—all we have to do is substitute μ for $\bar{x}$ and N (the size of the population) for n.

Exercises

1. The following are the wind velocities reported by an airport at 6 P.M. on five consecutive days: 14, 8, 17, 11, and 10 mph. Use the basic formula which defines s^2 on page 194 to find the variance of these figures.

2. While on a special weight-reducing diet, eight persons lost, respectively, 15, 23, 9, 11, 17, 20, 6, and 19 pounds. Use the basic formula which defines s on page 194 to find the standard deviations of these data.

3. In Exercise 2 on page 34 we stated that the records of 15 persons convicted of various crimes showed that, respectively, 4, 3, 0, 0, 2, 4, 4, 3, 1, 0, 2, 0, 2, 1, and 4 of their grandparents were born in the United States. Find the standard deviation of these figures using

- **a.** the basic formula which defines s on page 194;
- **b.** the shortcut formula.

4. On a field trip, ten students spotted, respectively, 12, 9, 16, 8, 17, 10, 15, 12, 14, and 12 different kinds of rocks. Find the standard deviation of these figures using

- **a.** the basic formula which defines s on page 194;
- **b.** the shortcut formula.

5. Use the shortcut formula to calculate s^2 for the I.Q.'s of the 16 jurors of Exercise 3 on page 34.

6. In the years 1964 through 1970, there were, respectively, 169, 117, 138, 77, 98, 101, and 104 new cases of morphine addiction in the United States. Use the shortcut formula to calculate the standard deviation of these data reported by the *Bureau of Narcotics and Dangerous Drugs*.

7. Repeat Example 10 on page 197 *twice*, once after subtracting 2 from each of the figures, and once after subtracting 9 from each of the figures. *What simplification does this suggest for the calculation of a standard deviation?*

8. Calculate the standard deviation of the distribution of absences obtained in Example 4 on page 14.

9. The following is the distribution of the percentage of students belonging to a certain ethnic group in 40 schools:

Percentage	*Frequency*
0–4	15
5–9	11
10–14	7
15–19	5
20–24	2

Calculate the standard deviation of this distribution.

10. Calculate the standard deviation of the service-time distribution of Exercise 8 on page 35.

11. **CODING** If we use the coding of Exercise 11 on page 35, the shortcut formula for the standard deviation becomes

$$s = c\sqrt{\frac{n(\sum u^2f) - (\sum uf)^2}{n(n-1)}}$$

where all the symbols are as explained on page 36—the u's are the class marks in the new scale, the f's are the corresponding class frequencies, n is the total number of observations, and c is the class interval of the distribution. For the distribution of the emission data of Example 7 on page 193 we would thus get

Class Mark x	u	f	uf	u^2f
7	−2	2	−4	8
12	−1	5	−5	5
17	0	13	0	0
22	1	14	14	14
27	2	5	10	20
32	3	1	3	9
		Totals	18	56

so that

$$s = 5\sqrt{\frac{40(56) - (18)^2}{40 \cdot 39}}$$

$$= 5\sqrt{1.23}$$

$$= 5.55 \text{ tons}$$

The difference between this answer and the one obtained on page 198 (namely, 5.54 tons) is due to rounding.

a. Use this method to rework Exercise 8.

b. Use this method to rework Exercise 9.

c. Use this method to rework Exercise 10.

12. THE SAMPLE RANGE When we are dealing with a very small sample, a good estimate of the population standard deviation can often be obtained on the basis of the sample range, namely, *the largest value minus the smallest.* Such estimates of the population standard deviation σ are given by the sample range divided by the divisor d, which depends on the size of the sample; for random samples from populations having roughly the shape of a normal distribution its values are

Sample size n	2	3	4	5	6	7	8	9	10
Divisor d	1.13	1.69	2.06	2.33	2.53	2.70	2.85	2.97	3.08

For instance, if we take the bacteriological data of Example 9 on page 196, we find that the smallest value is 5 and the largest value is 15, so that the sample range is $15 - 5 = 10$. Thus, we estimate the standard deviation of the population from which this sample of size 6 was obtained as

$$\frac{10}{2.53} = 3.95$$

and it should be observed that this figure is *very close* to the sample standard deviation $s = 4$ obtained for these data on page 196.

a. Use this method to estimate σ for the population which led to the data of Exercise 1, and compare the result with the square root of the value obtained for s^2 in that exercise.

b. Use this method to estimate σ for the population which led to the data of Exercise 2, and compare the result with the value of s obtained in that exercise.

c. In a chemistry experiment, a group of students determined the ignition temperature of yellow phosphorus *four times*, and obtained 94.2, 90.8, 95.3, and 91.6 degrees Fahrenheit. Use the method of this exercise to estimate σ for the population which led to these data, and explain what it measures.

d. Use this method to estimate σ for the population which led to the data (SAT scores) of Exercise 2 on page 50.

13. Show that $\sum (x - \bar{x}) = 0$ for any set of x's whose mean is $\bar{x}$ by writing out in full what is meant by $\sum (x - \bar{x})$.

14. Use the binomial expansion $(a - b)^2 = a^2 - 2ab + b^2$ to show that

$$\sum (x - \bar{x})^2 = \sum [(x - \mu) - (\bar{x} - \mu)]^2 = \sum (x - \mu)^2 - n(\bar{x} - \mu)^2$$

which proves that $\sum (x - \bar{x})^2$ *is always less than or equal to* $\sum (x - \mu)^2$. (*Hint:* If necessary, write each of the expressions in full, that is, without summation signs.)

The purpose of the last section has been to introduce the sample standard deviation as an estimate of the standard deviation of the population from which the sample was obtained; the exercises which follow deal with some other applications of the standard deviation of a set of data.

15. STANDARD UNITS If Mary and Joan, who attend different colleges, got grades of 68 and 76 in an English examination, it is difficult to judge which one is really doing better than the other. However, if we knew that Mary's class averaged 50 with a standard deviation of 12, while Joan's class averaged 64 with a standard deviation of 10, we could say that Mary's grade is $\frac{68 - 50}{12} = 1.5$ standard deviations above average while Joan's grade is only $\frac{76 - 64}{10} = 1.2$ standard deviations above average. Thus, even though Joan got a higher grade than Mary, Mary's performance *relative to her class* was better than Joan's. What we have done here is precisely what we did on page 157 when we converted measurements into *standard units* by means of the formula $z = \frac{x - \mu}{\sigma}$; for observed data the formula becomes $z = \frac{x - \bar{x}}{s}$, and it can be of great help in the comparison of measurements or observations belonging to different sets of data.

a. In a city in the Southwest, restaurants charge on the average \$6.35 for a steak dinner (with a standard deviation of \$0.40), \$3.35 for a chicken dinner (with a standard deviation of \$0.25), and \$8.55 for a lobster dinner (with a standard deviation of \$0.30). If a restaurant in this city charges \$6.95 for a steak dinner, \$3.75 for a chicken dinner, and \$8.95 for a lobster dinner, which of the three dinners is *relatively* most overpriced?

b. The applicants to one state university have an average ACT mathematics score of 20.4 with a standard deviation of 3.1, while the applicants to another state university have an average ACT mathematics score of 21.1 with a standard deviation of 2.8. If a student gets 25 on this test, with respect to which of these two universities is he or she relatively speaking in a *better position*? How about a student who gets 30 on this test?

16. RELATIVE VARIATION If a set of weights has a standard deviation of 0.13 ounce, we cannot really tell whether these weights are very *precise*. They would be very precise, indeed, if we repeatedly weighed a full-grown elephant, but they would be far from precise if we repeatedly weighed a loaf of bread. Thus, we cannot tell whether measurements are really precise unless we know something about the actual magnitude of the quantity being measured, and it is for situations like this that we need a measure of **relative variation**, such as the **coefficient of variation:**

$$V = \frac{s}{\bar{x}} \cdot 100$$

With this formula, we simply express the standard deviation as a percentage of the (average) size of the quantity being measured.

a. If, over the last few years, one golfer has averaged 82 for eighteen holes with a standard deviation of 3.4, while another golfer has averaged 96 with a standard deviation of 3.9, which of the two is *relatively more consistent*?

b. Use the result of Exercise 1 to determine the coefficient of variation of the wind velocities.

c. Use the result of Exercise 2 to determine the coefficient of variation of the weight losses.

d. In order to compare the precision of two micrometers, a laboratory technician studies recent measurements made with both instruments. The first micrometer was recently used to measure the diameter of a ball bearing and the measurements had a mean of 4.93 mm and a standard deviation of 0.019 mm; the second was recently used to measure the unstretched length of a spring and the measurements had a mean of 2.57 in. with a standard deviation of 0.011 in. Which of the two micrometers is relatively more precise?

17. SKEWNESS We say that the distribution of Figure 7.3 is **symmetrical**, for we can picture its histogram folded along the dashed line so that the two halves will more or less coincide. The distribution of Figure 7.4, on the other hand, is said to be **skewed**, because it has more of a "tail" on one side than it has on the other. If the "tail" is on the right, as in Figure 7.4,

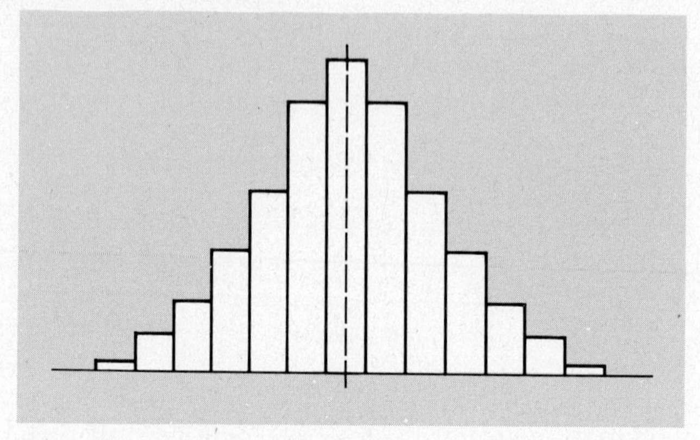

Figure 7.3 Symmetrical distribution.

the mean of the distribution will generally exceed its median (and if the "tail" is on the left, the order will be reversed). Based on this difference, the **Pearsonian coefficient of skewness** measures the skewness of a distribution by means of the formula

$$\frac{3(\bar{x} - \tilde{x})}{s}$$

where $\tilde{x}$ denotes the median. Note that for symmetrical distributions the mean and the median coincide and the Pearsonian coefficient of skewness is equal to *zero*.

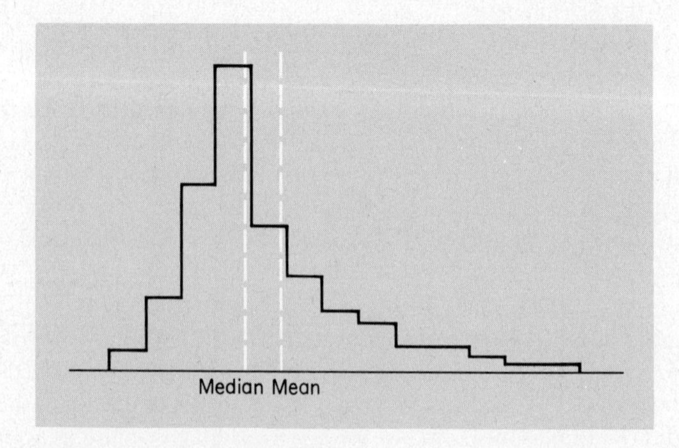

Figure 7.4 Skewed distribution.

a. For the air pollution data of Example 7 we showed that the mean is 19.6 tons and that the standard deviation is 5.5 tons. Find the median on the basis of the ungrouped data on page 191, and then calculate the Pearsonian coefficient of skewness.
b. For the data on absences of Example 4 of Chapter 2, we showed that the mean is 15.7 and the median is 15.2. Use the result of Exercise 8 or calculate s for the distribution of absences obtained on page 14, and then determine the Pearsonian coefficient of skewness.

The sampling distribution of the mean

When we calculate the mean of a sample, this mean is a value of a random variable. Clearly, the values we get for $\bar{x}$ will differ from sample to sample (as they did for the ten samples on page 191), and if we knew enough about the population from which we are sampling, we could actually determine the distribution of $\bar{x}$, or at least we could make some probability statement about its variability. For instance, on page 192 we judged on the basis of very meager information that in the given example *the odds were 6 to 4 that the mean of a random sample (of size 40) would differ from the true mean by less than one ton.* This argument was based on the means of 10 samples (taken from one and the same population), and it stands to reason that if we took 100 samples like this or even 1,000, we should be able to get a pretty good picture about the distribution (probability function or probability density) of $\bar{x}$.

Of course, in actual practice we do not take 1,000 samples from the same population, nor do we take 100 or even 10, and we must therefore rely on other methods to obtain information about the **sampling distribution of the mean**. [This is what we call the distribution of $\bar{x}$, for it tells us how the means of random samples (from one and the same population) would vary from sample to sample.] In most practical applications *in which we have only one sample,* we have no choice but to base our argument about the sampling distribution of the mean on two theorems: *one theorem, given in the paragraph which follows, about the mean and the standard deviation of this sampling distribution, and the other, explained on page* 208, *about its overall shape.*

The mean of the sampling distribution of $\bar{x}$ will be denoted by $\mu_{\bar{x}}$ (which reads "μ sub $\bar{x}$"), and the first of the two theorems states (among other things) that $\mu_{\bar{x}}$ equals the mean of the population from which the sample is obtained. This should not come as a surprise, for it stands to reason that the means of samples from one and the same population should *on the average* equal the mean of the population itself. The first theorem also provides a formula for the standard deviation $\sigma_{\bar{x}}$ of the sampling distribution of the mean; formally it states that for random samples of size n from a population having the mean μ and the standard deviation σ,

The sampling distribution of $\bar{x}$ has the mean

$$\mu_{\bar{x}} = \mu$$

and the standard deviation*

$$\sigma_{\bar{x}} = \frac{\sigma}{\sqrt{n}}$$

It is customary to refer to $\sigma_{\bar{x}}$, the standard deviation of the sampling distribution of the mean, as the **standard error of the mean**. Its role in statistics is fundamental, for *it measures the extent to which sample means vary, or fluctuate, due to chance*, and this is *The Problem* with which we have been concerned in this chapter.

The quotient $\sigma/\sqrt{n}$ demonstrates two things: *If σ is large and there is considerable variation in the population from which we are sampling, we can expect a proportionally large variation in the sampling distribution of the mean. So far as n is concerned, it is apparent from the formula that the larger the size of the sample, the smaller will be the chance fluctuations of the mean and the closer we can expect the mean of a sample to be to μ, the mean of the population.* Both of these properties make sense, intuitively, as we already indicated on page 192.

Example 7 (Continued). Let us now take another look at the example in which we were concerned with the average daily emission of sulfur oxides of an industrial plant. On page 195 we showed that the standard deviation of the 40 sample values is $s = 5.5$, and if we now use this figure as an estimate of σ, the standard deviation of the

*When the population from which we are sampling is *finite*, the formula for σ must be modified by multiplying $\sigma/\sqrt{n}$ by the **finite population correction factor** $\sqrt{\frac{N-n}{N-1}}$, where N is the size of the population. In actual practice, this modification is used only when the sample constitutes a substantial portion of the population (say, 5% or more), for otherwise the effect of the correction factor is negligible (see Exercise 2 on page 211).

population from which the sample was obtained, we can argue that for $n = 40$ the standard error of the mean is approximately

$$\frac{\sigma}{\sqrt{n}} = \frac{5.5}{\sqrt{40}} = 0.87$$

Having obtained this result by means of the formula for $\sigma_{\bar{x}}$, it would be interesting to see how close we will come if we actually calculate the standard deviation of the 10 sample means on page 191; after all, their fluctuations are also indicative of the variability of the sampling distribution of the mean, and their standard deviation constitutes an estimate of $\sigma_{\bar{x}}$ (for random samples of size 40 from the given population). If we add the ten means on page 191 and also their squares, we get 203.0 and 4,182.62, and if we substitute these figures together with $n = 10$ into the shortcut formula for s, we get

$$s = \sqrt{\frac{10(4{,}128.62) - (203.0)^2}{10 \cdot 9}} = \sqrt{0.86} = 0.93$$

This is fairly close, about as close as we can expect in an example like this, to 0.87, our first estimate of $\sigma_{\bar{x}}$. What we have done here does not *prove* anything, but it provides *supporting evidence* for the formula $\sigma_{\bar{x}} = \dfrac{\sigma}{\sqrt{n}}$.

To demonstrate how the formula for the standard error of the mean is used in actual practice, we really need the second theorem referred to on page 205, but even without it we can get some idea by referring to *Chebyshev's theorem* (see page 147), according to which we can now say that

The probability of getting a sample mean which differs from the population mean by less than k times $\sigma_{\bar{x}}$ is at least

$$1 - \frac{1}{k^2}$$

This application of Chebyshev's theorem is illustrated by the following example:

Example 11. If we take a random sample of size 100 from an infinite (or very large) population with $\sigma = 20$, then the probability is *at least* $1 - \dfrac{1}{2^2} = 0.75$ that we will get a sample mean within $k = 2$ standard errors of the mean of the population, namely, that we will get a sample mean which differs from the *true* mean by less than

$$k \cdot \frac{\sigma}{\sqrt{n}} = 2 \cdot \frac{20}{\sqrt{100}} = 4$$

Similarly, if we take a random sample of size 36 from an infinite (or very large) population with $\sigma = 10$, then the probability is *at least* $1 - \frac{1}{3^2} = 0.89$ that we will get a sample mean within $k = 3$ standard errors of the mean of the population, namely, that we will get a sample mean which differs from the *true* mean by less than

$$3 \cdot \frac{10}{\sqrt{36}} = 5$$

If we want to be more specific about probabilities like these, we shall have to refer to the second of the two theorems mentioned on page 205, namely, the **Central Limit theorem**, which states that

> **If n is large, the sampling distribution of the mean can be approximated closely with a normal distribution.**

It is difficult to say precisely how large n must be before the theorem applies, but unless the population has a very unusual shape, the approximation will be good even when n is relatively small—certainly when n is 30 or more. (Incidentally, when the population itself has the shape of a normal distribution, the theorem applies regardless of the size of the sample.)

Example 12. In case the reader still finds it difficult to picture sampling distributions and the significance of the Central Limit theorem, suppose that on a certain day hundreds upon hundreds of sportsmen are fishing on a large well-stocked lake, so that none of them has any trouble catching the legal limit of, say, three dozen trout. Well then, if we are interested in the *true* average length of the trout with which the lake has been stocked, the lengths of all the trout in this lake constitute the population from which the fishermen are "sampling," and it will have a certain mean μ and a certain standard deviation σ. If each fisherman were to calculate the mean length of his catch of 36 fish, we would have hundreds of sample means which could be grouped into a distribution, say, the one pictured in Figure 7.5. It is this kind of distribution, namely, a distribution of sample means,

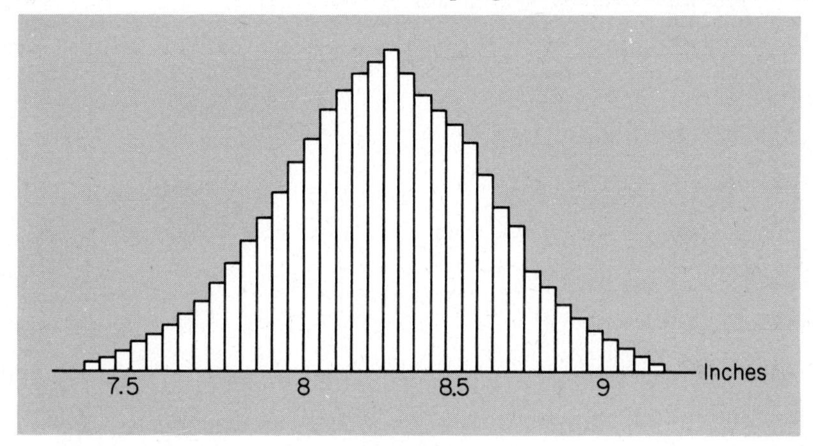

Figure 7.5 Distribution of sample means.

which according to the Central Limit theorem should have roughly the shape of a normal distribution (provided the sample size n is sufficiently large).

If the mean and the standard deviation of *all* the trout in the lake happened to be $\mu = 8.32$ inches and $\sigma = 2.45$ inches, we could use Table I to determine all sorts of probabilities about the average length of an angler's catch (assuming that he catches 36 trout and does not throw back any of those which he considers too short). For instance, the probability that the average length of his catch will exceed 8.5 inches is given by the area of the shaded region of Figure 7.6, namely, the area to the right of

$$z = \frac{8.5 - \mu_{\bar{x}}}{\sigma_{\bar{x}}} = \frac{8.5 - 8.32}{2.45/\sqrt{36}} = 0.44$$

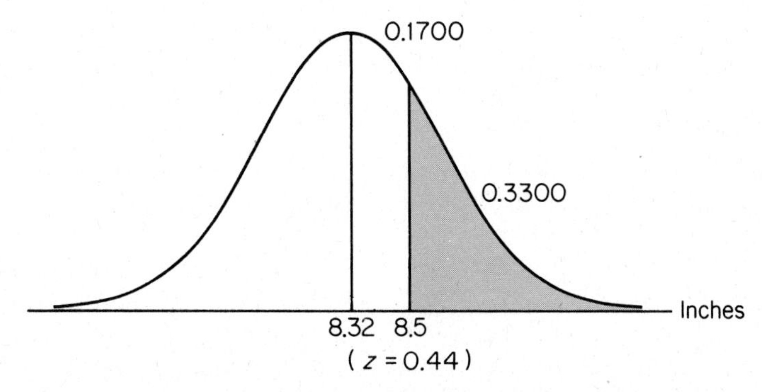

Figure 7.6 Diagram for Example 12.

where we did not bother to use the finite population correction factor in the formula for $\sigma_{\bar{x}}$, for as we said, the lake is very well stocked. Looking up the entry corresponding to $z = 0.44$ in Table I, we find that the desired probability is $0.5000 - 0.1700 = 0.3300$. In Exercise 8 on page 213 the reader will be asked to use the same kind of argument to find the probability that an angler's catch will average less than 7.4 inches and that it will average anywhere from 8.0 to 9.0 inches.

Example 11 (Continued). Let us return to the first of the two illustrations in which we used Chebyshev's theorem to study the chance fluctuations of means. We can now say that if a random sample of size $n = 100$ is taken from an infinite population with $\sigma = 20$, the probability of getting a sample mean which differs from the *true* (population) mean by less than 4 is given by the area of the tinted region of Figure 7.7. Since this is the normal curve area between $z = \dfrac{-4}{20/\sqrt{100}} = -2$ and $z = 2$, we find that the answer is *twice* 0.4772, namely, 0.9544 or roughly 0.95. Thus, whereas Chebyshev's theorem enabled us only to say that this probability is *at least* 0.75, we can now say that it is just about 0.95.

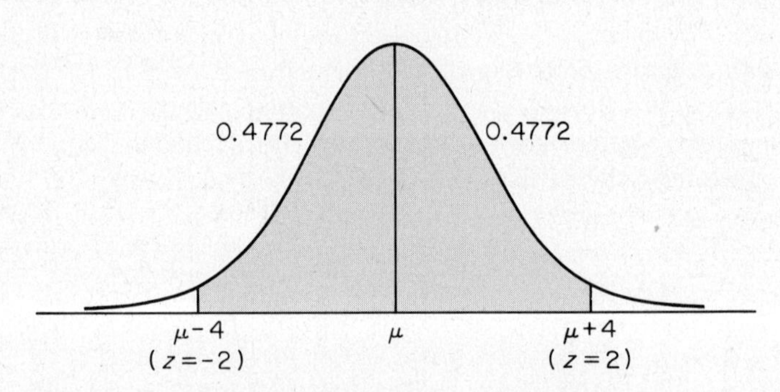

Figure 7.7 Diagram for continuation of Example 11.

The main purpose of this chapter has been to introduce the concept of a *sampling distribution*, and the one which we chose for this purpose was the sampling distribution of the mean. Note, however, that instead of the mean we could have studied the median or some other statistics and investigated its chance fluctuations. So far as corresponding theory is concerned, we would, of course, have obtained different formulas for the standard errors (that is, for the standard deviations of the respective sampling distributions); the one for the median, for example, is given in

Exercise 11 on page 213, and the one for the sample standard deviation is given on page 221. One thing which helps considerably is that most of these sampling distributions can be approximated with normal curves when the size of the samples is sufficiently large.

Exercises

1. What happens to the standard error of the mean for random samples from an infinite population when

a. the sample size is increased from 60 to 240;
b. the sample size is increased from 50 to 450;
c. the sample size is decreased from 750 to 30;
d. the sample size is increased from 100 to 225?

2. What is the value of the *finite population correction factor* in the formula for $\sigma_{\bar{x}}$ when

a. $n = 10$ and $N = 400$; **b.** $n = 50$ and $N = 1{,}000$?

3. Random samples of size 2 are taken *with replacement* (see page 181) from the population which consists of the numbers 1, 3, 5, and 7.

a. Assigning each number the probability $\frac{1}{4}$ and using the formulas on pages 140 and 144, show that the mean and the standard deviation of this population are $\mu = 4$ and $\sigma = \sqrt{5}$.

b. List the 16 possible samples of size $n = 2$ that can be drawn *with replacement* from this population (3 *and* 5, for example, is counted separately from 5 *and* 3), calculate their respective means, and, assigning each of these values the probability $\frac{1}{16}$, verify that we get the following *sampling distribution of the mean*:

$\bar{x}$	*Probability*
1	$\frac{1}{16}$
2	$\frac{2}{16}$
3	$\frac{3}{16}$
4	$\frac{4}{16}$
5	$\frac{3}{16}$
6	$\frac{2}{16}$
7	$\frac{1}{16}$

c. Use the formulas on pages 140 and 144 to calculate the mean and the standard deviation of the sampling distribution obtained in part (b), and verify the results with the use of the theorem on page 206.

4. Suppose that in Exercise 3 sampling had been *without replacement*, namely, that a random sample of size 2 is taken without replacement from the finite population which consists of the numbers 1, 3, 5, and 7.

a. List the six possible samples of size $n = 2$ that can be drawn without replacement from this population, calculate their respective means, and, assigning each of these values the probability $\frac{1}{6}$, verify that we get the following *sampling distribution of the mean*:

$\bar{x}$	*Probability*
2	$\frac{1}{6}$
3	$\frac{1}{6}$
4	$\frac{2}{6}$
5	$\frac{1}{6}$
6	$\frac{1}{6}$

b. Use the formulas on pages 140 and 144 to calculate the mean and the standard deviation of the sampling distribution obtained in part (a), and verify the value obtained for the standard deviation with the use of the results of part (a) of Exercise 3 and the formula given in the footnote on page 206.

5. The mean of a random sample of size $n = 25$ is used to estimate the mean of a very large population (consisting of the attention spans of persons over 65), which has a standard deviation of $\sigma = 2.5$ minutes. What can we say about the probability that the value of $\bar{x}$ we get will be "off" either way by less than 1.25 minutes, using

a. Chebyshev's theorem; **b.** the Central Limit theorem?

6. The mean of a random sample of size $n = 100$ is used to estimate the mean of a very large population (consisting of the incomes of persons ten years after they graduated from college), which has a standard deviation of $\sigma = \$12,000$. What can we say about the probability that the value of $\bar{x}$ we get will be "off" either way by less than $2,500, using

a. Chebyshev's theorem; **b.** the Central Limit theorem?

7. In the continuation of Example 11 on page 210 we showed that if a random sample of size $n = 100$ is taken from an infinite population with $\sigma = 20$, the probability of getting a sample mean which differs from the population mean by less than 4 is approximately 0.95. What would this probability be if the population were finite and of size $N = 500$? Explain why it stands to reason that this probability should be greater than 0.95.

8. With reference to Example 12 on page 208, find the probability that an angler's catch will average

a. less than 7.4 inches;
b. anywhere from 8.0 to 9.0 inches.

9. In the second part of Example 11 on page 208 we used Chebyshev's theorem to show that if a random sample of size 36 is taken from an infinite population with $\sigma = 10$, the probability that a sample mean will be "off" by less than 5 is at least 0.89. What value would we get for this probability if we used the Central Limit theorem?

10. The mean of a random sample of size $n = 64$ is used to estimate the mean of a very large population (consisting of the lifetimes of certain light bulbs) which has a standard deviation of $\sigma = 60$ hours. If we use the Central Limit theorem, what can we say about the probability that our estimate will be "off" either way by

a. less than 16.5 hours; **b.** less than 6.0 hours?

11. STANDARD ERROR OF THE MEDIAN As we indicated in Chapter 2 (see, for example, Exercise 7 on page 51), the sample mean is generally *more reliable* than the sample median; that is, it is subject to smaller chance fluctuations. Theoretically, this is expressed by the fact that for random samples from very large populations having roughly the shape of normal distributions, the **standard error of the median** (namely, the standard deviation of the sampling distribution of the median) is approximately $1.25 \cdot \dfrac{\sigma}{\sqrt{n}}$.

a. Verify that the mean of a random sample of size 64 is just about as reliable an estimate of a population mean as the median of a random sample of size 100.
b. How large a random sample would we have to take so that its mean is just about as reliable an estimate of the population mean as the median of a random sample of size 625?

12. PROBABLE ERROR OF THE MEAN In part (a) of Exercise 6 on page 165 the reader was asked to show that the probability is just about $\frac{1}{2}$ that a random variable having the standard normal distribution will take on a value between $z = -0.675$ and $z = 0.675$. Making use of this result, we can say that if the mean of a large random sample of size n is used to estimate the mean of an infinite population with the standard deviation σ, there is a *fifty-fifty chance* that the error is less than $0.675 \cdot \dfrac{\sigma}{\sqrt{n}}$.

It has been the custom to refer to this quantity, or more precisely $0.6745 \cdot \frac{\sigma}{\sqrt{n}}$, as the **probable error of the mean.**

a. If a random sample of size $n = 60$ is taken from a very large population (consisting of the I.Q.'s of army inductees) which has the standard deviation $\sigma = 12.8$, determine the probable error of the mean and explain its significance.

b. If a random sample of size $n = 120$ is taken from a very large population (consisting of the distances by which expert marksmen miss the center of a target) which has the standard deviation $\sigma = 8.3$ mm, determine the probable error of the mean and explain its significance.

8

The analysis of measurements

Introduction

In the beginning of Chapter 1 we explained that statistical inference is the science of basing decisions on numerical data; now we might say more specifically that statistical inference is *the science of making generalizations about populations on the basis of samples.* Traditionally, statistical inference has been divided into **problems of estimation** and **tests of hypotheses**—in problems of estimation we try to determine the parameters (statistical descriptions) of populations, and in tests of hypotheses we face the task of having to accept or reject specific assertions about populations.

Problems of estimation can be found everywhere, in science, in business, as well as in everyday life. *In science,* a psychologist may want to determine the average (mean) time it takes an adult to react to a given visual stimulus, or a biologist may want to find out what proportion of a certain kind of insect is born physically defective; *in business,* a retailer may want to know something about the average income of families living within 2 miles of his store, or a finance company may wish to estimate what percentage of all loan payments will be late; finally, *in everyday life,* we may want to know how long it takes (on the average) to iron a shirt, or what proportion of all automobile accidents are due to faulty brakes.

In each of the examples of the preceding paragraph somebody was interested in determining the "true" value of some quantity, so that they were all problems of estimation. They would have been *tests of hypotheses,* however, if the psychologist had wanted to decide on the basis of a sample whether the average time it takes an adult to react to the stimulus is really 0.44 second, if the biologist had wanted to check another scientist's claim that 0.073 (or 7.3%) of the given insects are born with defects, if the retailer had wanted to find out whether the true average income of all families living within 2 miles is at least $9,000, . . . , or if we wanted to check the claim that 0.026 (or 2.6%) of all automobile accidents are due to faulty brakes. Now it must be decided in each case whether to accept or reject a hypothesis (namely, an assertion or claim) about the parameter of a population.

Note that in each pair of examples (those from science, those from business, and those from everyday life) the first concerned **measurements,** while the other concerned **count data**. The reaction times, incomes, and ironing times are all quantities one has to measure in some way, while the insects born with defects, the late loan payments, and the number of accidents due to faulty brakes are all quantities one has to count. Since the statistical treatment of measurements differs in most instances from that of count data, we shall study the two separately—this chapter will be devoted to methods which apply mostly to measurements, while Chapter 9 will be devoted to methods which apply mostly to counts.

The estimation of means

In the analysis of measurements we are most often concerned with the *mean* of the population (or the *means* of the populations) from which our data are obtained.

Example 1. To illustrate some of the problems we face in the estimation of means, let us refer to a study in which a doctor wants to estimate the true average increase of the pulse rate of a person performing a certain strenuous task. The following are the data (increases in pulse rate in beats per minute) which the doctor obtained for 32 persons who performed the given task:

27	25	19	28	35	23	24	22
14	30	32	34	23	26	29	27
27	24	31	22	23	38	25	16
32	29	26	25	28	26	21	28

The mean of this sample is $\bar{x} = 26.2$ beats per minute, and in the absence of any other information this figure may well have to serve as an estimate of μ, the *true* average increase in the pulse rate of persons performing the given task.

An estimate like this is called a **point estimate**, as it consists of a single number, namely, a single point on the real number scale. Although this is the most common way of expressing an estimate, point estimates have the bad feature that they do not tell us how good they are—*they do not tell us* *how close we can expect them to be to the quantities they are supposed to estimate.* For instance, if an advertising agency claimed "on the basis of scientific evidence" that 80% of all women prefer a certain fabric softener to all others, this would not mean very much if their claim were based on interviews of five women among whom four happen to prefer the given kind of fabric softener. On the other hand, it might be meaningful and significant if the claim were based on interviews of 100 women, or perhaps 500 or more.

This demonstrates that *point estimates should always be accompanied by some information which makes it possible to judge their merits.* So far as the mean is concerned, we already saw in Chapter 7 that its chance fluctuations (and, hence, its reliability and its merit) depend on two things—the size of the sample and the variability (standard deviation) of the population.

Example 1 (Continued). Thus, we might supplement the above estimate of the true average increase in the pulse rate (namely, $\bar{x} = 26.2$ beats per minute) with the information that it is the mean of a sample of size $n = 32$, whose standard deviation is $s = 5.15$ (as can easily be verified by calculating s in accordance with the formula on page 195). Although this does not tell us the *exact* value of the standard deviation σ of the population, $s = 5.15$ is at least an estimate.

Scientific reports often present sample means together with the corresponding values of n and s, but this does not supply the reader with a coherent picture unless he has had some formal training in statistics. To make the supplementary information meaningful even to the layman, let us refer briefly to the two theorems of Chapter 7 concerning the sampling distribution of the mean (namely, the ones on pages 206 and 208), and to the result of part (b) of Exercise 7 on page 165, according to which 95% of the area under the standard normal curve falls between $z = -1.96$ and $z = 1.96$. Thus, making use of the fact that the sampling distribution of the mean can be approximated closely with a normal distribution having the mean μ and the standard deviation $\sigma_{\bar{x}} = \dfrac{\sigma}{\sqrt{n}}$, we find from Figure 8.1 that a sample mean will differ from the population mean μ by less than $1.96 \cdot \dfrac{\sigma}{\sqrt{n}}$ just about 95% of the time. In other words,

The probability that a sample mean will be "off" either way by less than $1.96 \cdot \dfrac{\sigma}{\sqrt{n}}$ is 0.95.

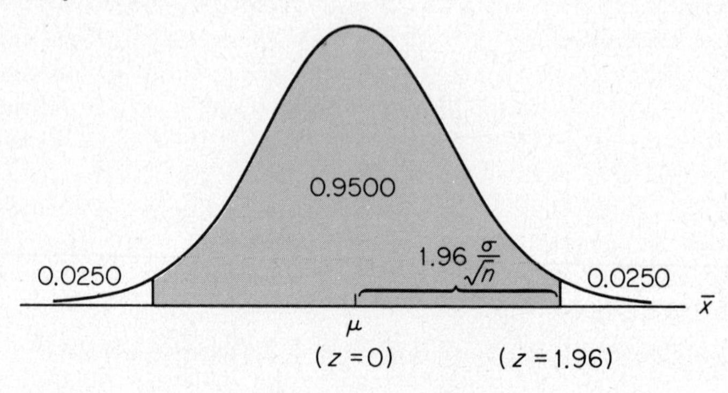

Figure 8.1 Sampling distribution of the mean.

The application of this result (which, incidentally, applies only to samples from infinite, or very large, populations) involves one difficulty—in order to judge the size of the error we might make when we use the mean of the sample to estimate the mean of a population, *we must know σ, the population standard deviation.* In actual practice, this is seldom the case, but we can get around this difficulty by replacing σ with an estimate, usually the sample standard deviation s. Generally, this is reasonable provided the sample size is not too small (that is, when n is 30 or more).

Example 1 (Continued). Returning now to our numerical example and using 26.2 beats per minute (the sample mean from page 216) as an estimate of the true average increase in the pulse rate of persons performing the given task, we can assert with a probability of 0.95 that the error of this estimate is less than

$$1.96 \cdot \frac{5.15}{\sqrt{32}} = 1.78$$

beats per minute. Of course, the error of the estimate is either less than 1.78 or it is not, *and we really do not know which,* but if we had to bet, 95 to 5 (or 19 to 1) would be *fair odds* that the error is less than 1.78.

If we had wanted to be "more certain" in this example, we could have substituted, say, 0.99 for 0.95 and correspondingly 2.58 for 1.96 [see the result of part (f) of Exercise 7 on page 165].

Example 1 (Continued). Thus, we could have asserted with a probability of 0.99 (and given odds of 99 to 1) that the sample mean is "off" by less than

$$2.58 \cdot \frac{5.15}{\sqrt{32}} = 2.35$$

beats per minute.

The error which we make when we use a sample mean to estimate the mean of a population is given by the difference $\bar{x} - \mu$, and the fact that the *magnitude* of this error is less than $1.96 \cdot \frac{\sigma}{\sqrt{n}}$ can be expressed in the following form:

$$-1.96 \cdot \frac{\sigma}{\sqrt{n}} < \bar{x} - \mu < 1.96 \cdot \frac{\sigma}{\sqrt{n}}$$

(In case the reader is not familiar with **inequality signs**, $a < b$ means "a is less than b," while $a > b$ means "a is greater than b"; also, $a \leq b$ means "a is less than or equal to b," and $a \geq b$ means "a is greater than or equal to b.") Applying some relatively simple algebra, we can rewrite the above inequality as

$$\bar{x} - 1.96 \cdot \frac{\sigma}{\sqrt{n}} < \mu < \bar{x} + 1.96 \cdot \frac{\sigma}{\sqrt{n}}$$

and we can assert with a probability of 0.95 that it will hold for any given sample. In other words, if $\bar{x}$ is the mean of a random sample of size n from an infinite (or very large) population with the standard deviation σ:

The probability is 0.95 that the interval from $\bar{x} - 1.96 \cdot \frac{\sigma}{\sqrt{n}}$ to $\bar{x} + 1.96 \cdot \frac{\sigma}{\sqrt{n}}$ will contain the true mean of the population.

An interval like this is called a **confidence interval**, its endpoints are called **confidence limits**, and the probability that such an interval will "do its job," namely, that it will contain the quantity we are trying to estimate, is called the **degree of confidence**. Note that if we had wanted the degree of confidence to be 0.99 instead of 0.95, we would have substituted 2.58 for 1.96, as in the example at the top of this page.

When σ is unknown (as is the case in most practical applications) and n is 30 or more, we proceed as before, substituting for σ the sample standard deviation s, and refer to the interval as a **large-sample confidence interval**.

Example 1 (Continued). Substituting $\bar{x} = 26.2$, $n = 32$, and $s = 5.15$, we thus get the 0.95 large-sample confidence interval

$$26.2 - 1.96 \cdot \frac{5.15}{\sqrt{32}} < \mu < 26.2 + 1.96 \cdot \frac{5.15}{\sqrt{32}}$$

$$24.4 < \mu < 28.0$$

for the true average increase in the pulse rate of persons performing the given task. Had we wanted to calculate a 0.99 large-sample confidence interval for this example, we would have obtained

$$26.2 - 2.58 \cdot \frac{5.15}{\sqrt{32}} < \mu < 26.2 + 2.58 \cdot \frac{5.15}{\sqrt{32}}$$

$$23.8 < \mu < 28.6$$

The two results which we obtained in this example illustrate the important fact that *if we increase the degree of certainty (namely, the degree of confidence), the confidence interval will become wider and, hence, tell us less about the quantity we are trying to estimate.*

When we estimate the mean of a population with the use of a confidence interval, we refer to this kind of estimate as an **interval estimate**. In contrast to point estimates, interval estimates require no further elaboration about their reliability—this is taken care of indirectly by their width and the degree of confidence.

The estimation of σ

So far we have concerned ourself only with the estimation of means, but *in principle* the methods that are used to estimate other population parameters are very much the same. By studying the sampling distributions of appropriate statistics, mathematicians have been able to develop formulas for confidence intervals for population standard deviations, population medians, population quartiles, . . . , and, as can well be imagined, the corresponding theory becomes quite involved. However, as we already indicated on page 211, the whole problem is greatly simplified by the fact that *for large samples most of these sampling distributions can be approximated closely with normal curves.*

For instance, if we are dealing with large samples, the sampling distribution of the standard deviation s can be approximated closely with a normal distribution having the mean σ and the standard deviation

$$\sigma_s = \frac{\sigma}{\sqrt{2n}}$$

called the **standard error of s** in accordance with the terminology introduced on page 206. Now, if we reason as on page 219, we arrive at the result that 95% of the time a sample value of s will fall on the interval from $\sigma - 1.96 \cdot \frac{\sigma}{\sqrt{2n}}$ to $\sigma + 1.96 \cdot \frac{\sigma}{\sqrt{2n}}$, and fairly straightforward algebra leads to the following **0.95 large-sample confidence interval for σ:**

$$\frac{s}{1 + \frac{1.96}{\sqrt{2n}}} < \sigma < \frac{s}{1 - \frac{1.96}{\sqrt{2n}}}$$

(To obtain a corresponding 0.99 large-sample confidence interval for σ we have only to substitute 2.58 for 1.96.)

Example 1 (Continued). Referring again to the pulse rates, and substituting $n = 32$ and $s = 5.15$ into the above formula, we get

$$\frac{5.15}{1 + \frac{1.96}{\sqrt{64}}} < \sigma < \frac{5.15}{1 - \frac{1.96}{\sqrt{64}}}$$

$$4.14 < \sigma < 6.82$$

Thus, we can assert with a probability of 0.95 that the interval from 4.14 to 6.82 contains σ, the *true* value of the standard deviation which measures the variability of the increase in the pulse rates. This may well make us think twice before we substitute $s = 5.15$ for σ (as on page 220); all we can really say is that 95 to 5 or 19 to 1 would be fair odds if we had to bet that the interval from 4.14 to 6.82 contains the true value of σ.

The method which we have described in this section applies only to *large samples*; there exist small-sample techniques for the estimation of σ, but they are not discussed in this text.

Exercises

1. To estimate the time it takes on the average to fill in a new income tax form, a research worker timed 49 persons (randomly selected from a large group), and got a mean of 46.8 minutes and a standard deviation of 5.6 minutes. What can we say with a probability of 0.95 about the research worker's error (namely, the amount by which he may be "off") if he estimates the true average time it takes to fill in the new tax form as 46.8 minutes?

2. Suppose that the principal of the elementary school referred to in Example 4 of Chapter 1 uses the mean of $\bar{x} = 15.6$ as an estimate of the true average number of students that are absent from school on any one day. Making use of the fact that the standard deviation of the 80 figures is 1.08, what can we say with a probability of 0.95 about the possible size of his error?

3. Suppose that in the air pollution study referred to in Example 7 of Chapter 7 the mean $\bar{x} = 19.6$ tons is used as an estimate of the plant's average daily emission of sulfur oxides. Making use of the fact that $n = 40$ and $s = 5.51$ tons, what can we say about the possible size of the error of this estimate

a. with a probability of 0.95;
b. with a probability of 0.99?

4. In a study of automobile collision insurance costs, a random sample of 60 body repair costs on a particular kind of damage had a mean of $467.25 and a standard deviation of $57.85. If $467.25 is used as an estimate of the true average cost of such repairs, what can we say about the possible size of the error

a. with a probability of 0.95;
b. with a probability of 0.99?

5. DETERMINATION OF SAMPLE SIZE An interesting by-product of the theory discussed on page 217 is that it sometimes enables us to determine the sample size which is required to attain a desired degree of reliability. If we want to use the mean of a random sample to estimate the mean of a population and we want to be able to assert with a probability of 0.95 that this estimate will be "off" either way by less than (or at most) some quantity E, we can write

$$E = 1.96 \cdot \frac{\sigma}{\sqrt{n}}$$

and, upon solving this equation for n, we get

$$n = \left(\frac{1.96 \cdot \sigma}{E}\right)^2$$

To illustrate this technique, suppose that a college dean wants to determine the average time it takes a student to get from one class to the next. He wants to be able to assert with a probability of 0.95 that his estimate, the mean of a suitable random sample, will not be "off" by more than 0.25 minute. Suppose also that preliminary studies have shown that it is reasonable to let $\sigma = 1.5$ minutes. Substituting $E = 0.25$ and $\sigma = 1.5$ into the above formula for n, he gets

$$n = \left(\frac{1.96 \cdot 1.50}{0.25}\right)^2 = 139$$

rounded *up* to the nearest whole number, and he concludes that a random sample of size 139 will be enough for the job. In other words, the dean will have to use 139 students to be able to assert with a probability of 0.95 that his estimate of the time it takes a student to get from one class to the next is not "off" by more than 0.25 minute. *Note that this method can be used only when we know (at least approximately) the value of the standard deviation of the population whose mean we are trying to estimate.*

a. In a study of television viewing habits, it is desired to estimate the average number of hours per week that teenagers spend watching television. Assuming that it is reasonable to use $\sigma = 3.4$ hours, how large a sample would be needed if one wants to be able to assert with a probability of 0.95 that the sample mean will be "off" by a quarter of an hour or less?

b. An efficiency expert wants to determine the average time it takes an adult to eat breakfast. If preliminary studies have shown that it is reasonable to let $\sigma = 2.1$ minutes, how large a sample will he need to be able to assert with a probability of 0.95 that his estimate, the mean of the sample, will be "off" by at most 0.2 minute?

c. Suppose that we want to estimate the average score that adults living in a large rural area should get in a current events test, and we want to be "99% sure" that our estimate, the mean of a suitable sample, will be "off" by at most 3.5 points. How large a sample will we need, assuming that it is known from similar studies that the standard deviation of such scores is 10? (*Hint:* Substitute 2.58 for 1.96 in the formula for n.)

d. Using the hint given in part (c), rework part (a) with the probability changed from 0.95 to 0.99.

6. Use the information given in Exercise 1 to construct a 0.95 confidence interval for the true average time it takes to fill in the new income tax form.

7. Use the information given in Exercise 2 to construct a 0.95 confidence interval for the true average number of students that are absent from the given elementary school on any one day.

8. Referring to Exercise 3, construct

a. a 0.95 confidence interval for the plant's average daily emission of sulfur oxides;

b. a 0.99 confidence interval for the plant's average daily emission of sulfur oxides.

9. Referring to Exercise 4, construct

a. a 0.95 confidence interval for the average cost of such body repairs;

b. a 0.99 confidence interval for the average cost of such body repairs.

10. A study of the annual growth of saguaro cacti showed that 64 of them grew on the average 48.2 mm with a standard deviation of 5.8 mm. Construct a 0.95 confidence interval for the true average annual growth of this kind of cactus.

11. To estimate the true average speed of cars traveling on a highway where the maximum speed is 55 mph, 225 cars were timed at an average speed of 61.5 mph with a standard deviation of 10.4 mph. Construct a 0.99 confidence interval for the true average speed of cars traveling on this highway.

12. A study of the time between the call for an ambulance (in a certain city) and the patient's arrival at the hospital showed that in 50 cases it took on the average 23.2 minutes with a standard deviation of 6.1 minutes. Construct

a. a 0.95 confidence interval for the true average time between the call for an ambulance (in this city) and the patient's arrival at the hospital;

b. a corresponding 0.99 confidence interval.

13. SMALL-SAMPLE CONFIDENCE INTERVALS FOR μ To be able to construct confidence intervals for μ on the basis of small samples (namely, when n is less than 30), we shall have to assume that the population from which we are sampling has roughly the shape of a normal distribution. Then, we can base confidence intervals for μ on the **t distribution**, a continuous distribution which is in many respects very similar to the normal distribution (see Figure 8.2). It is also symmetrical and has a zero

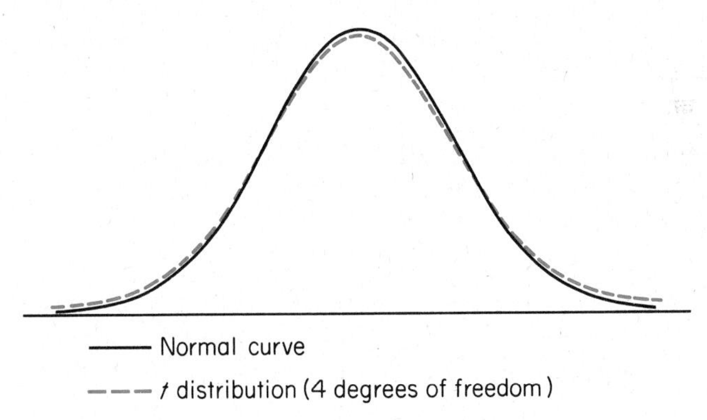

Figure 8.2 t distribution.

mean, but its shape depends on a parameter called the **number of degrees of freedom**, or simply the **degrees of freedom**, which, as the distribution will be used here, is given by *the sample size minus 1*. (In other applications of the t distribution, for instance, in Exercise 5 on page 245, the number of degrees of freedom will be given by a different expression.) Based on the t distribution and reasoning similar to that on page 219, we obtain the following **0.95 small-sample confidence interval for μ**:

$$\bar{x} - t_{0.025} \cdot \frac{s}{\sqrt{n}} < \mu < \bar{x} + t_{0.025} \cdot \frac{s}{\sqrt{n}}$$

which differs from the one on page 219 only insofar as σ is replaced by s (which we did anyhow) and that 1.96 is replaced by the quantity $t_{0.025}$, which is such that 95% of the area under the curve of the t distribution falls between $-t_{0.025}$ and $t_{0.025}$ (see Figure 8.3). Since the values of $t_{0.025}$ depend on the number of degrees of freedom, they will have to be looked up in Table II (which, incidentally, also contains the values of $t_{0.005}$ which must be used instead of $t_{0.025}$ when the degree of confidence is 0.99). To illustrate the calculation of a small-sample confidence interval for μ, suppose that the average amount of time it took 12 fuses (randomly

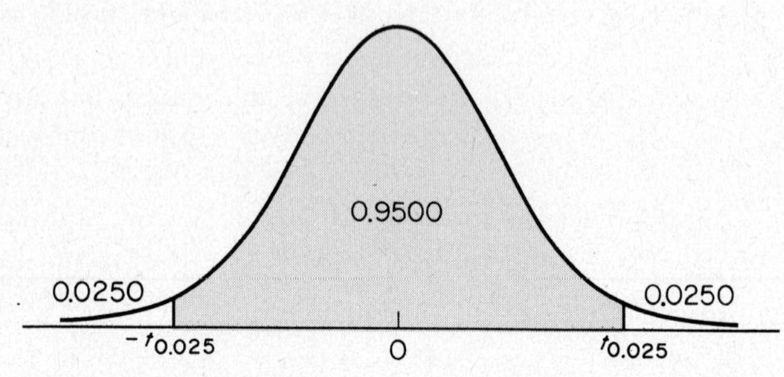

Figure 8.3 *t* distribution.

selected from a large lot) to blow with a 20% overload was 14.43 minutes with a standard deviation of 2.6 minutes. Since $t_{0.025}$ for $12 - 1 = 11$ degrees of freedom equals 2.201, substitution into the formula yields

$$14.43 - 2.201 \cdot \frac{2.6}{\sqrt{12}} < \mu < 14.43 + 2.201 \cdot \frac{2.6}{\sqrt{12}}$$

or

$$12.78 < \mu < 16.08$$

Information like this may be important in accepting or rejecting the lot of fuses, comparing them with other fuses, and so on.

a. In an air pollution study, an experiment station obtained an average of 2.18 micrograms of suspended benzene-soluble organic matter per cubic meter with a standard deviation of 0.54 for eight different samples of air. Construct a 0.95 confidence interval for the corresponding true mean.

b. In a survey conducted in a retirement community, it was found that ten "senior citizens" visited a physician on the average 6.6 times per year with a standard deviation of 1.7. Construct a 0.95 confidence interval for the true average number of times that such persons visit a physician per year.

c. In order to test the durability of a new paint, a highway department had test strips painted across heavily traveled roads in 15 different locations. If on the average the test strips disappeared after they had been crossed by 146,692 cars and the standard deviation is 14,380 cars, construct a 0.99 confidence interval for the number of cars it will actually take *on the average* to wear off this paint.

d. In setting type for a book, a compositor makes, respectively, 10, 11, 14, 8, 12, and 17 mistakes in six galleys. Verify that $\bar{x} = 12$ and $s = 3.16$ for these data, and construct a 0.99 confidence interval for the number of mistakes which this compositor actually averages per galley.

14. With reference to Exercise 1, construct a 0.95 confidence interval for the true standard deviation of the amount of time it takes to fill in the new income tax form.

15. Use the information of Exercise 2 to construct a 0.95 confidence interval for the true standard deviation of daily absences at the given elementary school.

16. With reference to Exercise 4, construct a 0.99 confidence interval for the true standard deviation of the costs of such body repairs.

Tests concerning means

In this section we shall study methods which will enable us to decide whether to accept or reject hypotheses (namely, assumptions or claims) about the means of populations. As we already pointed out on page 215, this includes such problems as deciding whether the average time it takes an adult to react to a certain visual stimulus is really 0.44 second, or whether the true average income of families living within 2 miles of a given store is at least $9,000.

In practice, we base most decisions like this on the *difference* between the mean of a sample and the assumed value of the population from which the sample was obtained—*if the difference is small, the sample mean supports the assumption about the mean of the population, and if the difference is large, the sample mean tends to refute it.*

Example 2. For instance, if the psychologist (who is studying such reaction times) takes a random sample of 55 adults, measures their reaction time to the given visual stimulus, and obtains a mean of 0.438 second, he may well consider the small difference of $0.44 - 0.438 = 0.002$ second as *supporting evidence* for the contention that the true mean is 0.44. On the other hand, if he obtained 0.363 or 0.526 (namely, a value which is much less or much greater than 0.44), he may well feel that the hypothesis $\mu = 0.44$ *cannot be correct.*

If the difference between a sample mean and the assumed value of the mean of the population is *very small or very large,* we really do not need high-powered statistical techniques to decide whether the sample evidence supports or refutes the hypothesis about μ. However, what can we do when the difference between a sample mean and the assumed mean of the population is *neither very small nor very large*?

Example 2 (Continued). What if the psychologist obtained a mean of 0.454 second, which differs from the assumed value by 0.454 − 0.44 = 0.014, or what if he obtained a mean of 0.416 second, which differs from the assumed value by 0.416 − 0.44 = −0.024?

All this really boils down to the question: Where do we draw the line? To answer this question we shall again have to study the chance fluctuations of means of random samples from the given population; in other words, we shall have to refer to the sampling distribution of the mean.

Example 2 (Continued). If the hypothesis $\mu = 0.44$ second is correct and it can be assumed (perhaps, on the basis of previous studies) that such reaction times can be expected to have the standard deviation $\sigma = 0.078$, we can picture this sampling distribution as in Figures 8.4

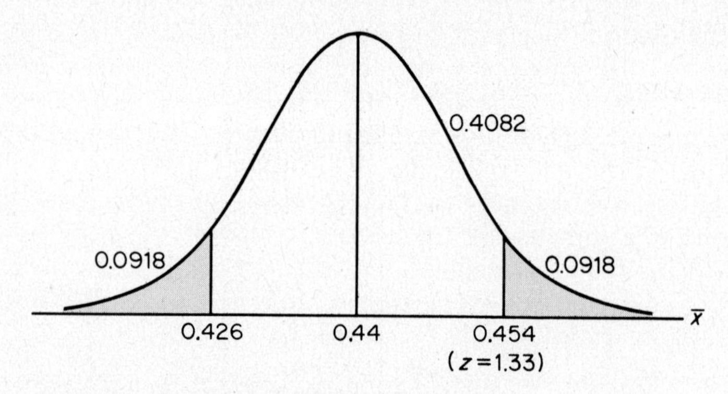

Figure 8.4 Sampling distribution of the mean.

and 8.5. According to the theorem on page 206, the mean of this sampling distribution equals the mean of the population, namely, $\mu = 0.44$ (if the assumption is correct), and its standard deviation is given by

$$\sigma_{\bar{x}} = \frac{\sigma}{\sqrt{n}} = \frac{0.078}{\sqrt{55}} = 0.0105$$

provided the population is so large that it can be treated as if it were infinite. Also, since the sample size is $n = 55$, we can use the Central Limit theorem to approximate the sampling distribution of the mean with a normal curve.

Suppose now that the psychologist actually obtains a sample mean of 0.454, which is "off" by 0.014, and wants to know the probability of *being "off" either way by that much or more*, namely the probability of getting a sample mean which is *at least* 0.454 or *at most* 0.44 − 0.014 = 0.426. The answer to this question is given by the total area of the tinted regions of Figure 8.4, which (owing to the symmetry of the normal distribution) is *twice* the area under the curve to the right of 0.454. Making use of the result of the preceding paragraph and converting to standard units, we get

$$z = \frac{0.454 - 0.44}{0.0105} = 1.33$$

and since the corresponding entry in Table I is 0.4082, we find that the desired probability is 2(0.5000 − 0.4082) = 2(0.0918) = 0.1836. Thus, the psychologist may well conclude that the sample mean of 0.454 does *not* constitute any real evidence against the assumption that the true mean is $\mu = 0.44$, for he could expect such a sample mean to be "off" by 0.014 *or more* better than 18% of the time.

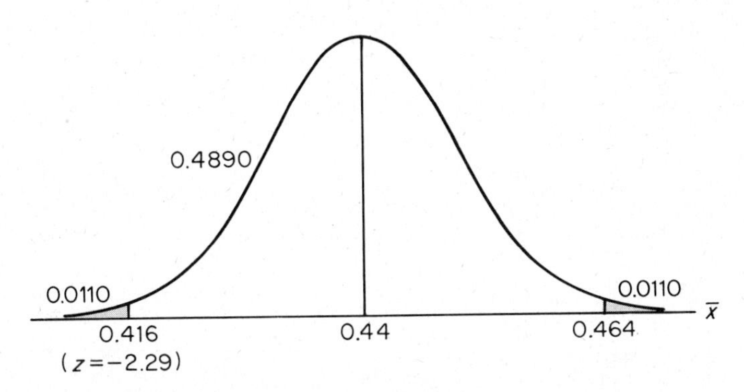

Figure 8.5 Sampling distribution of the mean.

If we use the same method to see how the psychologist might react if he obtained a sample mean of 0.416, we find that the probability of the mean being "off" either way by 0.024 *or more* is given by the total area of the tinted regions of Figure 8.5, namely, *twice* the area under the curve to the left of 0.416. Converting to standard units, we get

$$z = \frac{0.416 - 0.44}{0.0105} = -2.29$$

and since the entry in Table I corresponding to $z = 2.29$ is 0.4890, we find that the desired probability is $2(0.5000 - 0.4890) = 2(0.0110) = 0.0220$. In this case the psychologist may well conclude that the true average reaction time is *not* 0.44 second, for the odds against his being "off" by 0.024 *or more* are better than 44 to 1 (0.9780 to 0.0220 to be more exact).

We can now answer the question posed on page 228: Where do we draw the line? In practice, we proceed as is illustrated in Figure 8.6: *We specify the probability with which we are willing to risk rejecting the hypothesis about*

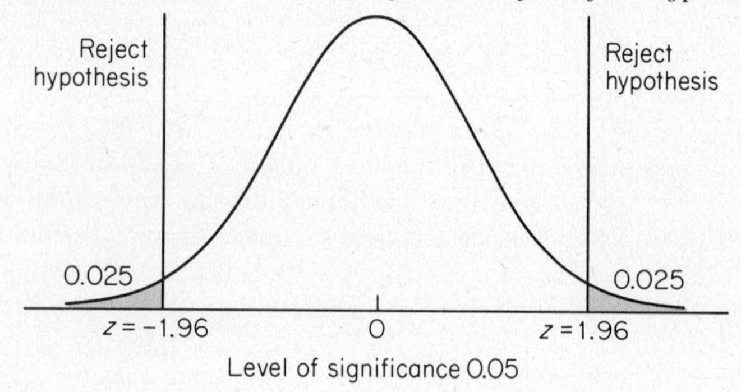

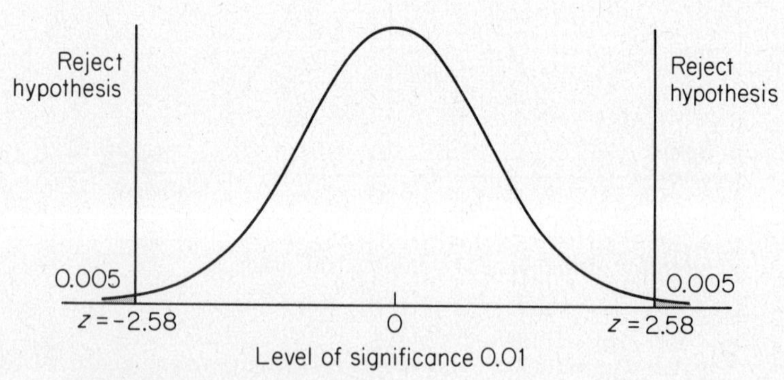

Figure 8.6 Two-tail test criteria.

the population mean even though it is true, usually, 0.05 or 0.01. Then we convert the sample mean into standard units (as in our two illustrations) and reject the hypothesis about the population mean if z is less than -1.96 or greater than 1.96, or if z is less than -2.58 or greater than 2.58. This procedure is called a **test of significance**, for it enables us to decide whether a difference between a sample mean and the assumed value of a population mean can be attributed to chance, or whether it is **statistically significant**, namely, too large to be "reasonably" attributed to chance. Correspondingly, the probability with which we are willing to risk rejecting the hypothesis about the population mean even though it is true is called the **level of significance**, and it is usually denoted by the Greek letter α (*alpha*). The choice of α, usually 0.05 or 0.01, depends in each case on the consequences (risks or penalties) of erroneously rejecting a hypothesis which is really true.

Example 3. To give another example, suppose that an oceanographer wants to check whether the average depth of the ocean in a certain area is 62.3 fathoms, as had previously been recorded. Deciding to use a level of significance of 0.05 and to take soundings at 40 random locations, she performs the necessary measurements, getting a mean of 64.8 fathoms and a standard deviation of 5.1. Then, using $s = 5.1$ as an estimate of σ, she finds that the standard error of the mean is

$$\sigma_{\bar{x}} = \frac{5.1}{\sqrt{40}} = 0.81$$

so that 64.8 fathoms, converted into standard units, becomes

$$z = \frac{64.8 - 62.3}{0.81} = 3.09$$

Since this value exceeds 1.96 (see Figure 8.6), it follows that the hypothesis $\mu = 62.3$ will have to be rejected, and the oceanographer may well decide to substitute her own value of 64.8 fathoms.

It is customary to refer to the criteria of Figure 8.6 as **two-sided tests** or as **two-tail tests**, because the hypothesis about the population mean is rejected for values of $\bar{x}$ falling into either "tail" of the sampling distribution. When the hypothesis about the population mean is rejected only for values of $\bar{x}$ falling into one of the "tails" of the sampling distribution, we refer to such a criterion as a **one-tail test**.

Example 4. To give an example of a one-tail test, let us refer to the problem on page 215, where a retailer wanted to determine whether the true average income of all families living within 2 miles of his store is *at least* \$9,000. Evidently, he will reject the hypothesis that the average income of these families is *at least* \$9,000 only when the sample mean is too small, for a large value of $\bar{x}$ would support rather than refute the claim. As we cannot perform a significance test without assigning a *specific* value to the population mean μ, we handle problems like this by testing the hypothesis $\mu = \$9{,}000$ against the **alternative hypothesis** that μ is *less than* \$9,000. This means that we will either accept the hypothesis $\mu = \$9{,}000$ or the **one-sided alternative** $\mu < \$9{,}000$. Returning for a moment to Example 2, where we were concerned with the reaction times to a visual stimulus, note that we were, in fact, testing the hypothesis $\mu = 0.44$ second against the **two-sided alternative** that the true average reaction time is either less than or greater than 0.44 second, namely, that $\mu \neq 0.44$.

The general procedure which we use to test the hypothesis that a population mean μ *equals a given value* against the one-sided alternative that it is *less than that value* is illustrated in Figure 8.7. If the level of significance is $\alpha = 0.05$, we reject the hypothesis (and accept the one-sided alternative) when z, the sample mean converted into standard units, is less than -1.64; if the level of significance is $\alpha = 0.01$, we reject the

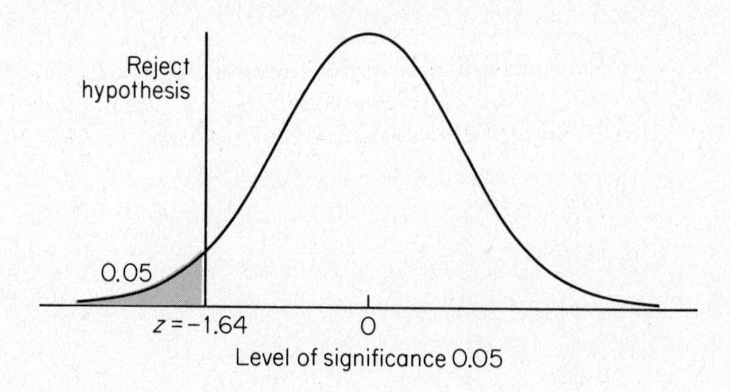

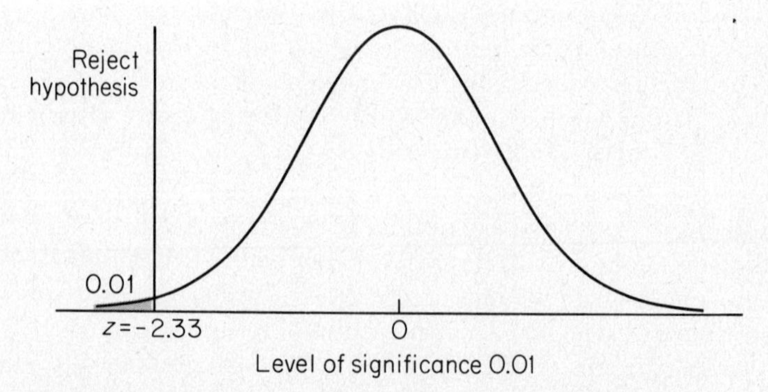

Figure 8.7 One-tail test criteria (alternative hypothesis μ less than given value).

hypothesis (and accept the one-sided alternative) when z is less than -2.33. These dividing lines, or **critical values**, of the criterion were obtained by approximating the sampling distribution of the mean with a normal curve; in fact, they are the results of parts (a) and (e) of Exercise 7 on page 165.

Example 4 (Continued). Returning to our example, suppose that the retailer who is concerned about the average income of families living within 2 miles of his store has 80 families interviewed, and gets $\bar{x} = \$8{,}862$ and $s = \$945$. Since $n = 80$ is large enough to justify substituting s for σ, we get $\frac{\sigma}{\sqrt{n}} = \frac{945}{\sqrt{80}} = 105.7$ for the standard error of the mean, and, hence,

$$z = \frac{8{,}862 - 9{,}000}{105.7} = -1.31$$

Since this value is *not* less than -1.64, the hypothesis $\mu = \$9{,}000$ *cannot be rejected* at the level of significance $\alpha = 0.05$. Even though the sample mean *is* less than \$9,000, the difference between \$8,862 and \$9,000 is **not statistically significant**; in other words, *the difference may well be attributed to chance.* Note that the statement "The difference is not statistically significant" does not necessarily imply that the hypothesis $\mu = \$9{,}000$ *must* be accepted; the retailer may still have some doubts, and he may prefer to continue the investigation with a larger sample.

The criteria of Figure 8.7 apply only for the one-sided alternative that the population mean μ is *less than* the value assumed under the hypothesis. To take care of the alternative hypothesis that the population mean μ is *greater than* the value assumed under the hypothesis, we have only to turn the diagrams around as in Figure 8.8. Then, if the level of significance is $\alpha = 0.05$, we reject the hypothesis (and accept the one-sided alternative) when z, the sample mean converted into standard units, is greater than 1.64; correspondingly, for $\alpha = 0.01$, the diving line of the criterion is 2.33.

Example 5. To illustrate, suppose that we want to investigate a vacuum cleaner manufacturer's claim that the average noise level of his latest model is *at most* 75 decibels, and that we have at our disposal readings of the sound intensity produced by 48 of his machines. If the mean of the readings is 76.4 and their standard deviation is 3.6, let us see whether we can reject the hypothesis $\mu = 75$ and accept the one-sided alternative $\mu > 75$ (namely, that the true average sound intensity exceeds 75 decibels) at the level of significance $\alpha = 0.01$.

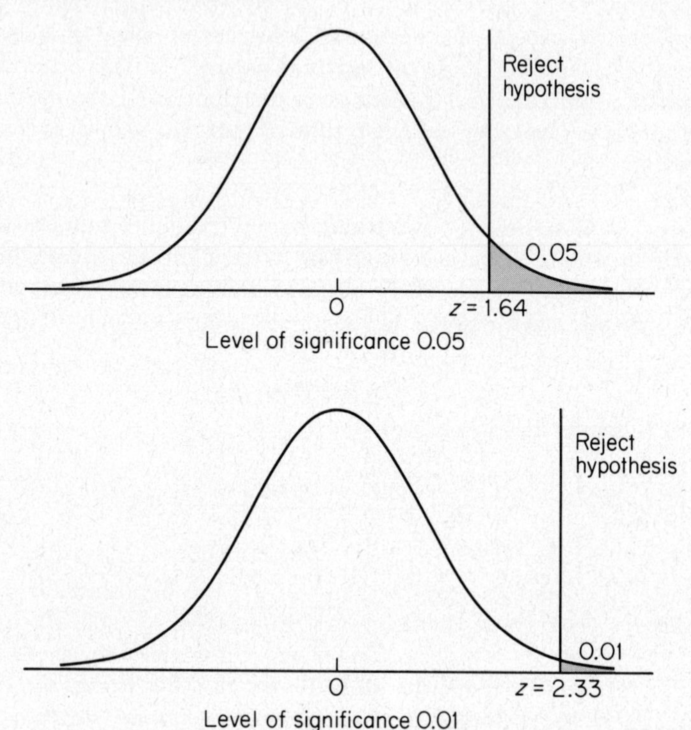

Figure 8.8 One-tail test criteria (alternative hypothesis μ greater than given value).

Since $n = 48$ is large enough to justify substituting s for σ, we get $\frac{\sigma}{\sqrt{n}} = \frac{3.6}{\sqrt{48}} = 0.52$ for the standard error of the mean, and, hence,

$$z = \frac{76.4 - 75}{0.52} = 2.69$$

Since this value exceeds 2.33, the "critical value" for a level of significance of 0.01, we conclude that the hypothesis $\mu = 75$ decibels will have to be rejected. In other words, the vacuum cleaners are on the average *noisier than claimed.*

Since the general procedure of testing hypotheses and constructing statistical decision criteria often tends to confuse the beginner, it helps to proceed systematically as outlined in the following steps:

1. **We formulate the hypothesis to be tested in such a way that the probability of erroneously rejecting it can be calculated.**

To follow this first step, we often have to assume the exact opposite of what we may want to prove. For instance, if it had been our purpose in Example 5 to show that on the average the vacuum cleaners are *noisier than claimed*, we would nevertheless have assumed that $\mu = 75$ decibels, namely, that on the average they are only as noisy as claimed. Had we assumed that $\mu > 75$, we would not have known what value to substitute for μ in the formula for z, and we would not have been able to perform the test. Similarly, to prove that a new drug will lower a heart patient's blood pressure more effectively than another, we assume that the new drug is only as good as the other drug (for which the average decrease in blood pressure is presumably known); and to show that a new programmed textbook is actually better than the one which has been in use, we assume that it is only as good as the old one (for which we know how much a student can be expected to learn). Since we assumed in each case that there is *no difference* (between the vacuum cleaner's performance and the manufacturer's claim, in the effectiveness of the two drugs, and in the merits of the two textbooks), we refer to hypotheses like these as **null hypotheses**. Nowadays, the term "null hypothesis" is used for any hypothesis *which is set up primarily to see whether it can be rejected*; in fact, this is why the symbol μ_0 is used to denote the value of μ assumed in a test of significance—the subscript "0" stands for "zero," "null," or "nought." For the same reason, null hypotheses are usually denoted H_0, and in contrast, alternative hypotheses are denoted H_A.

Actually, the idea of setting up a null hypothesis is not uncommon even in non statistical thinking. It is precisely what is done in criminal court procedures, where the accused is assumed to be innocent unless his guilt can be established beyond any reasonable doubt. The assumption that the accused is *not guilty* is a null hypothesis; if it cannot be rejected the accused will go free, but this does not necessarily imply that he is really innocent. Note also that the question of what we mean here by "reasonable doubt" is precisely the problem of assigning a level of significance.

2. We formulate the alternative hypothesis which is to be accepted when the null hypothesis must be rejected.

In Example 2, the one dealing with the reaction times to a visual stimulus, the null hypothesis was $\mu = 0.44$ second and, even though we did not say so specifically, the alternative hypothesis was $\mu \neq 0.44$ second (namely, the hypothesis that the true average reaction time does not equal 0.44 second). *Thus, we use the two-sided alternative* $\mu \neq \mu_0$ *if we want to reject the null hypothesis* $\mu = \mu_0$ *regardless of whether* μ_0 *happens to be too small or too large.*

The choice of an appropriate one-sided alternative depends usually on what we hope to be able to show, or better, perhaps, *where we want to put the burden of proof.*

Example 6. Suppose, for instance, that a manufacturer of dresses, whose sewing machine averages 115 dresses per work shift, is considering the purchase of a new sewing machine. If he does not want to buy the new machine unless it is definitely proved superior, he would test the null hypothesis $\mu = 115$ against the alternative hypothesis $\mu > 115$, and buy the new machine only if the null hypothesis can be rejected. If he wants to buy the new machine (which has some other nice features) unless it is actually slower than the old one, he would test the null hypothesis $\mu = 115$ against the alternative $\mu < 115$, and buy the new machine unless the null hypothesis can be rejected. *In the first case he would thus be putting the burden of proof on the new sewing machine, and in the second case he would be putting the burden of proof on the old one.*

Having formulated a null hypothesis and a suitable alternative, we then specify the level of significance (namely, the probability of erroneously rejecting the null hypothesis) and proceed with the following steps:

3. **Using the sample data, we calculate the value of the statistic**

$$z = \frac{\bar{x} - \mu_0}{\sigma/\sqrt{n}}$$

that is, we convert the sample mean into standard units, replacing σ (if necessary) with the sample standard deviation s.

4. **Finally, we base our decision on the appropriate criterion of Figure 8.6, 8.7, and 8.8, depending on the choice of the alternative hypothesis and the level of significance.**

As we already pointed out on page 231, the choice of the level of significance depends on the consequences (risks or penalties) of erroneously rejecting a null hypothesis which is actually true. Generally, we use $\alpha = 0.05$ or $\alpha = 0.01$, and preferably the smaller of the two values when the consequences of this kind of error are really serious.

If we *cannot reject* the null hypothesis, we have the option of either accepting it or reserving judgment. In the latter case we say that "the difference between $\bar{x}$ and μ_0 is *not statistically significant*." In other words, we admit that we are unable to *disprove* the null hypothesis, but by the same token we do not have to accept it either. This is the possibility which we suggested in Example 4 on page 233.

To use the methods of this section, the population standard deviation σ must be known, or the sample must be large enough so that we can substitute for σ the sample standard deviation s. If σ is unknown and n is *small* (less than 30), we can use a small-sample technique which will be explained in Exercise 9 on page 238, or, if the necessary assumptions cannot be met, an alternative method, based on the **sign test**, which will be explained in Exercise 11 on page 268.

Exercises

1. A law student, who wants to check a professor's claim that convicted embezzlers spend on the average 12.3 months in jail, takes a random sample of 36 such cases from court files. Using his results, namely, a mean of 10.8 months and a standard deviation of 3.9 months, test the null hypothesis $\mu = 12.3$ against the alternative hypothesis $\mu \neq 12.3$ at the level of significance $\alpha = 0.05$.

2. In a labor–management discussion it was brought up that workers (employed in a very large plant) take on the average 52.6 minutes to get to work. Is this figure substantiated by a survey in which a random sample of 60 workers took on the average 53.2 minutes with a standard deviation of 8.1 minutes? Use the level of significance $\alpha = 0.05$.

3. According to the norms established for a reading comprehension test, eighth graders should average 83.2 with a standard deviation of 8.6. If 45 randomly selected eighth graders from a certain school district averaged 86.9, test the null hypothesis $\mu = 83.2$ against the alternative hypothesis $\mu > 83.2$ at the level of significance $\alpha = 0.01$, and thus check the district superintendent's claim that his eighth graders are "above average."

4. A random sample of the boots worn by 50 soldiers in a desert region had an average (useful) life of 1.69 years with a standard deviation of 0.62 year. Under standard conditions, such boots are known to have an average (useful) life of 1.85 years. Can one conclude at the level of significance $\alpha = 0.05$ that use in the desert will *in general* decrease the (useful) life of such boots?

5. In a study of new sources of food, it is reported that a pound of a certain kind of fish will yield on the average 2.41 ounces of FPC (fish-protein concentrate), which is used to enrich various food products (including flour). Is this figure supported by a study in which 32 samples of this kind of fish yielded on the average 2.38 ounces of FPC (per pound of fish) with a standard deviation of 0.07 ounce, if we use

- **a.** the level of significance $\alpha = 0.05$;
- **b.** the level of significance $\alpha = 0.01$?

6. Investigating an alleged unfair trade practice, a statistician working for the Federal Trade Commission takes a random sample of 80 "8-ounce" packages of marshmallows, getting a mean of 7.98 ounces and a standard deviation of 0.13 ounce. Using an appropriate one-sided alternative and the level of significance $\alpha = 0.05$, test whether this constitutes evidence on which to base a finding of unfair practice.

7. Suppose that we want to use the data of Example 7 of Chapter 7 to test the hypothesis that *in general* the plant's emission of sulfur oxides averages 18 tons per day. Thus, using $\bar{x} = 19.6$, $s = 5.51$, and $n = 40$, what can we conclude at the level of significance $\alpha = 0.05$, if we let $\mu = 18$ and choose the alternative hypothesis $\mu > 18$ tons?

8. The average drying time of a manufacturer's paint is 20 minutes. Investigating the effectiveness of a modification in the chemical composition of his paint, the manufacturer wants to test the null hypothesis $\mu = 20$ minutes against a suitable alternative, where μ is the average drying time of the new paint.

a. What alternative hypothesis should the manufacturer use if he does not want to make the modification in the chemical composition of the paint unless it actually decreases the drying time.

b. What alternative hypothesis should the manufacturer use if the new process is actually cheaper and he wants to make the modification unless it increases the drying time of the paint?

9. SMALL-SAMPLE TESTS For small samples (that is, when n is less than 30), we proceed as in Exercise 13 on page 225, assume that the population from which we are sampling has roughly the shape of a normal distribution, and base our argument on the t distribution. (If this assumption about the shape of the population is unreasonable, we have to use instead one of the **nonparametric** tests mentioned in Chapter 11, for instance, the one of Exercise 11 on page 268.) So far as the outline on pages 234 through 236 is concerned, the first two steps remain the same, but in the third step we calculate the statistic

$$t = \frac{\bar{x} - \mu_0}{s/\sqrt{n}}$$

instead of z (which is really not much of a change, for even in the large-sample tests we replaced σ with s when the population standard deviation was unknown). So far as the fourth step is concerned, we base our decision on the value of t and we still refer to the criteria of Figures 8.6, 8.7, and 8.8, but in Figure 8.6 we substitute $t_{0.025}$ for 1.96 and $t_{0.005}$ for 2.58, while in Figures 8.7 and 8.8 we substitute $t_{0.05}$ for 1.64 and $t_{0.01}$ for 2.33. As in the small-sample confidence interval for μ, the number of *degrees of freedom* is $n - 1$ and the required values of $t_{0.05}$, $t_{0.025}$, $t_{0.01}$, and $t_{0.005}$ are given in Table II. To illustrate this technique, suppose that someone wants to check whether the fat content of a certain kind of ice cream is less than 12%. Taking a random sample of size $n = 5$ (say, samples of the ice cream produced on five different days), he gets a mean of 11.3 and a standard deviation of 0.38, so that substitution into the formula for t yields

$$t = \frac{11.3 - 12.0}{0.38/\sqrt{5}} = -4.12$$

Since this is less than -3.747, the value of $-t_{0.01}$ for $5 - 1 = 4$ degrees of freedom, he finds that the null hypothesis $\mu = 12.0$ will have to be rejected at the level of significance $\alpha = 0.01$. In other words, he can conclude that the average fat content of the ice cream *is* less than 12%.

a. The specifications for a certain kind of ribbon call for a mean breaking strength of 175 pounds. If six pieces of the ribbon (randomly selected from different rolls) have a mean breaking strength of 173.1 pounds with a standard deviation of 7.8 pounds, test the null hypothesis $\mu = 175$ against the alternative hypothesis $\mu < 175$ at the level of significance $\alpha = 0.05$.

b. Advertisements claim that the average nicotine content of a certain kind of cigarette is 0.30 milligram. If a random sample of fifteen cigarettes (chosen from different production lots) have a mean nicotine content of 0.32 milligram with a standard deviation of 0.016 milligram, test the null hypothesis $\mu = 0.30$ against the alternative hypothesis $\mu > 0.30$ at the level of significance $\alpha = 0.01$.

c. In an experiment with a new tranquilizer, the pulse rate of 18 patients was measured before they were given the tranquilizer and again 5 minutes later, and their pulse rate was found to be reduced on the average by 7.8 beats with a standard deviation of 1.9. Using the level of significance $\alpha = 0.05$, what can we conclude about the claim that this tranquilizer will reduce the pulse rate on the average by at least 9 beats in 5 minutes?

d. Test runs with six models of an experimental engine showed that they operated, respectively, for 25, 29, 22, 24, 33, and 23 minutes with 1 gallon of a certain kind of fuel. Is this evidence at the level of significance $\alpha = 0.05$ that *in general* the engine will not operate at the desired standard (average) of 28 minutes per gallon?

10. TYPE I AND TYPE II ERRORS The error which we make when we *reject a true null hypothesis* is called a **Type I error**, and the error which we make when we *accept a false null hypothesis* is called a **Type II error**. For instance, in Example 2 on page 227 the psychologist would be committing a Type I error if his sample data led him to *reject* the null hypothesis that the mean reaction time is 0.44 second even though this figure is correct, and he would be committing a Type II error if his sample data led him to *accept* the null hypothesis that the mean reaction time is 0.44 second even though this figure is not correct.

a. With reference to Example 3 on page 231, where the oceanographer is checking whether the average depth of the ocean in a certain area is 62.3 fathoms, explain under what conditions she would be committing a Type I error and under what conditions she would be committing a Type II error.

b. With reference to Example 4 on page 231, where the retailer tests the hypothesis that the average income of families living within 2 miles of his store is $9,000 (against the alternative hypothesis that their average income is less), explain under what conditions he would be committing a Type I error and under what conditions he would be committing a Type II error.

c. Suppose that a team of doctors is examining applicants to see whether they are physically fit to become astronauts. What type of error would they commit if they erroneously accepted the hypothesis that an applicant is physically fit to become an astronaut? What type of error would they commit if they erroneously rejected the hypothesis that an applicant is physically fit to become an astronaut?

d. Suppose we want to test the hypothesis that an antipollution device for cars is effective. Explain under what conditions we would be committing a Type I error and under what conditions we would be committing a Type II error.

11. THE PROBABILITY OF A TYPE II ERROR The probability of a Type I error is what we have been referring to as the level of significance, and it has always been *specified* in the test criteria which we have discussed. So far as the probability of a Type II error is concerned, the situation is much more complicated, as the reader will discover from the following example. Referring to Example 4 on page 233, let us first determine the value of $\bar{x}$ which corresponds to the dividing line of the $\alpha = 0.05$ criterion of the one-tail test of Figure 8.7, namely, $z = -1.64$. Substituting $n = 80$, $\sigma = 945$, $\mu_0 = 9{,}000$, and $z = -1.64$ into the formula for z on page 236, we get

$$-1.64 = \frac{\bar{x} - 9{,}000}{945/\sqrt{80}}$$

and, hence,

$$\bar{x} = 9{,}000 - 1.64 \cdot \frac{945}{\sqrt{80}} = 9{,}000 - 173 = \$8{,}827$$

rounded to the nearest dollar. Now we can calculate the probability of a

Type II error, say, when the true average income of the families is \$8,800. We would be committing this kind of error if we got a sample mean greater than \$8,827 when $\mu =$ \$8,800, and the answer is therefore given by the area of the tinted region of Figure 8.9. Converting 8,827 into standard units, we get

$$z = \frac{8{,}827 - 8{,}800}{945/\sqrt{80}} = 0.26$$

and since the corresponding entry in Table I is 0.1026, we find that the probability of a Type II error is $0.5000 - 0.1026 = 0.3974$ (or approximately 0.40) when the true average income of the families is \$8,800.

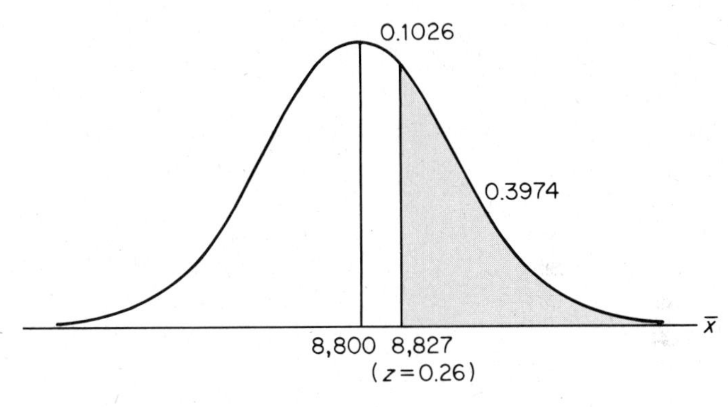

Figure 8.9 Probability of Type II error when $\mu =$ \$8,800.

a. Duplicating the last part of this argument, show that if $\mu =$ \$8,700, the probability of a Type II error is about 0.12 in this example; if $\mu =$ \$8,600, the probability of a Type II error is about 0.02; if $\mu =$ \$8,500, the probability of a Type II error is about 0.001; and if $\mu =$ \$8,900, the probability of a Type II error is about 0.75.

b. Draw a system of perpendicular axes on which μ is measured along the horizontal axis and the probability of a Type II error is measured along the vertical axis, plot the points which correspond to $\mu = 8{,}900$, $\mu = 8{,}800$, $\mu = 8{,}700$, $\mu = 8{,}600$, and $\mu = 8{,}500$, and join them by means of a smooth curve. This kind of curve is called an **operating characteristic curve**, and it provides us with an overall picture of the probabilities of Type II errors to which we are exposed when the true average income of the families is not \$9,000. *Of course, all this applies only to this particular example, and in most practical situations we are satisfied to specify the probability of a Type I error (namely, the level of significance), keeping our fingers crossed that the probabilities of serious Type II errors are "within reason."*

12. PROBABILITIES OF TYPE I AND TYPE II ERRORS Unfortunately, the relationship between the probabilities of Type I errors and Type II errors is such that *if we reduce one this automatically increases the other* (provided everything else remains unchanged). To illustrate, let us refer again to Example 4 on page 233 and *reduce* the probability of a Type I error from 0.05 to 0.01. If we proceed as in Exercise 11, we first obtain $\bar{x}$, the dividing line of the criterion, by solving the equation

$$-2.33 = \frac{\bar{x} - 9{,}000}{945/\sqrt{80}}$$

which yields $\bar{x} = \$8{,}754$. Then, if $\mu = \$8{,}800$, the probability of a Type II error is given by the normal curve area to the right of

$$z = \frac{8{,}754 - 8{,}800}{945/\sqrt{80}} = -0.44$$

and since the entry in Table I corresponding to $z = 0.44$ is 0.1700, we find that the probability of a Type II error has *increased* from 0.40 to $0.5000 + 0.1700 = 0.6700$ or 0.67. Use the same kind of argument to recalculate the probabilities of Type II errors for $\alpha = 0.01$ and the alternative hypotheses $\mu = \$8{,}500$, $\mu = \$8{,}600$, $\mu = \$8{,}700$, and $\mu = \$8{,}900$, and compare the results with those obtained in part (a) of Exercise 11.

13. PROBABILITIES OF TYPE I AND TYPE II ERRORS, Continued To illustrate the effect that changes in the sample size have on the probabilities of Type I and Type II errors, suppose that in Example 4 on page 233 the sample size had been $n = 100$ instead of $n = 80$. Duplicating the work done in the illustration of Exercise 11, show that if the probability of a Type I error is kept at $\alpha = 0.05$, the probability of a Type II error is *reduced* from 0.40 to 0.32.

Differences between means

There are many situations in which we must decide whether an observed difference between two sample means can be attributed to chance, or whether it is indicative of the fact that the two samples come from populations with unequal means. For instance, if tests performed with two kinds of compact cars showed that one averaged 23.6 miles per gallon while the other averaged 25.8 miles per gallon, we may want to decide (or have to decide) whether there really is a difference in the performance of the two kinds of cars, or whether the difference between the two average mileages can be attributed to chance. Similarly, we may want to decide on the basis of samples whether men can perform a certain task faster than women, whether one kind of ceramic insulator is more brittle than another, whether the average diet in one country is more nutritious than that in another country, and so on.

The test which we shall use to decide whether an observed difference between two sample means can be attributed to chance is based on the following theory: *If* $\bar{x}_1$ *and* $\bar{x}_2$ *are the means of independent random samples from two populations with the means* μ_1 *and* μ_2 *and the standard deviations* σ_1 *and* σ_2, *then the sampling distribution of the statistic* $\bar{x}_1 - \bar{x}_2$ *has the mean* $\mu_1 - \mu_2$ *and the standard deviation*

$$\sigma_{\bar{x}_1-\bar{x}_2} = \sqrt{\frac{\sigma_1^2}{n_1} + \frac{\sigma_2^2}{n_2}}$$

appropriately called the **standard error of the difference between two means.*** Also, for large samples (namely, when neither n_1 nor n_2 is less than 30), the sampling distribution of $\bar{x}_1 - \bar{x}_2$ can be approximated closely with a normal distribution, and, if necessary, the sample standard deviations s_1 and s_2 can be substituted for σ_1 and σ_2. Thus, we can base the test of the null hypothesis $\mu_1 = \mu_2$ on the statistic

$$z = \frac{\bar{x}_1 - \bar{x}_2}{\sqrt{\dfrac{s_1^2}{n_1} + \dfrac{s_2^2}{n_2}}}$$

which has approximately the standard normal distribution as it was obtained by *subtracting* from $\bar{x}_1 - \bar{x}_2$, the mean of its sampling distribution (which under the null hypothesis equals $\mu_1 - \mu_2 = 0$), and then *dividing* by the expression for the standard error of the difference between two means with the sample standard deviations s_1 and s_2 substituted for σ_1 and σ_2. Depending on whether the *alternative hypothesis* is $\mu_1 - \mu_2 \neq 0$, $\mu_1 - \mu_2 < 0$, or $\mu_1 - \mu_2 > 0$, we can thus base the test of the null hypothesis on the criteria of Figures 8.6, 8.7, and 8.8.

*By "independent" samples we mean that the selection of one sample should in no way affect the selection of the other. Thus, the theory does *not* apply to "before and after" kinds of comparisons, nor does it apply, for example, if we want to compare the ages of husbands and wives. Special methods for handling situations like these are given in Exercise 6 on page 246 and Exercise 13 on page 270.

Example 7. To illustrate this *large-sample* technique, suppose that we want to investigate whether there *is* a difference between the average heights of adult females born in two different countries. The sample data which we have at our disposal show that 120 adult females born in one country have an average height of 62.7 inches with a standard deviation of 2.50 inches, and that 150 adult females born in the other country have an average height of 61.8 inches with a standard deviation of 2.62 inches. To test the null hypothesis $\mu_1 = \mu_2$ (where μ_1 and μ_2 are the corresponding true average heights of adult females born in the two countries) against the alternative hypothesis $\mu_1 \neq \mu_2$, say, at the level of significance $\alpha = 0.01$, we have only to substitute all the given values into the formula for z, getting

$$z = \frac{62.7 - 61.8}{\sqrt{\frac{(2.50)^2}{120} + \frac{(2.62)^2}{150}}} = 2.87$$

Since this exceeds 2.58, the dividing line of the $\alpha = 0.01$ criterion of Figure 8.6, we find that *the null hypothesis will have to be rejected*; in other words, the sample data show that there *is* a difference between the true average heights of adult females in the two given countries.

Exercises

1. A company claims that its light bulbs are superior to those of a competitor on the basis of a study which showed that a sample of 40 of its bulbs had an average "lifetime" of 628 hours of continuous use with a standard deviation of 27 hours, while a sample of 30 bulbs made by the competitor had an average "lifetime" of 619 hours of continuous use with a standard deviation of 25 hours. Check, at the level of significance $\alpha = 0.05$, whether this claim is justified.

2. To compare high school seniors' knowledge of current events in two different school districts, samples of 60 seniors from each district were given a special test. If those of the first district obtained an average score of 78.2 with a standard deviation of 5.0, while those of the second district obtained an average score of 74.4 with a standard deviation of 4.2, test at the level of significance $\alpha = 0.05$ whether the difference between these two sample means is significant.

3. Suppose that we want to investigate whether male and female file clerks in a large city earn comparable wages. If sample data show that 60 male file clerks earn on the average $158.50 per week with a standard deviation of \$18.20, while 60 female file clerks earn on the average \$141.60 per week with a standard deviation of \$20.60, test the null hypothesis $\mu_1 = \mu_2$ against the alternative hypothesis $\mu_1 > \mu_2$ at the level of significance $\alpha = 0.05$.

4. A sample study was made of the number of business lunches that executives claim as deductible expenses in a given month. If 40 executives in the insurance industry average 9.7 such deductions with a standard deviation of 1.8, while 50 bank executives average 8.3 such deductions with a standard deviation of 2.2, test at the level of significance $\alpha = 0.01$ whether the difference between these two sample means is significant.

5. SMALL-SAMPLE TESTS For small samples (that is, when either sample size is less than 30), we can base our argument on the t distribution as in Exercise 9 on page 238, provided the two populations from which we are sampling have roughly the shape of normal distributions and *equal standard deviations.* So far as the null hypothesis and the alternative hypothesis are concerned, they are the same as in the large-sample test discussed in the text, and the decision criteria are identical with those of Exercise 9 on page 238, but the formula for the statistic on which we base our decision is now

$$t = \frac{\bar{x}_1 - \bar{x}_2}{\sqrt{\frac{(n_1 - 1)s_1^2 + (n_2 - 1)s_2^2}{n_1 + n_2 - 2}} \sqrt{\frac{1}{n_1} + \frac{1}{n_2}}}$$

Also, the number of *degrees of freedom* (needed to look up the appropriate value of $t_{0.05}$, $t_{0.025}$, $t_{0.01}$, or $t_{0.005}$ in Table II) is now $n_1 + n_2 - 2$. To illustrate, suppose that a professional bowler averaged 207.3 with a standard deviation of 8.2 in five games with a relatively light ball, and 210.9 with a standard deviation of 10.1 in six games with a somewhat heavier ball. Can we conclude at the level of significance $\alpha = 0.05$ that the difference in weight actually has an effect on the bowler's game? To test the null hypothesis $\mu_1 = \mu_2$ against the alternative hypothesis $\mu_1 \neq \mu_2$, we substitute into the above formula for t and get

$$t = \frac{207.3 - 210.9}{\sqrt{\frac{4(8.2)^2 + 5(10.1)^2}{9}} \sqrt{\frac{1}{5} + \frac{1}{6}}} = -0.64$$

Since $t_{0.025}$ for $5 + 6 - 2 = 9$ degrees of freedom is 2.262 and -0.64 is *not* less than -2.262, the difference between the two sample means is *not significant*; in other words, the null hypothesis that on the average the bowler performs equally well with both balls cannot be rejected.

a. Six guinea pigs injected with 0.5 mg of a tranquilizer took on the average 12.4 seconds to fall asleep with a standard deviation of 2.2 seconds, while six guinea pigs injected with 1.5 mg of the tranquilizer took on the average 8.2 seconds to fall asleep with a standard deviation of 1.8 seconds. Use the level of significance $\alpha = 0.05$ to test the null hypothesis that the difference in dosage has no effect.

b. If 15 randomly selected citrus trees in one orchard have a mean height of 12.3 feet with a standard deviation of 1.2 feet, while 12 randomly selected citrus trees in another orchard have a mean height of 13.4 feet with a standard deviation of 1.4 feet, check on the claim that the trees in the second orchard are *taller on the average* than those in the first orchard. Use the alternative hypothesis that the claim is true and the level of significance $\alpha = 0.01$.

c. With reference to Example 15 on page 49, use the level of significance $\alpha = 0.05$ to check whether the difference between the average mileages obtained with cars A and C is significant.

6. DEPENDENT SAMPLES As we pointed out on page 243, the methods of this section do not apply unless the two samples are *independent.* Suppose, then, that we want to study the effectiveness of new traffic controls installed at dangerous intersections and that during the first month of their use at eight of these intersections (chosen at random) the number of accidents decreased from 18 to 14, decreased from 9 to 2, increased from 5 to 6, decreased from 19 to 10, decreased from 13 to 10, remained 6, decreased from 12 to 1, and decreased from 14 to 7. Thus, the average number of accidents in the month *before* the installation of the new controls was

$$\frac{18 + 9 + 5 + 19 + 13 + 6 + 12 + 14}{8} = 12$$

in the month *after* the installation of the new controls it was

$$\frac{14 + 2 + 6 + 10 + 10 + 6 + 1 + 7}{8} = 7$$

and to test whether the difference between these two means is significant we proceed as follows. We calculate the *change* in the number of accidents at the eight intersections, getting -4, -7, $+1$, -9, -3, 0, -11, and -7, and *then we test the null hypothesis that these differences (changes) constitute a random sample from a population with the mean* $\mu = 0$. Since $n = 8$ in this example, use the small-sample test of Exercise 9 on page 238 to test this null hypothesis against the alternative $\mu < 0$ (which implies that the new traffic controls are effective) at the level of significance $\alpha = 0.05$.

7. **DEPENDENT SAMPLES, Continued** To study the effectiveness of physical exercise in weight reduction, a group of 16 persons was asked to engage in a prescribed program of physical exercise for two months. The results are shown in the following table:

Weight Before (pounds)	*Weight After (pounds)*	*Weight Before (pounds)*	*Weight After (pounds)*
209	196	170	164
178	171	153	152
169	170	183	179
212	207	165	162
180	177	201	199
192	190	179	173
158	159	243	231
180	180	144	140

Use the method suggested in Exercise 6 to check, at the level of significance $\alpha = 0.01$, whether the prescribed program of exercise is effective.

Differences among *k* means (analysis of variance)*

Let us now generalize the work of the preceding section and consider the problem of deciding whether observed differences among *more than two* sample means can be attributed to chance, or whether they are indicative of actual differences among the means of the corresponding populations. For instance, we may want to decide on the basis of sample data whether there really is a difference in the effectiveness of three methods of teaching computer programming, we may want to compare the average yield per acre of several varieties of wheat, we may want to see whether there really is a difference in the average mileage obtained with different kinds of gasoline, we may want to judge whether there really is a difference in the performance of several kinds of fishing reels, and so on.

*The material in this section is somewhat more advanced, and it may be omitted without loss of continuity.

Example 8. To illustrate, suppose that we want to compare the effectiveness of three methods of teaching the programming of a certain electronic computer—method A, which is straight teaching-machine instruction; method B, which involves the personal attention of an instructor and some direct experience working with the computer; and method C, which involves the personal attention of an instructor but no work with the computer itself. Now suppose that random samples of size 4 are taken from large groups of students taught by the three methods and that these students obtained the following scores in an appropriate achievement test:

Method A	*Method B*	*Method C*
71	90	72
75	80	77
65	86	76
69	84	79

The means of these three samples are, respectively, 70, 85, and 76, and what we would like to know is whether the differences among these means are actually significant; after all, the size of the samples is *very small.*

In general, in a problem like this, if we let $\mu_1, \mu_2, \ldots,$ and μ_k denote the means of the k populations from which the respective samples are obtained, we shall want to decide whether it is reasonable to say that these μ's are all equal; in other words, we shall want to test the null hypothesis $\mu_1 = \mu_2 = \ldots = \mu_k$ against the alternative hypothesis that the μ's are *not all equal.* Evidently, this null hypothesis would be supported by the data if the sample means were all nearly the same size, and the alternative hypothesis would be supported by the data if the differences among the sample means were large. Thus, we need a measure of the discrepancies among the $\bar{x}$'s, and the most obvious, perhaps, is their standard deviation (or their variance), which we can calculate according to the formula on page 194.

Example 8 (Continued). For our three samples the mean of the three $\bar{x}$'s is $\frac{70 + 85 + 76}{3} = 77$, so their variance is

$$s_{\bar{x}}^2 = \frac{(70 - 77)^2 + (85 - 77)^2 + (76 - 77)^2}{3 - 1} = 57$$

where we used the subscript $\bar{x}$ to indicate that this variance measures the variability of the sample means.

Of course, we still have to decide whether the value which we obtain for $s_{\bar{x}}^2$ is "small" or "large," that is, whether or not it supports the null hypothesis that the population means are all equal, and to this end we make the following assumption, which is critical to the method of analysis we shall employ:

> **We assume that the populations from which we are sampling can be approximated closely with normal distributions having the same standard deviation σ.**

If the null hypothesis $\mu_1 = \mu_2 = \ldots = \mu_k$ is true, we can then look upon our samples as samples from *one and the same population*, and, hence, upon $s_{\bar{x}}^2$ as an estimate of $\sigma_{\bar{x}}^2$, the *square* of the standard error of the mean. Now, if we make use of the theorem on page 206 according to which $\sigma_{\bar{x}} = \frac{\sigma}{\sqrt{n}}$ and, hence, $\sigma_{\bar{x}}^2 = \frac{\sigma^2}{n}$, we can look upon $s_{\bar{x}}^2$ as an estimate of $\sigma_{\bar{x}}^2 = \frac{\sigma^2}{n}$ and upon $n \cdot s_{\bar{x}}^2$ as an estimate of σ^2.

Example 8 (Continued). For our example we thus get $n \cdot s_{\bar{x}}^2 = 4 \cdot 57 = 228$ as an estimate of σ^2, the common variance of the three populations.

If σ^2 were known, we could compare the value which we obtained for $n \cdot s_{\bar{x}}^2$ with σ^2 and *reject* the null hypothesis (that the population means are all equal) if this value is much larger than σ^2. However, in most practical problems σ^2 is not known and we have no choice but to estimate it on the basis of the given data. Having assumed, in fact, that the samples all come from identical populations, we could use any one of their variances (s_1^2, s_2^2, $\ldots$, or s_k^2) as an estimate of σ^2 and, hence, we can also use their *mean*.

Example 8 (Continued). For our example we, thus, estimate σ^2 as

$$\frac{s_1^2+s_2^2+s_3^2}{3} = \frac{1}{3}\left[\frac{(71-70)^2+(75-70)^2+(65-70)^2+(69-70)^2}{4-1}\right.$$
$$+\frac{(90-85)^2+(80-85)^2+(86-85)^2+(84-85)^2}{4-1}$$
$$\left.+\frac{(72-76)^2+(77-76)^2+(76-76)^2+(79-76)^2}{4-1}\right]$$
$$= \tfrac{130}{9} \text{ (or } 14\tfrac{4}{9}\text{)}$$

Observe that we now have the following two estimates of σ^2,

$$n \cdot s_{\bar{x}}^2 = 228 \qquad \text{and} \qquad \frac{s_1^2 + s_2^2 + s_3^2}{3} = 14\tfrac{4}{9}$$

and if the *first estimate* (which is based on the variation among the sample means) is much larger than the *second estimate* (which is based on the variation within the samples and, hence, measures variation that is due to chance), it stands to reason that the null hypothesis should be rejected. After all, *in that case the variation among the sample means would be greater than it should be if it were due only to chance.*

To put comparisons such as this on a rigorous basis, we use the following statistic, which is appropriately called a **variance ratio**:

$$F = \frac{\text{estimate of } \sigma^2 \text{ based on the variation among the } \bar{x}\text{'s}}{\text{estimate of } \sigma^2 \text{ based on the variation within the samples}}$$

If the null hypothesis is true, the sampling distribution of this statistic is the F **distribution**, an example of which is shown in Figure 8.10. Since the null hypothesis will be rejected only when F is *large* (namely, when the variability of the $\bar{x}$'s is too great to be attributed to chance), we ultimately base our decision on the criterion of Figure 8.10. Here $F_{0.05}$ is such that the area under the curve to its right is equal to 0.05, and it provides the dividing line of the criterion when the level of significance is $\alpha = 0.05$; correspondingly, $F_{0.01}$ provides the dividing line of the criterion when the

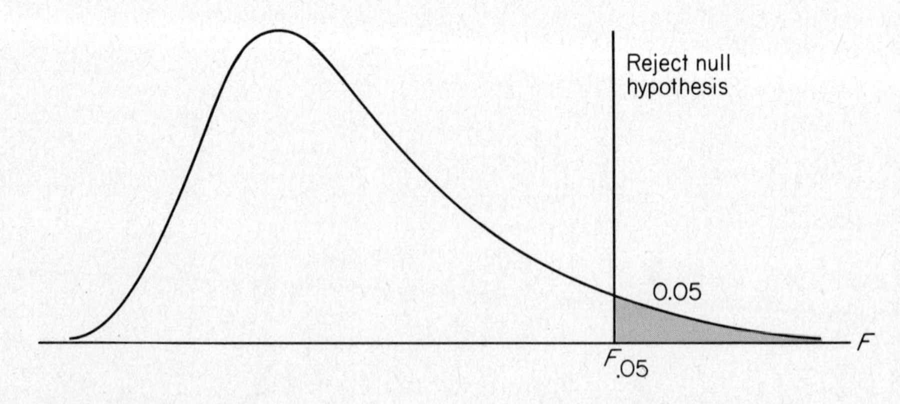

Figure 8.10 F distribution.

level of significance is $\alpha = 0.01$. These quantities, which depend on two parameters called, respectively, the **numerator** and **denominator degrees of freedom**, are given in Tables IVa and IVb. When we compare the means of k samples of size n, the number of degrees of freedom for the numerator of F is $k - 1$, and the number of degrees of freedom for the denominator of F is $k(n - 1)$.*

Example 8 (Continued). Returning now to our numerical example, we find that $F_{0.05} = 4.26$ for $k - 1 = 3 - 1 = 2$ and $k(n - 1) = 3(4 - 1) = 9$ degrees of freedom, and since this value is exceeded by

$$F = \frac{228}{\frac{130}{9}} = 15.8$$

the value of the F statistic which we obtained in our example, the null hypothesis will have to be *rejected*. In other words, we conclude that there *is* a difference in the effectiveness of the three methods of teaching the programming of the given computer. (Incidentally, had we used the level of significance $\alpha = 0.01$, we would have found that $F_{0.01} = 8.02$ for 2 and 9 degrees of freedom, and the result would have been the same.) Actually, the result should not come as a surprise since the superiority of method B was quite apparent from the sample data; but as we already indicated on page 248, the samples were very small and we just wanted to make sure that the differences among the means are significant. This will not be quite so obvious in some of the exercises at the end of this section.

If we have k random samples of size n and let x_{ij} denote the ith observation of the jth sample (for instance, x_{31} is the *third* observation of the *first* sample and for the data on page 248 it equals 65, x_{42} is the *fourth* observation of the *second* sample and for the data on page 248 it equals 84, and x_{23} is the *second* observation of the *third* sample and for the data on page 248 it equals 77), we can write the formula for F directly as

$$F = \frac{k(n - 1)[n \cdot \sum (\bar{x}_j - \bar{x})^2]}{(k - 1)[\sum \sum (x_{ij} - \bar{x}_j)^2]}$$

*So far as the formula for the numerator degrees of freedom is concerned, note that the numerator is $n \cdot s_{\bar{x}}^2$, where $s_{\bar{x}}^2$ is the variance of k means and, hence, has $k - 1$ degrees of freedom in accordance with the terminology introduced in the footnote to page 194. So far as the formula for the denominator degrees of freedom is concerned, note that the denominator is the mean of k sample variances, with each having $n - 1$ degrees of freedom.

where $\bar{x}_j$ is the mean of the jth sample and $\bar{x}$ is the **grand mean** of all the data. The expression bracketed in the *numerator* is n times the sum of the squares of the differences between the sample means and the grand mean, and in actual practice it is usually calculated by means of the shortcut formula

$$n \cdot \sum (\bar{x}_j - \bar{x})^2 = \frac{1}{n} \cdot \sum T_j^2 - \frac{T^2}{kn}$$

where T_j is the total of the observations in the jth sample and T is the grand total of all the data. The expression bracketed in the *denominator* is the sum of the squares of the differences between the individual observations and the means of the respective samples; in fact, it is a **double summation**, as we indicated by means of the $\sum \sum$, for we are summing on i as well as j (in other words, it applies to *all* the data). In actual practice, this quantity is usually calculated by means of the shortcut formula

$$\sum \sum (x_{ij} - \bar{x}_j)^2 = \sum \sum x_{ij}^2 - \frac{1}{n} \cdot \sum T_j^2$$

where $\sum \sum x_{ij}^2$ is the sum of the squares of all the given observations, and the quantity which we subtract was already determined in connection with the shortcut formula for the expression bracketed in the numerator of F.

Example 8 (Continued). Had we used these formulas in our numerical example, we would have obtained $T_1 = 280$, $T_2 = 340$, $T_3 = 304$, $T = 924$, and $\sum \sum x_{ij}^2 = 71{,}734$, so that

$$n \cdot \sum (\bar{x}_j - \bar{x})^2 = \tfrac{1}{4}(280^2 + 340^2 + 304^2) - \frac{924^2}{3 \cdot 4} = 456$$

$$\sum \sum (x_{ij} - \bar{x}_j)^2 = 71{,}734 - \tfrac{1}{4}(280^2 + 340^2 + 304^2) = 130$$

and, hence,

$$F = \frac{3(4-1)456}{(3-1)130} = \frac{4{,}104}{260} = 15.8$$

which is identical with the result obtained before. The advantages of the shortcut technique may not be apparent from this example, but this is due to the fact that all the sample means happened to be whole numbers.

The method which we have studied in this section belongs to a very important branch of statistics called **analysis of variance**. We, so to speak, analyze what part of the total variation of our data might be attributed to specific "causes," or *sources of variation* (for instance, the three teaching methods of our example), and we then compare it with that part of the variation of the data which can be attributed to chance.

Exercises

1. The following are the yields of four varieties of wheat (in pounds per plot) which an agronomist obtained for three test plots of each variety:

Variety A:	52,	45,	44
Variety B:	51,	51,	48
Variety C:	48,	48,	45
Variety D:	53,	54,	49

Use the level of significance $\alpha = 0.05$ to test the null hypothesis that there is no difference in the true average yields of the four varieties of wheat; first calculate the value of F without the shortcut formulas on page 252, and then use them to check the result.

2. To find the best arrangement of instruments on a control panel of an airplane, three different arrangements were tested by simulating an emergency condition and observing the reaction time required to correct the condition. The reaction times (in *tenths* of a second) of twelve pilots (randomly assigned to the different arrangements) were as follows:

Arrangement 1:	7,	14,	9,	10
Arrangement 2:	15,	10,	13,	18
Arrangement 3:	11,	6,	12,	7

Use the level of significance $\alpha = 0.05$ to test the null hypothesis that the true average reaction time is the same for all three arrangements; first calculate the value of F without the shortcut formulas on page 252, and then use them to check the result.

3. Mr. Cooper can drive to a friend's house along four different routes, and the following are the number of minutes in which he timed himself on five different occasions for each route:

Route 1:	15,	17,	16,	19,	19
Route 2:	17,	20,	24,	21,	24
Route 3:	14,	16,	17,	15,	18
Route 4:	11,	15,	14,	14,	20

Use the shortcut formula to calculate F, and test the null hypothesis that there is no difference in the true average time it takes Mr. Cooper to drive to his friend's house along the four different routes. Let $\alpha = 0.05$.

4. The following are the number of mistakes made in three successive days by four technicians working for a photographic laboratory:

Technician A	*Technician B*	*Technician C*	*Technician D*
4	5	12	5
12	13	20	11
10	8	17	12

Test at the level of significance $\alpha = 0.01$ whether the differences among the four sample means can be attributed to chance.

5. With reference to Example 15 on page 49, check (at the level of significance 0.05) whether the differences among the average mileages are significant. Use the shortcut formulas on page 252 to calculate F.

9

The analysis of count data

Introduction

Many problems in science, business, and everyday life deal with **count data** (namely, data obtained by counting rather than measuring) that are used to estimate, or test hypotheses about, proportions, percentages, and probabilities. In principle, the work of this chapter will be very similar to that of Chapter 8. In *problems of estimation* we shall again construct confidence intervals and worry about the possible size of our error, and in the *testing of hypotheses* we shall again have to be careful in formulating null hypotheses and their alternatives, in deciding between one-tail tests and two-tail tests, in choosing the level of significance, and so on. The main difference is that the parameters with which we will now be concerned are "true" proportions (percentages, or probabilities) instead of population means, and this is why we shall base our methods on counts instead of measurements.

The estimation of proportions

The kind of information that is usually available for the estimation of a "true" proportion is the **relative frequency** with which an event has occurred, namely, a **sample proportion**. If an event has occurred x times out of n, the relative frequency of its occurrence is x/n, and we generally use this sample proportion to estimate the corresponding "true" proportion, which we shall denote p.

Example 1. For instance, if 69 of 150 adults (interviewed in a sample survey) say that television watching is their favorite way of spending an evening, then $\frac{x}{n} = \frac{69}{150} = 0.46$, and we can use this figure as an estimate of the *true* proportion of adults who will respond in the same way. Note that since a *percentage* is simply a proportion multiplied by 100 and a *probability* can be interpreted as a proportion "in the long run," we could also say that we are thus estimating that 46% of all adults consider television watching as their favorite way of spending an evening, or that we are estimating as 0.46 the probability that an adult will consider television watching as his or her favorite way of spending an evening. We have made this point to impress upon the reader that *the problem of estimating a true percentage or a true probability is the same as that of estimating a true proportion.*

Throughout this section it will be assumed that the situations with which we are dealing satisfy (at least approximately) the conditions underlying the *binomial distribution.* Our information will always tell us how many successes there are in a given number of independent trials, and that in each trial the probability of a success has the same value p. Thus, the sampling distribution of the *number of successes* (namely, the sampling distribution of the *count* on which our methods will be based) is the binomial distribution for which we indicated on pages 141 and 146 that the mean and the standard deviation are given by the formulas $\mu = np$ and $\sigma = \sqrt{np(1 - p)}$.

The fact that this formula for σ involves the quantity p (which we want to estimate) leads to some difficulties, but we can avoid them, at least for the moment, by constructing confidence intervals for p with the use of Tables Va and Vb. (So far as the construction of these special tables is concerned, let us merely point out that the statisticians who made them used methods which are very similar to what we did on page 219, except that they referred to very detailed tables of binomial probabilities instead of areas under the normal curve.) Tables Va and Vb are very easy to use and require no calculations. If a sample proportion x/n is less than or equal to 0.50, we begin by marking its value on the *bottom scale*; then we go up vertically until we reach the two contour lines (curves) which correspond to the size of the sample, and finally we read the confidence limits for p off the *left-hand scale,* as is indicated in Figure 9.1. If the sample proportion exceeds 0.50, we mark its value on the scale which is *at the top,* go down vertically until we reach the two contour lines (curves) which correspond to the size of the sample, and then we read the confidence limits for p off the *right-hand scale,* as is indicated in Figure 9.2.

Example 2. To illustrate the use of these tables in a situation where the sample proportion is less than or equal to 0.50, suppose that in a random sample of 400 high school students given a flu vaccine only 136 experienced any discomfort. Marking $\frac{136}{400} = 0.34$ on the bottom scale of Table Va and proceeding as in Figure 9.1, we find that

$$0.29 < p < 0.39$$

is a 0.95 confidence interval for the true proportion of high school students who will experience any discomfort from the flu vaccine.

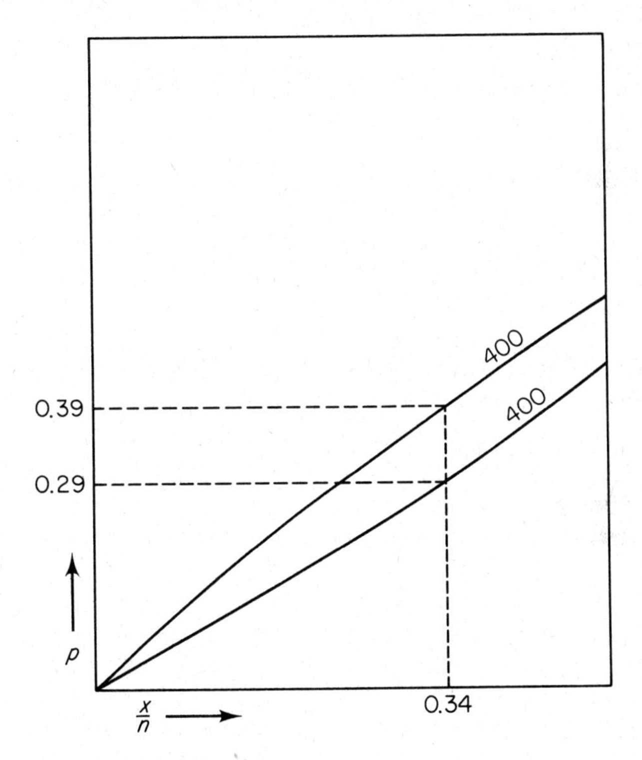

Figure 9.1 Confidence limits for p.

Had we wanted to use a degree of confidence of 0.99 instead of 0.95, Table Vb would have yielded the confidence interval

$$0.28 < p < 0.41$$

The two confidence intervals of Example 2 demonstrate again that *an increase in the degree of confidence will lead to a wider interval and, hence, to less specific information about the quantity we are trying to estimate.*

Example 3. To give an example where the sample proportion exceeds 0.50, suppose that 108 of 150 persons interviewed in a large western

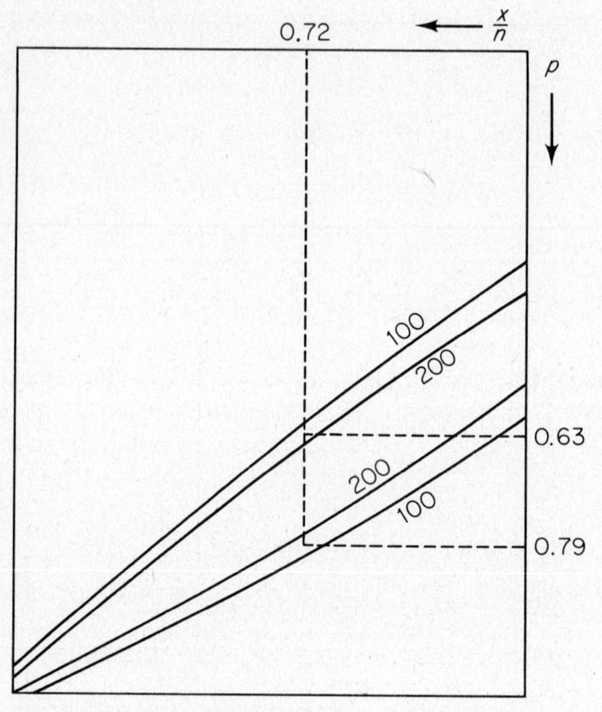

Figure 9.2 Confidence limits for p.

city opposed the construction of any more freeways. The corresponding sample proportion is $\frac{108}{150} = 0.72$, and the first thing we discover is that there are no curves for $n = 150$ in Tables Va and Vb. Reading between the lines, however, as in Figure 9.2, we find that a 0.95 confidence interval for the corresponding true proportion (of persons in that city who oppose the construction of any more freeways) is given by

$$0.63 < p < 0.79$$

Had we not wanted to "read between the lines" in Example 3, we could have constructed an approximate confidence interval for p on the basis of the *normal curve approximation* of the binomial distribution which we discussed in Chapter 6. This approximate large-sample technique can also be used when tables like Tables Va and Vb are not available, when it is desired to use a different degree of confidence, or the sample size exceeds 1,000. Using the formulas for μ and σ on pages 141 and 146, the normal curve approximation to the binomial distribution enables us to assert with the probability 0.95 that x, the number of successes, will fall within 1.96 standard deviations of the mean, namely, on the interval from $np - 1.96\sqrt{np(1-p)}$ to $np + 1.96\sqrt{np(1-p)}$. This can be written as

$$np - 1.96\sqrt{np(1-p)} < x < np + 1.96\sqrt{np(1-p)}$$

which becomes

$$\frac{x}{n} - 1.96\sqrt{\frac{p(1-p)}{n}} < p < \frac{x}{n} + 1.96\sqrt{\frac{p(1-p)}{n}}$$

after some algebraic manipulation (which the reader will be asked to perform in Exercise 14 on page 264). This double inequality *could* serve as a confidence interval for p if it were not for the fact that the quantity p itself appears on both sides in the expression $\sqrt{\frac{p(1-p)}{n}}$. This expression is referred to as the **standard error of a proportion**, for it is, in fact, the standard deviation of the sampling distribution of x/n (see Exercise 15 on page 264).

There are several ways in which we can get around this difficulty. The easiest way is to substitute the sample proportion x/n for p in $\sqrt{\frac{p(1-p)}{n}}$, which leads to the following 0.95 *large-sample confidence interval for* p:

$$\frac{x}{n} - 1.96\sqrt{\frac{\frac{x}{n}\left(1-\frac{x}{n}\right)}{n}} < p < \frac{x}{n} + 1.96\sqrt{\frac{\frac{x}{n}\left(1-\frac{x}{n}\right)}{n}}$$

If we want to change the degree of confidence to 0.99, we have only to substitute 2.58 into this formula for 1.96.

Example 2 (Continued). To illustrate this large-sample technique, let us refer again to Example 2, where 136 of 400 high school students experienced some discomfort from the flu vaccine. Substituting $x = 136$ and $n = 400$ into the above 0.95 confidence interval formula for p, we obtain

$$\frac{136}{400} - 1.96\sqrt{\frac{\frac{136}{400}(1-\frac{136}{400})}{400}} < p < \frac{136}{400} + 1.96\sqrt{\frac{\frac{136}{400}(1-\frac{136}{400})}{400}}$$

or

$$0.294 < p < 0.386$$

which is *very close* to the interval obtained previously with the use of Table Va.

The large-sample theory which we have presented here can also be used to judge the possible size of the error we make when we use a sample proportion as an estimate of the corresponding "true" proportion p. Corresponding to what we said on page 217, the normal curve approximation to the binomial distribution permits us to assert that

The probability is 0.95 that the numerical difference between x/n and p, namely, the size of our error either way, is less than

$$1.96\sqrt{\frac{\frac{x}{n}\left(1-\frac{x}{n}\right)}{n}}$$

where we again substituted x/n for p in the formula for the standard error of a proportion. Of course, if we want the probability (with which we make this assertion about the error) to be 0.99, we have only to substitute 2.58 for 1.96.

Example 3 (Continued). To illustrate this technique, let us refer again to Example 3, where 108 of 150 persons opposed the construction of any more freeways. Now, if we use $\frac{108}{150} = 0.72$ as a *point estimate* of the true proportion of persons in the given city who oppose the construction of any more freeways, we can assert with a probability of 0.95 that this estimate is "off" either way by less than

$$1.96\sqrt{\frac{0.72(1-0.72)}{150}} = 0.07$$

rounded to two decimals. In Exercise 12 on page 262 the reader will be asked to show that 99 to 1 would be *fair odds* in this example that the estimate is not "off" by more than 0.09 (rounded to two decimals).

Exercises

1. In a sample survey, 280 of 1,000 persons interviewed in a large city said that they regularly use public transportation. Use Table Va to construct a 0.95 confidence interval for the true proportion of persons in this city who regularly use public transportation.

2. In a random sample of 200 adults in a midwestern city, 152 answered "yes" to the question: "Do you have occasion to use alcoholic beverages such as liquor, wine, or beer?" Use Table Vb to construct a 0.99 confidence interval for the corresponding true proportion.

3. In a random sample of 60 claims filed against an insurance company writing collision insurance on cars, 36 exceeded $800. Use Table Va to construct a 0.95 confidence interval for the actual *percentage* of claims filed against this company that exceed $800.

4. In a random sample of 300 high school seniors in a large school district, 234 said that they expect to continue their education. Use Table Vb to construct a 0.99 confidence interval for the probability that any one of the seniors in this school district will continue his or her education.

5. Among 200 fish caught in a certain lake, 32 were inedible as a result of the chemical pollution of their environment.

a. Use Table Va to construct a 0.95 confidence interval for the probability that a fish caught in this lake will be inedible for the given reason.

b. Repeat part (a) using the large-sample confidence interval formula on page 259.

6. In a sample of 100 supposed UFO sightings, 83 could easily be explained in terms of natural phenomena.

a. Use Table Vb to construct a 0.99 confidence interval for the corresponding true proportion of supposed UFO sightings that can easily be explained in terms of natural phenomena.

b. Repeat part (a) using the large-sample confidence interval formula on page 259.

7. Repeat Exercise 1 using the large-sample confidence interval formula on page 259 and compare the results.

8. Repeat Exercise 2 using the large-sample confidence interval formula on page 259 and compare the results.

9. In a random sample of 600 voters interviewed in a large city, only 342 felt that the President was doing a good job.

a. Use Table Va to construct a 0.95 confidence interval for the true proportion of voters in this city who feel the same way.

b. If we use $\frac{342}{600} = 0.57$ as an estimate of the true proportion of voters in this city who feel that the President is doing a good job, what can we say with a probability of 0.95 about the possible size of our error?

10. In a sample of 120 persons convicted in U.S. District Courts on narcotics charges, 45 received probation. What can we say with a probability of 0.95 about the possible size of our error, if we estimate the probability as 0.375 that a person convicted in a U.S. District Court on narcotics charges will be put on probation?

11. With reference to Exercise 5, what can we say with a probability of 0.99 about the possible size of our error, if we estimate as 0.16 the probability that a fish caught in the given lake will be inedible as a result of the chemical pollution of its environment?

12. Verify the statement concerning the construction of more freeways which we made on page 260, namely, that 99 to 1 would be fair odds that the estimate (of the true proportion of persons in the given city who oppose the construction of any more freeways) is not "off" by more than 0.09 (rounded to two decimals).

13. DETERMINATION OF SAMPLE SIZE As in the estimation of means (see Exercise 5 on page 222), the formula which enables us to judge the possible size of our error can also be used to determine *how large a sample is needed to attain a desired degree of reliability.* If we want to be able to assert with a probability of 0.95 that a sample proportion will be "off" either way by less than (or at most) some quantity E, we can write

$$E = 1.96\sqrt{\frac{p(1-p)}{n}}$$

and, upon solving this equation for n, we get

$$n = p(1-p)\left[\frac{1.96}{E}\right]^2$$

Unfortunately, this formula requires knowledge of p, the true proportion we are trying to estimate, which means that it cannot be used unless we have some prior (perhaps, indirect) information about the possible size of p. To get around this difficulty, we shall make use of the fact which the reader will be asked to verify in part (d) of this exercise, namely, that $p(1-p) = \frac{1}{4} - (p - \frac{1}{2})^2$ and, hence, that the *maximum* value of $p(1-p)$ is $\frac{1}{4}$. Thus, *when we have no idea about the possible size of p, we use the formula*

$$n = \frac{1}{4}\left[\frac{1.96}{E}\right]^2$$

and since this may make the sample size *unnecessarily large*, we can assert with a probability of *at least* 0.95 that the sample proportion we get will be "off" by less than E. To illustrate this technique, let us suppose that we want to estimate what proportion of heavy smokers will get lung cancer, and that we want to be "95% sure" that the error of our estimate will be less than $E = 0.04$. If we have no idea about the true value of p, we substitute the given value of E, namely, 0.04, into the last formula, getting

$$n = \frac{1}{4}\left[\frac{1.96}{0.04}\right]^2 = 601$$

rounded *up* to the nearest whole number. Thus, if the estimate is a sample proportion based on a random sample of size $n = 601$, we can assert with a probability of *at least* 0.95 that it will be "off" by less than 0.04. Had we known from other studies that the proportion we are trying to estimate is in the neighborhood of, say, 0.20, the first formula for n would have yielded

$$n = (0.20)(0.80)\left[\frac{1.96}{0.04}\right]^2 = 385$$

rounded up to the nearest whole number. This illustrates how prior information about the possible size of p can lead to an appreciable reduction in the required size of a sample.

a. Suppose that we want to estimate what proportion of the drivers are actually exceeding the 55-mph speed limit on a stretch of road between Indio and Blyth in California. How large a sample will we need to be able to assert with a probability of at least 0.95 that the error of our estimate (the sample proportion) will be less than 0.03? What would be the answer to this question if we guessed (or assumed) that the proportion we are trying to estimate is in the neighborhood of 0.70?

b. A private opinion poll is hired by a political leader to estimate what *percentage* of the voters in his state favor more lenient abortion laws. How large a sample will they have to take to be able to assert with a probability of at least 0.95 that their estimate, the sample *percentage*, will be within 5% of the correct value?

c. In a study of advertising campaigns, a national manufacturer wants to determine what proportion of shirts purchased for use by men are actually purchased by women. How large a sample will they need to be able to assert with a probability of 0.99 that their estimate (the sample proportion) will not be "off" by more than 0.02? If they have reason to believe that the proportion they want to estimate is about 0.40, how would this affect the required size of the sample?

d. Making use of the *binomial expansion* $(a + b)^2 = a^2 + 2ab + b^2$, verify that $p(1 - p) = \frac{1}{4} - (p - \frac{1}{2})^2$. Explain why the maximum value of $p(1 - p)$ is $\frac{1}{4}$, which occurs at $p = \frac{1}{2}$.

14. Show that if we add $1.96\sqrt{np(1 - p)}$ to the expressions on both sides of the inequality $np - 1.96\sqrt{np(1 - p)} < x$ and then divide by n, we get the inequality $p < \frac{x}{n} + \frac{1.96}{n}\sqrt{np(1 - p)}$ and, hence,

$$p < \frac{x}{n} + 1.96\sqrt{\frac{p(1 - p)}{n}}$$

Use a similar argument to change the other part of the double inequality on page 259.

15. Since the *proportion* of successes is simply the *number* of successes divided by n, the mean and the standard deviation of the sampling distribution of the *proportion* of successes may be obtained by *dividing* by n the mean and the standard deviation of the sampling distribution of the *number* of successes. Use this argument to obtain formulas for the mean and the standard deviation of the sampling distribution of the proportion of successes in n "binomial" trials.

Tests concerning proportions

Having learned in Chapter 8 what is involved in the test of a (statistical) hypothesis, the reader should find it easy to understand the various other tests which we shall study in this book. In this section we will be concerned with tests which enable us to decide, on the basis of sample data, *whether the "true" value of a proportion (percentage, or probability) equals a given constant.* These tests will enable us to check, say, whether the true proportion of high school seniors who can name the two senators of their state is 0.40, whether it is true that less than 30% of all taxpayers make some kind of mistake in their income tax returns, or whether the probability is greater than 0.20 that the "downtime" of a new computer will exceed 5 hours a week.

Questions of this kind are generally decided on the basis of the observed number, or proportion, of "successes" in n trials, and it will be assumed throughout this section that these trials are independent and that the probability of a success is the same for each trial. In other words, we shall assume that we can use the binomial distribution. When n is *small,* tests concerning the "true" proportion of "successes" (namely, tests concerning the parameter p of the binomial distribution) are usually based directly on tables of binomial probabilities, such as Table XI.

Example 4. To illustrate, let us refer to the third of the three examples of the preceding paragraph, and let us suppose that the computer was inoperative for more than 5 hours in 6 out of 14 weeks. To test the null hypothesis $p = 0.20$ against the alternative $p > 0.20$ at the level of significance $\alpha = 0.05$, we shall have to see whether the probability of "6 or more successes in 14 trials when $p = 0.20$" is 0.05 or less. Since Table XI shows that this probability is

$$0.032 + 0.009 + 0.002 = 0.043$$

which is less than 0.05, we find that *the null hypothesis will have to be rejected*; in other words, we conclude that the probability is greater than 0.20 that the "downtime" of the computer will exceed 5 hours a week. To picture this as in Figure 8.8 on page 234, we can say that $x = 6$ falls into the right-hand tail of the distribution of the number of successes. Note that if the level of significance had been $\alpha = 0.01$, the null hypothesis $p = 0.20$ could not have been rejected.

Example 5. To give an example where a two-tail test would be appropriate, suppose that we want to investigate the claim that 60% of the students attending a large university are opposed to a plan to increase student fees in order to build new parking facilities. Suppose, furthermore, that 4 of 12 students interviewed are opposed to the plan and that the level of significance is to be $\alpha = 0.05$. To test the null hypothesis $p = 0.60$ against the alternative $p \neq 0.60$, we shall have to see whether $x = 4$ falls into either tail of the distribution of the number of successes. Since the observed number of successes, 4, is less than the *expected* number of successes, $np = 12(0.60) = 7.2$, we shall have to worry only about the left-hand tail, that is, we shall only have to check whether the probability of "4 or fewer successes in 12 trials when $p = 0.60$" is less than 0.025. Thus, since Table XI shows that this probability is

$$0.002 + 0.012 + 0.042 = 0.056$$

which exceeds 0.025, we conclude that *the null hypothesis cannot be rejected.*

When n is *large*, tests concerning "true" proportions are usually based on the normal curve approximation to the binomial distribution, and (as we indicated on page 174) this approximation is satisfactory so long as np and

$n(1 - p)$ both exceed 5. Making use of the formulas on pages 141 and 146, according to which the mean and the standard deviation of the binomial distribution are given by $\mu = np$ and $\sigma = \sqrt{np(1 - p)}$, we base all large-sample tests of the null hypothesis $p = p_0$ on the statistic

$$z = \frac{x - np_0}{\sqrt{np_0(1 - p_0)}}$$

whose sampling distribution is approximately the *standard normal distribution*. Note that we are thus converting x, the observed number of "successes," into *standard units*; first we subtract from it the mean of its sampling distribution and then we divide by the standard deviation. The actual test criteria which we shall use are again those of Figures 8.6, 8.7, and 8.8. For the two-sided alternative $p \neq p_0$ we use the criteria of Figure 8.6 on page 230, for the one-sided alternative $p < p_0$ we use the criteria of Figure 8.7 on page 232, and for the one-sided alternative $p > p_0$ we use the criteria of Figure 8.8 on page 234.*

Example 6. To illustrate this kind of test, suppose that a nutritionist claims that at most 75% of the preschool children in a certain country have protein-deficient diets, and that a sample survey reveals that this is true for 244 out of 300 preschool children in that country. To test the null hypothesis $p = 0.75$ against the alternative hypothesis $p > 0.75$, say, at the level of significance $\alpha = 0.01$, we substitute $x = 244$, $n = 300$, and $p_0 = 0.75$, getting

$$z = \frac{244 - 300(0.75)}{\sqrt{300(0.75)(0.25)}} = 2.53$$

Since this value exceeds 2.33 (the dividing line of the $\alpha = 0.01$ criterion of Figure 8.8 on page 234), we conclude that *the null hypothesis will have to be rejected*. In other words, we conclude that more than 75% of the preschool children in the given country have protein-deficient diets.

Example 7. To give an example where it would be appropriate to use a two-tail test, suppose that we want to investigate the claim that 30% of all families who leave California move to Arizona, and that a

*To make the same *continuity correction* as on page 168, many statisticians substitute $x - \frac{1}{2}$ or $x + \frac{1}{2}$ instead of x into the formula for z, and use whichever value makes z smallest. However, in most cases the effect of this correction is negligible so long as n is large.

sample check of several large van lines reveals that the belongings of 163 of 600 families leaving California are being shipped to Arizona. To test the null hypothesis $p = 0.30$ against the alternative that this figure may be *too high or too low* at the level of significance $\alpha = 0.05$, we substitute $x = 163$, $n = 600$, and $p_0 = 0.30$ into the above formula for z, getting

$$z = \frac{163 - 600(0.30)}{\sqrt{600(0.30)(0.70)}} = -1.51$$

Since this is not less than -1.96 (the left-hand dividing line of the $\alpha = 0.05$ criterion of Figure 8.6 on page 230), we find that *the null hypothesis that 30% of all families leaving California move to Arizona cannot be rejected.*

Exercises

1. A physicist claims that at most 10% of all persons exposed to a certain amount of radiation will feel any effects. To test the hypothesis $p = 0.10$ against the alternative $p > 0.10$ at the level of significance $\alpha = 0.05$, 15 cases (where persons were exposed to the given amount of radiation) were pulled from extensive files, and it was found that in three of them there were some effects. What can we conclude from this about the physicists's claim?

2. The advertising manager of a chewing gum manufacturer claimed that 80% of the adult population recognize the company's jingle and associate it with the brand of chewing gum. To check whether this claim may be too low or too high, a market research organization interviewed 11 persons and found that only 5 of them recognized the jingle and associated it with the correct brand of chewing gum. What can they conclude at the level of significance $\alpha = 0.05$?

3. Suppose we want to investigate the claim that 30% of the graduates of a certain large university go on to graduate school. If in a sample of 14 graduates of this university only one says that he is going on to graduate school, what can we conclude about the claim at the level of significance $\alpha = 0.05$, if

a. we use the alternative hypothesis $p \neq 0.30$;

b. we use the alternative hypothesis $p < 0.30$?

4. Suppose we want to test the "honesty" of a coin on the basis of the number of heads we will get in 15 flips. How few or how many heads would we have to get so that we could reject the null hypothesis $p = 0.50$ against the alternative hypothesis $p \neq 0.50$ at the level of significance $\alpha = 0.05$? What is the *actual* level of significance we would be using with this criterion?

5. The manufacturer of a spot remover claims that his product removes at least 80% of all spots. What can we conclude about this claim at the level of significance $\alpha = 0.01$, if the spot remover removed only 154 of 200 spots chosen at random from spots on clothes brought to a dry cleaning establishment?

6. In a random sample of 300 industrial accidents, it was found that 183 were due at least in part to unsafe working conditions. Use the level of significance $\alpha = 0.05$ to check whether this supports the claim that 65% of such accidents are due at least in part to unsafe working conditions; that is, check whether this supports the hypothesis $p = 0.65$ or the alternative hypothesis $p \neq 0.65$.

7. To check on an ambulance service's claim that at least 40% of its calls are life-threatening emergencies, a random sample was taken from its files, and it was found that only 46 of 150 calls were life-threatening emergencies. Test the null hypothesis $p = 0.40$ against a suitable alternative at the level of significance $\alpha = 0.05$.

8. A fund-raising organization claims that it gets an 8% response to its mail solicitations. Test this claim against the alternatives that their figure is too high, if only 137 responses are received to 2,000 letters which they sent out to raise funds for several charities. Use the level of significance $\alpha = 0.05$.

9. With reference to Example 5 on page 265, suppose that a follow-up study shows that 156 of 250 students oppose the plan to increase student fees in order to build new parking facilities. Use these new data to retest the hypothesis of Example 5.

10. With reference to Exercise 1, suppose that a more extensive study shows that 19 of 120 persons exposed to the radiation felt any effects. Use these new data to retest the hypothesis of Exercise 1.

11. THE SIGN TEST The small-sample test concerning means, which we gave in Exercise 9 on page 238, is based on the assumption that the population from which we are sampling has roughly the shape of a normal distribution. When this assumption is untenable or difficult to justify, we can use many other tests, among them the **sign test**, which requires only that the population has a symmetrical distribution, so that the probability of getting a value less than the mean equals the probability of getting a value greater than the mean. *To test the null hypothesis* $\mu = \mu_0$ *against an appropriate alternative, we replace each sample value exceeding* μ_0

with a plus sign and each sample value less than μ_0 with a minus sign, discard values that actually equal μ_0, and we can then test instead the null hypothesis $p = 0.50$, where p is the probability of getting a plus sign or the probability of getting a minus sign. Depending on the size of the sample, the actual test is performed by either of the methods described in the preceding section. To illustrate the first of these techniques, let us refer again to the oceanographer of Example 3 on page 231, and let us suppose that in 15 soundings in a certain area she obtained the following ocean depths in fathoms: 46.4, 48.3, 51.9, 38.8, 46.5, 45.6, 52.1, 41.0, 54.2, 44.9, 52.3, 43.0, 48.7, 46.2, and 44.9. To test the null hypothesis $\mu = 43.0$ fathoms (the previously recorded ocean depth in that area) against the alternative $\mu \neq 43.0$ at $\alpha = 0.05$, we first replace each value less than 43.0 with a minus sign, each value greater than 43.0 with a plus sign, and discard the one value which actually equals 43.0. Thus, we get

$$+ \quad + \quad + \quad - \quad + \quad + \quad + \quad - \quad + \quad + \quad + \quad + \quad + \quad +$$

and it remains to be seen whether "12 successes in 14 trials" supports the null hypothesis $p = 0.50$ or the alternative hypothesis $p \neq 0.50$. Using Table XI, we find that the probability of "12 or more successes" is $0.006 + 0.001 = 0.007$ for $n = 14$ and $p = 0.50$, and since this is less than 0.025, the null hypothesis will have to be rejected—*the true average ocean depth in the given area is not 43.0 fathoms.*

a. The yield of alfalfa from nine test plots is 1.2, 2.2, 1.9, 1.1, 1.8, 1.4, 2.0, 1.3, and 1.6 tons per acre. Use the sign test and the level of significance $\alpha = 0.05$ to test the null hypothesis that the true average yield is 2.1 tons per acre against the two-sided alternative $\mu \neq 2.1$.

b. The following sample data are the amounts of money which fifteen persons spent on a visit to a certain amusement park: \$10.50, \$13.00, \$16.75, \$12.50, \$13.35, \$14.45, \$11.85, \$12.25, \$9.65, \$18.60, \$14.10, \$11.50, \$16.85, \$12.75, and \$13.15. Use the sign test and the level of significance $\alpha = 0.05$ to test the null hypothesis that on the average a person spends \$12.00 at the park against the alternative that this figure is too low.

c. The following data are the octane ratings obtained for 16 samples of a certain kind of gasoline: 101.0, 98.7, 101.0, 103.6, 97.1, 102.5, 100.0, 103.3, 99.4, 107.7, 105.4, 102.8, 100.0, 104.5, 98.2, and 100.0. Use the sign test and the level of significance $\alpha = 0.05$ to test the null hypothesis that the true mean octane rating of this kind of gasoline is 100.0 against the alternative that it is greater than 100.0.

12. THE SIGN TEST, Continued In the illustration of the preceding exercise we could have used the large-sample technique described on page 266, since np and $n(1 - p)$ both exceed 5 for $n = 14$ and $p = 0.50$.

Thus, substituting $x = 12$, $n = 14$, and $p = 0.50$ into the formula for z, we get

$$z = \frac{12 - 14(0.50)}{\sqrt{14(0.50)(0.50)}} = 2.67$$

and since this exceeds 1.96 (the dividing line of the $\alpha = 0.05$ criterion of Figure 8.6), we find, as before, that the null hypothesis will have to be rejected.

a. Suppose that the data of Example 1 on page 216 are to be used to test the hypothesis that in the performance of the given task the true average increase of a person's pulse rate is 23.5 beats per minute. Use the method described above to test this null hypothesis against the alternative hypothesis $\mu > 23.5$ at the level of significance $\alpha = 0.05$.

b. Suppose that the data of Example 7 on page 190 are to be used to test the hypothesis that the plant's true average daily emission of sulfur oxides is 22 tons. Use the method described above to test this null hypothesis against the alternative $\mu \neq 22$ tons at the level of significance $\alpha = 0.05$.

c. Suppose that the data of Exercise 14 on page 26 are to be used to test the hypothesis that the true average burning time of the given kind of rocket is 4.8 seconds. Use the method described above to test this null hypothesis against the alternative hypothesis $\mu < 4.8$ at the level of significance $\alpha = 0.01$.

It should be noted that the methods described here and in Exercise 11 are often used as shortcut techniques *even though the corresponding standard methods of Chapter 8 apply.*

THE SIGN TEST, Continued The sign test can also be used as an alternative to the method described in Exercise 6 on page 246, namely, when we are dealing with *dependent samples.* Thus, in the illustration of that exercise we replace each decrease in the number of accidents with a minus sign, each increase with a plus sign, and discard the one case where there is no change. Thus, we get $- - + - - - -$, and it remains to be seen whether the probability of "6 or more successes in 7 trials" is less than 0.05, namely, whether it enables us to reject the null hypothesis $p = 0.50$ against the alternative hypothesis that the probability of getting a minus sign exceeds 0.50. Table XI shows that this probability is $0.055 + 0.008 = 0.063$, so that *the null hypothesis cannot be rejected.*

a. The following are the number of speeding tickets issued by two policemen on 16 days: 7 and 10, 11 and 13, 14 and 14, 11 and 15, 12 and 9, 6 and 10, 9 and 13, 8 and 11, 10 and 11, 11 and 15, 13 and 11, 7 and 10, 8 and 8, 11 and 12, 9 and 14, and 10 and 9. Use the sign test at the level of significance $\alpha = 0.05$ to test the null hypothesis that on the average the two policemen issue equally many speeding tickets per day against the alternative hypothesis that on the average the second policeman issues more speeding tickets per day than the first.

b. Use the sign test to rework Exercise 7 on page 247.

When we are dealing with dependent samples whose size does not permit the use of Table XI, we can still use the sign test, but base it on the method described on page 266.

c. The following are the number of employees absent from two departments of a large firm on 25 days: 2 and 4, 6 and 3, 5 and 5, 7 and 2, 3 and 1, 4 and 3, 2 and 5, 3 and 1, 4 and 3, 5 and 6, 5 and 4, 3 and 8, 6 and 4, 5 and 2, 4 and 3, 3 and 0, 2 and 5, 6 and 4, 3 and 1, 2 and 4, 5 and 2, 3 and 2, 4 and 6, 6 and 3, and 4 and 3. Use the sign test at the level of significance $\alpha = 0.05$ to test the null hypothesis that on the average there are equally as many absences in the two departments against the alternative hypothesis that on the average there are more absences in the first department.

d. The following are the number of artifacts dug up by two archaeologists at an ancient cliff dwelling on 30 days: 1 and 0, 0 and 0, 2 and 1, 3 and 0, 1 and 2, 0 and 0, 2 and 0, 2 and 1, 3 and 1, 0 and 2, 1 and 0, 1 and 1, 4 and 2, 1 and 1, 2 and 1, 1 and 0, 3 and 2, 5 and 2, 2 and 6, 1 and 0, 3 and 2, 2 and 3, 4 and 0, 1 and 2, 3 and 1, 2 and 0, 0 and 1, 2 and 0, 4 and 1, and 2 and 0. Use the sign test at the level of significance $\alpha = 0.01$ to test the null hypothesis that the two archaeologists are equally good at finding artifacts against the alternative hypothesis that the first one is better.

Differences among proportions

If we were to follow the pattern of Chapter 8, we would continue our study of the analysis of count data by looking at methods which enable us to decide whether observed differences between *two* sample proportions are significant or whether they can be attributed to chance. This kind of problem would arise, for example, if 114 of 200 randomly selected housewives in one city say that they take empty aluminum cans to places where they are collected for recycling, but only 33 of 100 randomly selected housewives in another city make this claim. The corresponding sample proportions are $\frac{114}{200} = 0.57$ and $\frac{33}{100} = 0.33$, and it may be of interest to know whether the difference between these sample proportions is *significant*, namely, whether housewives in the first city are actually more ecology-minded than those in the other city. Similarly, if, during an epidemic,

51 of 150 persons who did not get flu shots caught the disease, while 42 of 50 persons who got flu shots did *not* catch it, it may be of interest to know whether the difference between the sample proportions $\frac{51}{150} = 0.34$ and $\frac{8}{50} = 0.16$ can be attributed to chance, or whether it is indicative of the effectiveness of the shots.

The method which we shall study in this section is actually *more general*—it applies to problems in which we have to decide whether differences among *two or more* sample proportions are significant, or whether they can be attributed to chance. Thus, we will be able to judge, for example, whether there really is a difference in the quality of four kinds of automobile tires, if 24 of 200 tires of brand A failed to last 30,000 miles, while the corresponding figures for 200 tires of brands B, C, and D, are, respectively, 21, 16, and 33. Similarly, the method will also apply if we added to the first example of the preceding paragraph that among 300 housewives interviewed in a third city there were 147 who say that they take empty aluminum cans to places where they are collected for recycling.

Example 8. Let us use this last example to illustrate the method of analysis which we shall employ, and let us begin by presenting all the sample data together in the following table:

	City 1	*City 2*	*City 3*
Housewives taking aluminum cans to be recycled	114	33	147
Housewives not taking aluminum cans to be recycled	86	67	153
Total	200	100	300

where the entries of the second row were obtained by subtraction from the corresponding totals. Using this table, we shall now investigate whether the differences among the three proportions of housewives who take empty aluminum cans to be recycled, namely, $\frac{114}{200} = 0.57$, $\frac{33}{100} = 0.33$, and $\frac{147}{300} = 0.49$, can be attributed to chance.

If we let p_1, p_2, and p_3 denote the *true* proportions of housewives in the three cities who take their empty aluminum cans to be recycled, the null hypothesis we shall want to test is

$$p_1 = p_2 = p_3$$

and the alternative hypothesis is that p_1, p_2, and p_3 are *not all equal*. If the null hypothesis is true and the differences *are* due to chance, we can then combine all the data and look upon the three samples as *one sample from one and the same population*, and estimate the "true" proportion of all housewives (in the three cities) who take their empty aluminum cans to be recycled as

$$\frac{114 + 33 + 147}{200 + 100 + 300} = \frac{294}{600} = 0.49$$

Then, using this estimate, we can say that we could have *expected* 200(0.49) = 98 of the 200 housewives from the first city to take their empty aluminum cans to be recycled, that we could have *expected* 100(0.49) = 49 of the 100 housewives from the second city to take their empty aluminum cans to be recycled, and that we could have *expected* 300(0.49) = 147 of the 300 housewives from the third city to take their empty aluminum cans to be recycled. To find the corresponding *expectations* for the number of housewives who do *not* take their empty aluminum cans to be recycled, we have only to subtract the figures we have just calculated from the respective totals, getting 200 − 98 = 102, 100 − 49 = 51, and 300 − 147 = 153. These results are summarized in the following table, where the **expected frequencies** are shown in parentheses below the ones that were actually observed:

	City 1	*City 2*	*City 3*
Housewives taking aluminum cans to be recycled	114 (98)	33 (49)	147 (147)
Housewives not taking aluminum cans to be recycled	86 (102)	67 (51)	153 (153)

To test the null hypothesis that the p's are all equal in an example like this, we compare the frequencies which were actually observed with those which we could have *expected* if the null hypothesis were true. If the two sets of frequencies are very much alike, it stands to reason that the null hypothesis should be *accepted*—after all, we would then have obtained almost exactly what we could have expected if the null hypothesis were true. On the other hand, if the discrepancies between the two sets of frequencies are large, the null hypothesis will probably have to be *rejected*, for in that case we did not get what we could have expected if the null hypothesis were true.

To judge *how large* the discrepancies between the two sets of frequencies must be before the null hypothesis can be rejected, we use the following statistic, where the observed frequencies are denoted by the letter f and the expected frequencies by the letter e:

$$\chi^2 = \sum \frac{(f - e)^2}{e}$$

This statistic is called **chi-square** (χ is the Greek letter *chi*), and it should be observed that the summation extends over all the "cells" of the table. In other words, χ^2 is the *sum* of the quantities which we obtain by dividing the squared difference $(f - e)^2$ by e *separately* for each cell of the table.

Example 8 (Continued). Returning to our numerical example, we thus obtain

$$\chi^2 = \frac{(114 - 98)^2}{98} + \frac{(33 - 49)^2}{49} + \frac{(147 - 147)^2}{147} + \frac{(86 - 102)^2}{102} + \frac{(67 - 51)^2}{51} + \frac{(153 - 153)^2}{153} = 15.37$$

and it still remains to be seen whether this value is *large enough* to reject the null hypothesis $p_1 = p_2 = p_3$, namely, the null hypothesis that the proportion of housewives taking their empty aluminum cans to be recycled is the same for all three cities.

If the null hypothesis that the p's are all equal is true, the sampling distribution of the χ^2 statistic is approximately the **chi-square distribution,** an example of which is shown in Figure 9.3. Since the null hypothesis will be rejected only when χ^2 is too large (namely, when the discrepancies between the f's and the e's are too great to be attributed to chance), we ultimately base our decision on the criterion of Figure 9.3. Here $\chi^2_{0.05}$ is such that the area under the curve to its right equals 0.05, and it provides the dividing line of the criterion when the level of significance is $\alpha = 0.05$; correspondingly, $\chi^2_{0.01}$ provides the dividing line of the criterion when the level of significance is $\alpha = 0.01$. These quantities, which depend on a parameter which (again) is called the **number of degrees of freedom**, or simply the

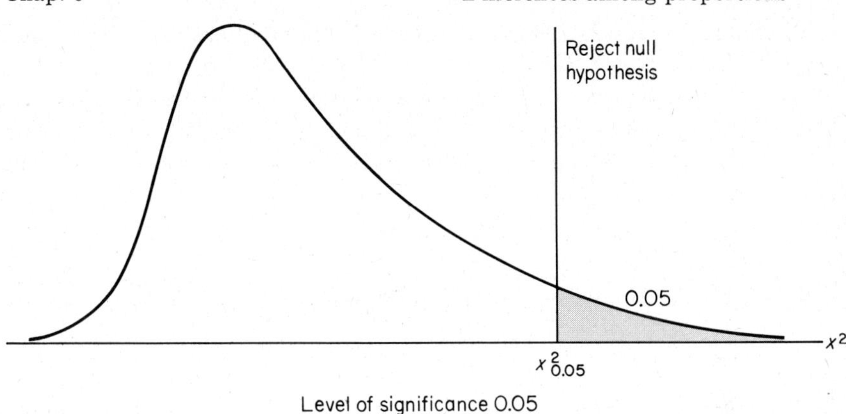

Figure 9.3 Chi-square distribution.

degrees of freedom, are given in Table III. It equals 2 in our example, and *in general it equals* $k - 1$ *when we compare* k *sample proportions.* Intuitively, we can justify this formula with the argument that once we have calculated any $k - 1$ of the expected frequencies in either row of the table, all of the other expected frequencies can be obtained by subtraction from the totals of the rows and columns (see Exercise 9 on page 280 and also the discussion on page 284).

Example 8 (Continued). Returning to our numerical example, we find from Table III that $\chi^2_{0.05} = 5.991$ for $3 - 1 = 2$ degrees of freedom. Since this is exceeded by 15.37, the value of the χ^2 statistic which we obtained in our example, we find that *the null hypothesis will have to be rejected.* In other words, we conclude that the true proportion of housewives who take their empty aluminum cans to be recycled is *not* the same for the three cities.

In general, if we have k sample proportions and want to test the null hypothesis $p_1 = p_2 = \ldots = p_k\ (= p)$ against the alternative that these "true" proportions are *not all equal,* we first calculate the expected frequencies as on page 273. Combining the data, we estimate the true value of the common proportion p as

$$\frac{x_1 + x_2 + \ldots + x_k}{n_1 + n_2 + \ldots + n_k}$$

where the n's are the sizes of the k samples and the x's are the corresponding number of "successes." (This is how we obtained the estimate of 0.49 on page 273.) Then, multiplying the n's by the above estimate of p we obtain the *expected frequencies* for the first row of the table, and subtracting these values from the sizes of the respective samples we obtain the expected frequencies for the second row of the table. (This is precisely how we obtained the expected frequencies shown in parentheses in the table on page 273.) Finally, we calculate χ^2 according to the formula on page 274, with $\frac{(f-e)^2}{e}$ determined separately for each of the $2k$ cells of the table, and we *reject* the null hypothesis that the differences among the sample proportions are due to chance if the value which we obtain for χ^2 *exceeds* $\chi^2_{0.05}$ or $\chi^2_{0.01}$ (depending on the level of significance) for $k - 1$ degrees of freedom.

When we calculate the expected frequencies, it is customary to round (if necessary) to the nearest integer or to one decimal. The entries of Table III are given to three decimals, but there is seldom any need to carry more than two decimals in calculating the value of the χ^2 statistic, itself. Let us also remind the reader that the test which we have been discussing is only *approximate*, and it is for this reason that it is best not to use the test when one of the *expected frequencies* is less than 5. (If this is the case, we can sometimes "salvage" the situation by combining some of the samples.)

Exercises

1. In a sample of the visitors to a famous tourist attraction (a cavern which had served as a hideout to a notorious outlaw almost 100 years ago), 86 of 250 men and 152 of 250 women bought souvenirs. Use the level of significance $\alpha = 0.01$ to test whether the difference between the corresponding sample proportions is significant.

2. A manufacturer of electronic equipment subjects samples of two competing brands of transistors to an accelerated performance test. If 45 of 180 transistors of the first kind and 34 of 120 transistors of the second kind fail the test, what can he conclude at the level of significance $\alpha = 0.05$ about the difference between the corresponding sample proportions?

3. To test the effectiveness of a new pain-relieving drug, 80 patients at a clinic are given a pill containing the drug while 80 others are given a placebo not containing any drug. If 52 in the first group and 38 in the second group report a beneficial effect, what can we conclude at the level of significance $\alpha = 0.01$ about the difference between the corresponding true proportions?

4. On page 272 we referred to a study which showed that 24 of 200 tires of brand A failed to last 30,000 miles, while the corresponding figures for 200 tires each of brands B, C, and D were, respectively, 21, 16, and 33. Use the level of significance $\alpha = 0.05$ to test the null hypothesis that there is no difference in the quality of the tires, namely, that there is no difference between the corresponding true proportions.

5. The following table shows the results of a sample survey in which the members of three large unions were asked whether they are for or against a certain piece of legislation:

	Union X	*Union Y*	*Union Z*
For the legislation	83	67	104
Against the legislation	37	33	76

Use the level of significance $\alpha = 0.05$ to test the null hypothesis that there is no difference between the corresponding true proportions.

6. A research organization, interested in investigating whether the proportions of sons taking up the occupations of their fathers are equal for a selected group of professions, took suitable samples and obtained the results shown in the following table:

	Doctors	*Lawyers*	*Engineers*
Same occupation	47	38	31
Different occupation	153	112	119

Use the level of significance $\alpha = 0.01$ to test the null hypothesis that there is no difference between the corresponding true proportions.

7. TWO SAMPLE PROPORTIONS (ONE-SIDED ALTERNATIVE)

When we compare two sample proportions, we are sometimes interested in testing the null hypothesis $p_1 = p_2$ against the one-sided alternative $p_1 < p_2$ or $p_1 > p_2$. This is really what we should have asked the reader to do in Exercise 3 to test the effectiveness of the new drug, and what we should do if we wanted to see whether an item in a true–false test, answered correctly by 164 of 200 good students and 102 of 200 poor students, *really discriminates between the two groups.* In situations like this we can base our decision on the following statistic provided that both samples are large:

$$z = \frac{\frac{x_1}{n_1} - \frac{x_2}{n_2}}{\sqrt{p(1-p)\left(\frac{1}{n_1} + \frac{1}{n_2}\right)}} \qquad \text{with } p = \frac{x_1 + x_2}{n_1 + n_2}$$

The sampling distribution of this statistic is approximately the *standard normal distribution* when the null hypothesis $p_1 = p_2$ is true, and the criteria on which we base the tests are again those of Figure 8.7 on page 232 when the alternative hypothesis is $p_1 < p_2$, and those of Figure 8.8 on page 234 when the alternative hypothesis is $p_1 > p_2$. For instance, in the *item analysis* referred to above, p_1 is the true proportion of good students who will answer the question correctly, p_2 is the true proportion of poor students who will answer the question correctly, and we shall want to test the null hypothesis $p_1 = p_2$ against the one-sided alternative $p_1 > p_2$, say, at the level of significance $\alpha = 0.01$. Substituting the given data, namely, $x_1 = 164$, $n_1 = 200$, $x_2 = 102$, and $n_2 = 200$, into the formula for p and then into the formula for z, we get

$$p = \frac{164 + 102}{200 + 200} = \frac{266}{400} = 0.665$$

and, hence,

$$z = \frac{\frac{164}{200} - \frac{102}{200}}{\sqrt{(0.665)(0.335)(\frac{1}{200} + \frac{1}{200})}} = \frac{\frac{62}{200}}{\sqrt{0.0022}} = \frac{0.31}{0.0469} = 6.61$$

Since this exceeds 2.33, the $\alpha = 0.01$ criterion of Figure 8.8, we conclude that the test item definitely discriminates between good students and poor students.

a. Repeat Exercise 3 using the alternative hypothesis $p_1 > p_2$, where p_1 is the probability that a person receiving the new drug will report a beneficial effect, and p_2 is the probability that a person receiving the placebo will report a beneficial effect.

b. If one method of producing rain by "seeding" clouds was successful in 16 of 50 attempts, while another method was successful in 29 of 80 attempts, can we conclude at the level of significance $\alpha = 0.05$ that the second method is better than the first?

c. If one mail solicitation for a charity brought 412 responses to 5,000 letters and another, more expensive, mail solicitation for the same charity brought 261 responses to 3,000 letters, can we conclude at the level of significance $\alpha = 0.05$ that *in general* the more expensive kind of solicitation is somewhat more effective?

8. THE MEDIAN TEST In Exercise 5 on page 245 we gave a *small-sample* significance test for the difference between two means, but it has the disadvantage that it is based on rather stringent assumptions. If the populations cannot be looked upon as having roughly the shape of normal distributions *or* their standard deviations cannot be considered equal, we may instead use the **median test**, in which we first find the median of the *combined data*, determine how many of the values in each sample fall above and below this median, and then analyze the resulting table by the method of the preceding section. (Values that are actually equal to the median of the combined data are omitted.) To illustrate, suppose that random samples of freshmen from two high schools obtained the following scores in a mathematics achievement test:

School 1: 75, 70, 56, 90, 85, 91, 95, 98, 89, 68, 96, 74, 78, 95, 88, 60, 67, 88, 90, 76, 94, 99, 82, 71, 94

School 2: 86, 61, 61, 86, 35, 61, 73, 76, 32, 73, 61, 66, 74, 87, 72, 43, 54, 69, 72, 98, 92, 86, 64, 60, 49

Since there is *much more variability* among the scores of the students from the second school, it would be appropriate here to use the median test in order to check whether it is reasonable to maintain that *in general* the freshmen from both schools are equally good in mathematics. Thus, we first determine the median of the combined data (see page 43), which is 74.5, and we arrive at the following table:

	School 1	*School 2*
Below median	7	18
Above median	18	7

a. Continue the illustration by showing that the expected frequencies are all 12.5, $\chi^2 = 9.68$, and, hence, that the null hypothesis (that *in general* the freshmen from both schools are equally good in mathematics) must be rejected at the level of significance $\alpha = 0.01$.

b. The following are data on the breaking strength (in pounds) of samples of two kinds of 6-inch cotton ribbon:

Type I ribbon: 143, 180, 199, 186, 168, 170, 185, 193, 198, 196, 175, 181, 132, 192, 196, 164, 179, 197, 176, 180

Type II ribbon: 176, 165, 173, 195, 177, 199, 155, 135, 172, 183, 170, 165, 186, 160, 162, 190, 171, 165, 189, 171, 180, 176, 193, 194

Verify that the median is 178 and use the median test to check at the level of significance $\alpha = 0.05$ whether the two kinds of ribbon may be regarded as equally strong.

c. For comparison, two kinds of feed are fed to samples of pigs, and the following are their gains in weight (in ounces) after a fixed period of time:

Feed C: 13.8, 13.4, 12.0, 12.9, 15.2, 14.2, 13.1, 14.6, 13.3, 12.5, 10.4, 14.1, 15.0, 12.4, 13.5, 14.8, 15.5, 13.2, 12.3, 12.0, 15.6, 14.1

Feed D: 11.4, 13.8, 11.7, 14.3, 12.6, 11.1, 13.6, 12.2, 13.8, 10.5, 12.8, 14.0, 10.2, 12.8, 12.7, 15.4, 11.3, 12.6, 11.8, 13.1, 10.8, 11.9, 14.0, 12.4, 12.7, 11.8, 13.1

Verify that the median is 12.9 and use the median test to check at the level of significance $\alpha = 0.05$ whether the two kinds of feed are equally good.

9. Making use of the fact that the expected number of successes for the k samples are obtained by multiplying $\frac{x_1 + x_2 + \ldots + x_k}{n_1 + n_2 + \ldots + n_k}$, respectively, by $n_1, n_2, \ldots$, and n_k show that *the sum of the expected number of successes for the k samples equals the sum of the observed number of successes.*

Contingency tables

The chi-square criterion which we introduced in the preceding section plays an important role in many different kinds of problems involving count data. In this section, we shall apply it to *two* kinds of problems, which differ conceptually but are analyzed the same way.

In the first kind of problem we deal with trials having *more than two possible outcomes*—for instance, the weather can get better, remain the same, or get worse, and an undergraduate in a four-year college can be a freshman, a sophomore, a junior, or a senior.

Example 8 (Continued). Also, in Example 8 on page 272, each housewife might have been asked whether she *always*, *sometimes*, or *never* takes her empty aluminum cans to be recycled, and this might have resulted in the following table:

	City 1	*City 2*	*City 3*
Always	58	18	86
Sometimes	56	15	61
Never	86	67	153
Total	200	100	300

In the above table, as in the example of the preceding section, *the column totals are fixed*, for they represent the sizes of the respective samples. On the other hand, *the row totals* ($58 + 18 + 86 = 162$, etc.) *depend on the responses of the housewives interviewed, and, hence, on chance.*

In the second kind of problem we shall study in this section, *the column totals also depend on chance*, as in the following example:

Example 9. Suppose, for instance, that a sociologist wants to know whether there really is a relationship between the intelligence of boys who have gone through a special training program and their subsequent success in their jobs. Suppose, furthermore, that a sample of 400 cases taken from very extensive files yielded the following results:

	Doing Well	*Doing Fair*	*Doing Poorly*	
Above-average I.Q.	37	23	10	70
Average I.Q.	56	76	42	174
Below-average I.Q.	25	64	67	556
	118	163	119	400

Tables such as this are called **two-way tables** or **contingency tables**, and it is common practice to describe them by giving their number of rows and columns. Thus, the table of Example 9 is a 3×3 table, and in general a table with r rows and k columns is referred to as an $\boldsymbol{r \times k}$ **table** (where $r \times k$ reads "r by k").

Before we demonstrate how $r \times k$ tables are analyzed, let us examine briefly what null hypotheses we shall actually want to test. In the first kind of problem, the one given above in which the column totals are fixed, we shall want to test the null hypothesis that the probabilities of choosing housewives who always, sometimes, or never take their empty aluminum cans to be recycled are the same for all three cities. In other words, we shall want to test the null hypothesis that housewives' behavior with regard to this problem of ecology is *independent* of the location where we conduct the poll. In the second kind of problem we are also concerned with a null hypothesis of *independence*, namely, the hypothesis that the success of a boy who has gone through the special training program does not depend on his I.Q.

Example 9 (Continued). To illustrate how an $r \times k$ table is analyzed, let us refer to the table of Example 9, and let us begin by calculating the *expected frequencies* as on page 273. If the null hypothesis of independence is true, the probability of randomly selecting a boy (from among those who have gone through the special training program) who has an above-average I.Q. *and* who is doing well in his job is the *product* of the probability that he has an above-average I.Q. and the probability that he is doing well in his job. (This is simply an application of the Special Multiplication Rule on page 112.) Since $37 + 23 + 10 = 70$ of the 400 boys had above-average I.Q.'s while $37 + 56 + 25 = 118$ are doing well in their jobs, we *estimate* the probability of choosing a boy with an above-average I.Q. as $\frac{70}{400}$ and the probability of choosing a boy who is doing well in his job as $\frac{118}{400}$. Hence, we *estimate* the probability of choosing a boy with an above-average I.Q. who is doing well in his job as $\frac{70}{400} \cdot \frac{118}{400}$, and in a sample of size 400 we would *expect* to find

$$400\left(\frac{70}{400} \cdot \frac{118}{400}\right) = \frac{70 \cdot 118}{400} = 20.6$$

boys who fit this description.

Note that this last result was obtained by evaluating $\dfrac{70 \cdot 118}{400}$, namely, by multiplying the total of the first row by the total of the first column and then dividing by the grand total for the entire table. Using precisely the same sort of reasoning, we can show that this rule applies to any cell of an $r \times k$ table, namely, that

The expected frequency of any cell of an $r \times k$ table is the product of the totals of the row and the column to which it belongs, divided by the grand total for the entire table.

Example 9 (Continued). Thus, we obtain an expected frequency of $\frac{70 \cdot 163}{400} = 28.5$ for the second cell of the first row, and expected frequencies of $\frac{174 \cdot 118}{400} = 51.3$ and $\frac{174 \cdot 163}{400} = 70.9$ for the first two cells of the second row.

Actually, it is not necessary to calculate all the expected cell frequencies in this way. If we use an argument similar to that of Exercise 9 on page 280, it can be shown that *the sum of the expected frequencies for any row or column must equal the sum of the corresponding observed frequencies*; hence, some of the expected cell frequencies can be obtained by subtraction.

Example 9 (Continued). For our example we, thus, find by subtraction that the expected frequency for the third cell of the first row is

$$70 - 20.6 - 28.5 = 20.9$$

that the expected frequency for the third cell of the second row is

$$174 - 51.3 - 70.9 = 51.8$$

and that the expected frequencies for the three cells of the third row are, respectively,

$$118 - 20.6 - 51.3 = 46.1$$

$$163 - 28.5 - 70.9 = 63.6$$

and

$$119 - 20.9 - 51.8 = 46.3$$

All these expected frequencies were rounded to one decimal, and they are summarized in the following table, where they are shown in parentheses below the corresponding frequencies which were actually observed:

	Doing Well	*Doing Fair*	*Doing Poorly*
Above-average I.Q.	37 (20.6)	23 (28.5)	10 (20.9)
Average I.Q.	56 (51.3)	76 (70.9)	42 (51.8)
Below-average I.Q.	25 (46.1)	64 (63.6)	67 (46.3)

From here on, the work is like that of the preceding section: Writing the observed and expected frequencies again as f and e, we calculate χ^2 according to the formula on page 274, namely,

$$\chi^2 = \sum \frac{(f - e)^2}{e}$$

where the quantity $\frac{(f - e)^2}{e}$ is again determined separately for each individual cell. Then we *reject* the null hypothesis of *independence* if the value which we obtain for χ^2 *exceeds* $\chi^2_{0.05}$ or $\chi^2_{0.01}$ (depending on the level of significance) for $(r - 1)(k - 1)$ degrees of freedom. As before, r is the number of rows and k is the number of columns, and to justify the formula for the degrees of freedom, intuitively at least, let us point out that after $(r - 1)(k - 1)$ of the expected frequencies have been calculated according to the rule on page 282, all the others can be obtained by subtraction from the totals of appropriate rows or columns. In our numerical example we had $r = 3$ and $k = 3$, and it should be observed that after we had determined $(r - 1)(k - 1) = (3 - 1)(3 - 1) = 4$ of the expected frequencies according to the rule on page 282, all the others were obtained by subtraction from the totals of rows or columns.

Example 9 (Continued). Returning now to our numerical example, we find that

$$\begin{aligned}\chi^2 &= \frac{(37 - 20.6)^2}{20.6} + \frac{(23 - 28.5)^2}{28.5} + \frac{(10 - 20.9)^2}{20.9} \\ &\quad + \frac{(56 - 51.3)^2}{51.3} + \frac{(76 - 70.9)^2}{70.9} + \frac{(42 - 51.8)^2}{51.8} \\ &\quad + \frac{(25 - 46.1)^2}{46.1} + \frac{(64 - 63.6)^2}{63.6} + \frac{(67 - 46.3)^2}{46.3} \\ &= 41.39\end{aligned}$$

and since $\chi^2_{0.01} = 13.277$ for $(3 - 1)(3 - 1) = 4$ degrees of freedom (see Table III), we find that the null hypothesis will have to be rejected. This means that there *is* a relationship between the intelligence of these boys and their success in the jobs for which they were trained in the special program. Does this suggest, perhaps, that not all of these boys were trained for the right kind of job?

The method by which we analyzed the 3×3 table of Example 9 applies also in first kind of problem described on page 281, namely, when the column totals are fixed sample sizes and do not depend on chance. The expected frequencies are determined in the same way, but the formula according to which we multiply the row total by the column total and then divide by the grand total has to be justified differently (see Exercise 8 on page 287). Let us also point out that, as in the preceding section, the chi-square criterion is only approximate and, hence, should be used only when each expected cell frequency is at least 5. If one or more of the expected cell frequencies is less than 5, we can sometimes "salvage" the situation by combining some of the cells and subtracting one degree of freedom for each cell which is thus eliminated.

Exercises

1. Analyze the 3×3 table of the continuation of Example 8 on page 281, and decide at the level of significance $\alpha = 0.05$ whether the pattern of the housewives' behavior with regard to the recycling of their empty aluminum cans is the same in all three cities.

2. The results of polls conducted two weeks and four weeks before a gubernatorial election are shown in the following table:

	Two Weeks before Election	*Four Weeks before Election*
For Republican candidate	79	91
For Democratic candidate	84	66
Undecided	37	43

Use the level of significance $\alpha = 0.05$ to decide whether there has been a change in opinion during the two weeks between the two polls.

3. The following sample data pertain to the shipments received by a large firm from three different vendors:

	Number Rejected	*Number Imperfect but Acceptable*	*Number Perfect*
Vendor A	12	23	89
Vendor B	8	12	62
Vendor C	21	30	119

Test at the level of significance $\alpha = 0.01$ whether the three vendors ship products of equal quality.

4. A large electronics firm which hires many handicapped workers wants to determine whether their handicaps affect such workers' performance. Use the level of significance $\alpha = 0.05$ to decide on the basis of the sample data shown in the following table whether it is reasonable to maintain that the handicaps have no effect on the workers' performance:

	Performance		
	Above Average	*Average*	*Below Average*
Blind	21	64	17
Deaf	16	49	14
No handicap	29	93	28

5. The following table is based on a study of the relationship between race and blood type in a Near Eastern country:

	Blood Type			
	O	A	B	AB
Race 1	176	148	96	72
Race 2	78	50	45	12
Race 3	15	19	8	7

Use the level of significance $\alpha = 0.01$ to test the null hypothesis that there is no relationship between race and blood type so far as the given Near Eastern country is concerned.

6. Suppose that a social scientist who wants to determine whether there is a relationship between success in life (as measured by means of a questionnaire) and a person's sense of humor (as measured by means of a special test) obtained the data shown in the following table:

		Sense of Humor		
		Low	*Average*	*High*
Success in Life	*Low*	53	51	46
	Average	117	162	91
	High	40	57	73

Use the level of significance $\alpha = 0.05$ to test the null hypothesis that there is no relationship.

7. The following data pertain to the intelligence and the standard of clothing of a sample of 800 school children:

	Intelligence		
	Low	*Average*	*High*
Very well dressed	62	258	240
Well dressed	70	84	38
Poorly dressed	18	23	7

Use the level of significance $\alpha = 0.01$ to test the null hypothesis that there is no relationship between school children's intelligence and their standard of clothing (in the city where this study was made).

8. Using an argument similar to that on pages 275 and 276, show that the rule on page 282 for calculating the expected cell frequencies applies also when the column totals are fixed sample sizes and do not depend on chance.

Goodness of fit

The chi-square criterion can also be used to compare observed frequency distributions with distributions which we might *expect* according to mathematical theory or assumptions.

Example 10. This kind of problem might arise, for example, if we want to check whether a die may be regarded as "honest." In that case the probability for each face of the die would be $\frac{1}{6}$, and if we rolled the die 240 times, we might get the results shown in the following table:

Face of Die	*Probability*	*Observed Frequency* f	*Expected Frequency* e
1	$\frac{1}{6}$	34	40
2	$\frac{1}{6}$	43	40
3	$\frac{1}{6}$	32	40
4	$\frac{1}{6}$	35	40
5	$\frac{1}{6}$	45	40
6	$\frac{1}{6}$	51	40
		240	240

The expected frequencies were obtained by multiplying the respective probabilities by 240, and it should be observed that there are substantial discrepancies between the f's and the e's.

To check whether such differences between observed and expected frequencies can be attributed to chance, we use the same chi-square statistics as in the two preceding sections, namely,

$$\chi^2 = \sum \frac{(f - e)^2}{e}$$

where $\frac{(f - e)^2}{e}$ is calculated separately for each class of the distribution. Then if the value of χ^2 is *too large* we reject the null hypothesis on which the expected frequencies are based—in Example 10 this would be the null hypothesis that the probability is $\frac{1}{6}$ for each face of the die. More specifically, we reject the null hypothesis at the level of significance $\alpha = 0.05$ if the value we get for χ^2 exceeds $\chi^2_{0.05}$ (or at the level of significance $\alpha = 0.01$ if the value we get for χ^2 exceeds $\chi^2_{0.01}$). The number of degrees of freedom for this kind of test of **goodness of fit** is $k - m$, where k is the number of terms $\frac{(f - e)^2}{e}$ which we have to add in the formula for χ^2, and m is *the number of quantities we have to obtain from the observed data in order to calculate the expected frequencies.* In Example 10 we had to know only the sum of the observed frequencies (or the number of times the die was to be rolled, namely, 240), so that $m = 1$ and the formula for the number of degrees of freedom is $k - 1$.

Example 10 (Continued). Returning now to our numerical example and substituting into the formula for χ^2 the f's and the e's of the table on page 287, we get

$$\begin{aligned}\chi^2 &= \frac{(34 - 40)^2}{40} + \frac{(43 - 40)^2}{40} + \frac{(32 - 40)^2}{40} \\ &\quad + \frac{(35 - 40)^2}{40} + \frac{(45 - 40)^2}{40} + \frac{(51 - 40)^2}{40} \\ &= 7.00\end{aligned}$$

Since this is less than 11.070, the value of $\chi^2_{0.05}$ for $k - 1 = 6 - 1 = 5$ degrees of freedom, we find that *the null hypothesis cannot be rejected*;

in other words, the data do *not* constitute evidence that there is anything wrong with the die.

The method we have illustrated in this section can be used quite generally to see how well a distribution which we *expect* on the basis of theory or assumptions fits observed data. It applies to all the distributions we studied in Chapter 5, and (although we shall not discuss it in this text) it can be used even to show how well a *continuous distribution curve* fits a distribution of observed data.

Exercises

1. In Exercise 1 on page 149 we claimed that the probabilities are 0.4, 0.3, 0.2, and 0.1 that there are, respectively, 0, 1, 2, or 3 bank robberies per month in a given western city. Test at the level of significance $\alpha = 0.05$ whether this claim is substantiated by ten years' data showing that there were no bank robberies in 62 months, one bank robbery in 28 months, two bank robberies in 21 months, and three bank robberies in 9 months.

2. A quality control engineer takes daily samples of ten electronic components and on 200 consecutive working days he obtained the data summarized in the following table:

Number of Defectives	*Number of Days*
0	56
1	77
2	40
3	16
4 *or more*	11
	200

a. Use Table XI to find the probabilities of getting 0, 1, 2, 3, and 4 *or more* defectives in a sample of size 10, when actually 10% of all the electronic components (from which the samples are obtained) are defective.
b. Find the corresponding expected frequencies for 200 samples.
c. Test at the level of significance $\alpha = 0.01$ whether it is reasonable to maintain that 10% of all the components are defective.

3. Using any five columns of Table XII (that is, a total of 250 random digits), construct a table showing how many times each of the digits 0, 1, 2, . . . , and 9 occurred. Compare these observed frequencies with the corresponding expected frequencies (which are all equal to 25) by means of the χ^2 statistic, and decide at the level of significance $\alpha = 0.05$ whether this **uniform distribution** provides a good fit to the observed distribution. Note that this is one of many criteria that are used to check on the *randomness* of tables of random numbers.

4. According to the Mendelian theory of heredity, if plants having round yellow seeds are crossbred with plants having wrinkled green seeds, in the second generation $\frac{9}{16}$ of the plants can be expected to have round yellow seeds, $\frac{3}{16}$ of the plants can be expected to have wrinkled yellow seeds, $\frac{3}{16}$ of the plants can be expected to have round green seeds, and $\frac{1}{16}$ of the plants can be expected to have wrinkled green seeds. Use the level of significance $\alpha = 0.05$ to check whether this theory is supported by an experiment which yielded the following results:

Type of Seed	*Number of Plants*
Round yellow	348
Wrinkled yellow	131
Round green	128
Wrinkled green	33

10

The analysis of paired data

Introduction

There are many statistical studies in which we deal with *paired* measurements or observations—the ages of husbands and wives, the supply and the demand for crude oil, rainfall and the per acre yield of wheat, water temperature and the amount of chemical that can be dissolved, the grades and the I.Q.'s of students in a class of European history, the dosage of a medication and its effect, sales and money spent on advertising, and so forth. In studies like these we are usually interested in one aspect or another of the following question:

> **Is there a relationship, or dependence, between the two variables under consideration, and if so, what is its strength and what is its "nature" or its form?**

To answer questions like this we usually try to express the relationship between the variables in terms of a mathematical equation. Among the many equations that can be used for this purpose, the simplest (and also the most widely used) is the **linear equation** in two variables,

$$y = a + bx$$

where x and y are the two variables and a and b are numerical constants. The name "linear" is accounted for by the fact that, when plotted on ordinary graph paper, all points whose x- and y-coordinates satisfy an equation of the form $y = a + bx$ will fall on a straight line.

Example 1. To illustrate, let us consider the equation

$$y = 0.23 + 4.42x$$

where x is a county's annual rainfall (in inches, measured from September through August) and y is its average annual yield of wheat (in bushels per acre). The graph of this equation is shown in Figure 10.1, and any pair of values of x and y which are such that $y = 0.23 + 4.42x$ form a point (x, y) that falls on the line. In most practical

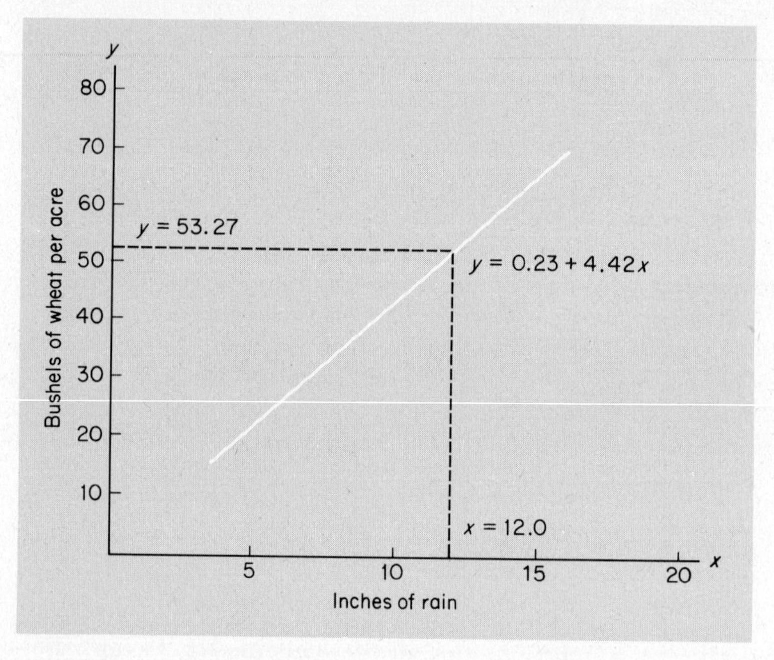

Figure 10.1 Graph of linear equation.

applications we use lines like this to make *predictions*, and in our example we might predict, say, that if there is an annual rainfall of 12.0 inches, the average annual yield of wheat can be expected to equal $y = 0.23 + 4.42(12.0) = 53.27$ bushels per acre.

Whenever we use paired measurements or observations to arrive at an equation which represents, or expresses, the relationship between the corresponding two variables, we face *three basic problems*. *First* we must decide what kind of equation (or curve) is to be used, and this question is usually answered by plotting the **data points** (x, y), which correspond to the given pairs of measurements or observations, on a piece of graph paper—hopefully, this will enable us to discern a definite pattern.* So far

*There exist methods for putting this kind of decision on a more *objective* basis, but they are fairly advanced and will not be treated in this text.

as the work of this chapter is concerned, this question is really academic, for linear equations are the only ones we shall discuss. This is not as confining as it may seem. *Linear equations are useful not only because many of the relationships which we meet in actual practice are of this form, but also because they often provide excellent approximations to relationships which would otherwise be difficult to describe in mathematical terms.*

Once we have decided to use a straight line, we are faced with the *second* kind of problem, namely, that of finding the equation of the particular line which in some sense provides the *best possible fit* to the paired data with which we are concerned.

Example 2. To illustrate, let us refer to the following data obtained in a study of the relationship between the number of years that applicants to certain foreign service jobs studied German in high school or college and the grades which they received in a proficiency test in that language:

Number of Years x	*Grade in Test* y
3	57
4	78
4	72
2	58
5	89
3	63
4	73
5	84
3	75
2	48

If we plot the points which correspond to these ten pairs of x's and y's as in Figure 10.2, it is apparent that although the points do not actually fall on a straight line, the overall pattern of the relationship is pretty well described by the dashed line. Consequently, it would seem *reasonable*, or *justifiable*, to represent the relationship between the number of years an applicant has studied German and his or her grade in the proficiency test by means of a linear equation.

Logically speaking, there is no end to the number of different lines which we could draw on the diagram of Figure 10.2. Some of these obviously would not fit the given data and they can immediately be ruled out, but we can draw many different lines which come fairly close to the 10 points and, hence, must be taken into account. In order to single out *one* line as *the line* which "best" fits the given set of paired data, we shall have to state explicitly what we mean here by "best"—in other words, we shall have to

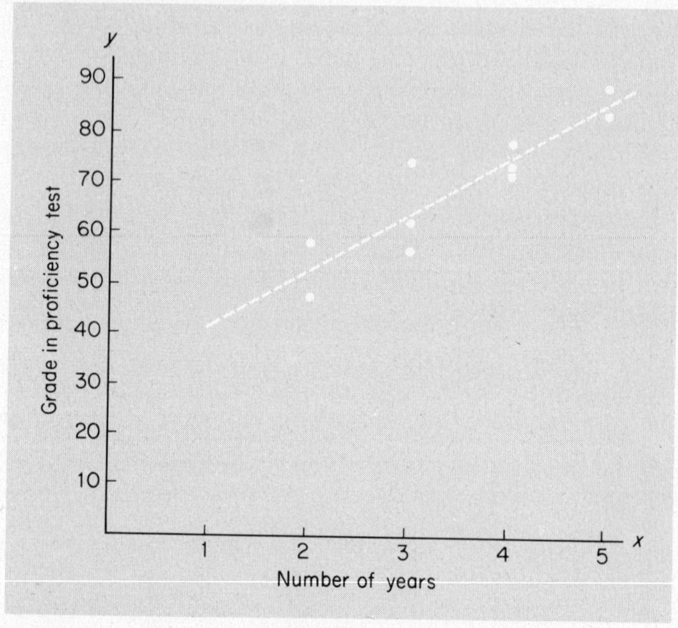

Figure 10.2 Data points of Example 2.

give a criterion, a "yardstick," which will enable us to choose the line which we thus specify as "best." (If all of the points fell on a straight line, *this would be it*, but this kind of situation is seldom, if ever, met in actual practice.)

The criterion which is most widely used for fitting straight lines to paired data is based on the **method of least squares**, which we shall study in the next section; it dates back to the early part of the nineteenth century and it has found many different important applications in various branches of statistics. Before we go into this, however, let us point out that after we have decided to fit a straight line and after we have decided to use the method of least squares, we still face a *third* kind of question. We must ask ourselves: How well does this line actually fit? After all, if the data points are scattered all over as in Figure 10.3, even the best-fitting line cannot provide a good fit. This third kind of question about the analysis of paired data will be discussed later in the section beginning on page 303.

The method of least squares

If the reader has ever had the occasion to work with paired data which were plotted as points on a piece of graph paper, he has probably felt the urge to take a ruler, move it around, and decide *by eye* on a line which presents a fairly good fit. There is no rule anywhere which says that this

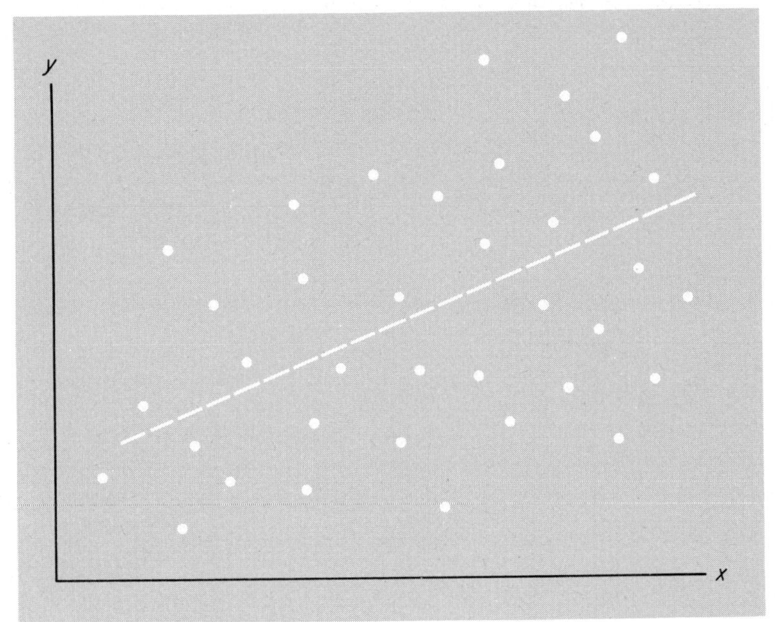

Figure 10.3 Poor fit.

cannot be done, but it would certainly not be very "scientific." More objective is the **least-squares criterion,** which requires that

> **The line which we fit to a set of data points must be such that the sum of the squares of the vertical deviations (distances) from the points to the line is as small as possible.**

With reference to the illustration of the preceding section, the least-squares criterion requires that the sum of the squares of the distances represented by the heavily drawn line segments of Figure 10.4 be a minimum.

Example 2 (Continued). To explain why this is done, let us refer to the data on page 293, and take one of the applicants to the foreign service jobs, say, the one who studied German for two years and obtained a grade of 48 in the test. If we mark $x = 2$ on the horizontal scale and read the corresponding value of y off the line of Figure 10.4, we find that the corresponding grade is about 53; thus, the *error* of this "prediction," *represented by the vertical distance from the point to the line,* is $48 - 53 = -5$. Altogether, there are ten such errors in our example, and the least-squares criterion requires that we minimize the sum of their squares.

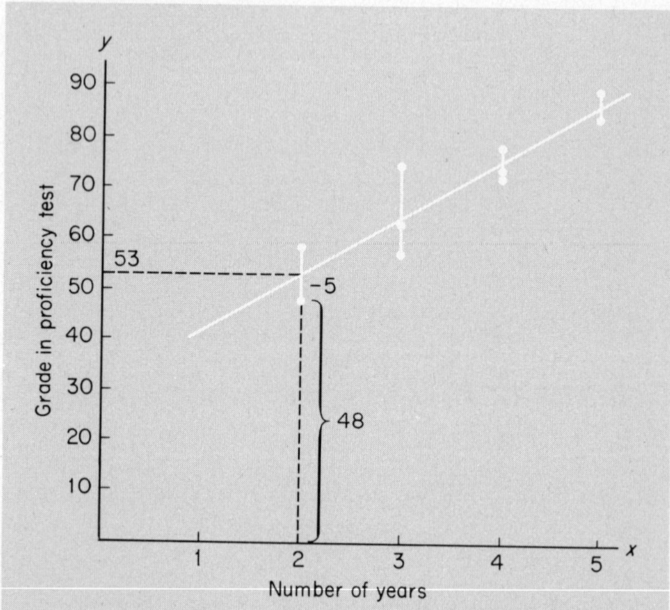

Figure 10.4 Error of prediction based on line fit to data of Example 2.

To explain why we minimize the sum of the *squares* of the deviations and not just the sum of the deviations themselves, let us point out that the deviations (namely, the differences between the observed values of y and the corresponding values on the line) are *positive* for points which fall above the line and *negative* for points which fall below the line, so that their sum can be very small (in fact, even zero), even though the points are widely scattered, as in Figure 10.3. By working with the squares of the deviations, which *cannot be negative*, we are making sure that we are paying attention to the *magnitude* of the deviations and not to their signs.

To demonstrate how a **least-squares line** is actually fit to a set of paired data, let us consider n pairs of numbers (x_1, y_1), (x_2, y_2), . . . , and (x_n, y_n), which might represent the heights and weights of n persons, the automobile registrations and annual gasoline sales in n different states, the I.Q.'s of n fathers and sons, the incomes and medical expenditures of n families, and so on. If we write the equation of the line which we want to fit to the corresponding n points as

$$y = a + bx$$

we shall have to work with *two* values of y for each given value of x—*the corresponding value of y which was actually observed and the value of y which was obtained by substituting the given value of x into the equation of the line.*

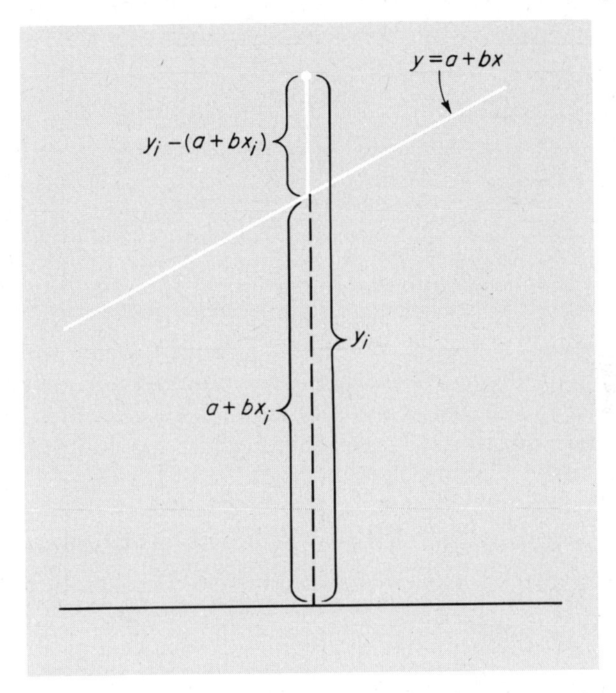

Figure 10.5 Vertical deviation from the point (x_i, y_i) to the line, where i can be 1, 2, 3, . . ., or n.

Thus, for x_1 the observed value of y is y_1 and the value obtained by substituting $x = x_1$ into the equation of the line is $a + bx_1$; correspondingly, for x_2 the two values are y_2 and $a + bx_2$, . . . , and for x_n the two values are x_n and $a + bx_n$. As can be seen from Figure 10.5, the corresponding differences (namely, the vertical deviations from the points representing the observed data to the line) are $y_1 - (a + bx_1)$, $y_2 - (a + bx_2)$, . . . , and $y_n - (a + bx_n)$, and the least-squares criterion requires that we minimize

$$\sum [y - (a + bx)]^2$$

namely, the sum of their squares.*

As it takes calculus (or a fairly tedious algebraic process called "completing the square") to find the values of a and b which will minimize this sum, let us merely state the result that these values are given by the

*In the notation of Exercise 20 on page 41, this sum of squares can be written more explicitly as

$$\sum_{i=1}^{n} [y_i - (a + bx_i)]^2$$

solution of the following system of linear equations in the two unknowns a and b:

$$\sum y = n \cdot a + b(\sum x)$$

$$\sum xy = a(\sum x) + b(\sum x^2)$$

Here n is the number of pairs of observations, $\sum x$ and $\sum y$ are, respectively, the sum of the given x's and the sum of the given y's, $\sum x^2$ is the sum of the *squares* of the given x's, and $\sum xy$ is the sum of the *products* obtained by multiplying each of the given x's by the corresponding observed value of y. It is customary to refer to the above system of equations as the **normal equations**, whose solution gives the desired least-squares values of a and b.

Example 2 (Continued). Returning to our numerical example and copying the first two columns from page 293, we obtain the sums needed for substitution into the normal equations by means of the calculations shown in the following table:

Number of Years x	*Test Grade* y	x^2	xy
3	57	9	171
4	78	16	312
4	72	16	288
2	58	4	116
5	89	25	445
3	63	9	189
4	73	16	292
5	84	25	420
3	75	9	225
2	48	4	96
35	697	133	2,554

(If this work is done with a desk calculator or more elaborate equipment, the sum of the x^2's and the sum of the xy's can be accumulated directly, and there is no need to fill in all the details.) Then, if we substitute $n = 10$ and the appropriate column totals into the two normal equations, we get

$$697 = 10a + 35b$$

$$2{,}554 = 35a + 133b$$

and all that remains to be done is to solve this system of equations for a and b. There are several ways in which this can be done—simplest, perhaps, is the **method of elimination**, which the reader may recall from elementary algebra. Using this method, let us eliminate a by dividing the expressions on both sides of the first equation by 10, dividing the expressions on both sides of the second equation by 35, and then subtracting "equals from equals." We thus get

$$69.70 = a + 3.5b$$

$$72.97 = a + 3.8b$$

and by subtraction

$$3.27 = 0.3b$$

Hence, $b = \frac{3.27}{0.3} = 10.9$, and if we substitute this value of b into the first of the original equations, we get $697 = 10a + 35(10.9)$, $697 = 10a + 381.5$, $10a = 315.5$, and finally $a = 31.55$ (or 31.6 rounded to one decimal). Thus, the equation of the least-squares line is

$$y = 31.6 + 10.9x$$

If the reader is not familiar with the method of elimination, he can instead use the following formulas for a and b, which were actually obtained by *symbolically* solving the two normal equations on page 298 by the method of elimination (see Exercise 6 on page 303):

$$a = \frac{(\sum y)(\sum x^2) - (\sum x)(\sum xy)}{n(\sum x^2) - (\sum x)^2}$$

$$b = \frac{n(\sum xy) - (\sum x)(\sum y)}{n(\sum x^2) - (\sum x)^2}$$

Example 2 (Continued). Had we used these formulas for a and b in our numerical example, we would have obtained

$$a = \frac{(697)(133) - (35)(2{,}554)}{10(133) - (35)^2} = \frac{3{,}311}{105} = 31.5$$

and

$$b = \frac{10(2{,}554) - (35)(697)}{105} = \frac{1{,}145}{105} = 10.9$$

where we did not have to calculate the denominator in the formula for b because it is the same as that in the formula for a. Actually, the results should have been the same as before, and the difference between $a = 31.5$ and $a = 31.6$ (obtained earlier by the method of elimination) is entirely due to rounding.

Now that we have determined the equation of the least-squares line for the data on page 293, suppose that we want to predict *more accurately than on page* 295 the grade of an applicant who has studied German in high school or college for two years. Substituting $x = 2$ into the equation, we get

$$y = 31.6 + 10.9(2) = 53.4$$

Rounded to the nearest whole number the prediction is 53, and this is the same as the value we read off the line of Figure 10.4.

When we make a prediction like this, we cannot really expect to hit the answer right on the nose—in fact, we cannot possibly be right when the answer has to be a whole number and our prediction is 53.4. Even if we predict a grade of 53, however, and not 51 or 54, *we should be delighted if it turned out that close.* After all, it would be very unreasonable in our example to expect that every applicant who has studied German for the same number of years will get the identical grade in the test; indeed, the data on page 293 show that this is not the case. Thus, *to make meaningful predictions based on least-squares lines, they will have to be interpreted as averages, namely, as mathematical expectations.* It is in this sense that we said in Example 1 on page 292 that "the yield of wheat can be *expected* to equal 53.27 bushels per acre."

Interpreted in this way as a line which enables us to read off (or calculate) the expected value of y for any given value of x, a least-squares line is also referred to as a **regression line** or, better, as an **estimated regression line**. This term is due to Sir Francis Galton, the nineteenth-century English scientist who first employed it in a study of the relationship between the heights of fathers and sons, in which he observed a "regression," or turning back from the heights of sons to the heights of their fathers.

We refer to the line $y = 31.6 + 10.9x$ of Example 2 only as an *estimated* regression line because it is based on sample data, and it must be evident that if we repeated the whole analysis with data pertaining to ten different applicants to the foreign service jobs, we would probably get different values for a and b. Thus, if the *true* relationship is given by the equation $y = \alpha + \beta x$, we must look upon the value we get for a as an estimate of α (*alpha*) and we must look upon the value we get for b as an estimate of β (*beta*), and this raises a virtual *Pandora's box* of questions: "*How good* are these estimates?" "*How good* are predictions based on the equations of estimated regression lines?" "*How good* are corresponding estimates of the *true* average value of y for a given value of x?" All these questions are answered in a fairly difficult branch of statistics called **regression analysis**; so far as the work of this book is concerned, a partial answer to the first of these questions will be given in Exercise 11 on page 314.

Exercises

1. The following data show the improvement (gain in reading speed) of eight students in a speed-reading program, and the number of weeks they have been in the program:

Number of Weeks	*Speed Gain (words per minute)*
3	85
5	117
2	48
8	192
6	163
9	231
3	72
4	108

a. Use the formulas on page 299 to calculate a and b for the least-squares line which will enable us to predict the gain in reading speed of a person who has been in the program for a given number of weeks.
b. Repeat part (a) by solving the two normal equations directly, and compare the results.
c. Use the equation of the least-squares line to predict the gain in reading speed which a person can expect if he has been in the program for six weeks.

2. The following data pertain to the demand for a product (in thousands of units) and its price (in cents) charged in five different market areas:

Price	*Demand*
9	177
12	109
10	135
15	81
6	218

a. Use the formulas on page 299 to calculate a and b for the least-squares line which will enable us to determine the expected demand which corresponds to any given price.
b. Repeat part (a) by solving the two normal equations directly, and compare the results.
c. Use the equation of the least-squares line to predict the demand for the product when it is priced at 14 cents.

3. The following are data on the amount of potassium bromide which a chemistry student was able to dissolve in 100 grams of water at various temperatures:

Temperature (degrees centigrade)	*Amount Dissolved (grams)*
10	61
20	65
30	72
40	77
50	85
60	90

a. Find the equation of the least-squares line which will tell the student how much potassium bromide he can expect to dissolve in 100 grams of water at any given temperature.
b. Draw a diagram showing the least-squares line together with the original data.
c. Use the line of part (b) to read off how much potassium bromide the student can expect to dissolve in 100 grams of water at 52 degrees centigrade, and check the result by substituting $x = 52$ into the equation obtained in part (a).

4. Verify that the equation of Example 1 on page 291 can be obtained by fitting a least-squares line to the following data:

Rainfall (inches)	*Yield of Wheat (bushels per acre)*
12.9	62.5
7.2	28.7
11.3	52.2
18.6	80.6
8.8	41.6
10.3	44.5
15.9	71.3
13.1	54.4

5. LINEAR TRENDS When business data are recorded at regularly spaced intervals of time (say, every month or every year) we refer to them as a **time series**, and when a straight line is fit to such data we refer to it as a **trend line** or as a **linear trend**. When dealing with *annual data* it is convenient to "code" the x's by assigning the years the values, . . . , -3, -2, -1, 0, 1, 2, 3, . . . , when n is *odd*, or the values . . . , -5, -3, -1, 1,3, 5, . . . , when n is *even*, making sure that this will make $\sum x = 0$. The formulas for a and b will then become

$$a = \frac{\sum y}{n} \qquad \text{and} \qquad b = \frac{\sum xy}{\sum x^2}$$

Of course, the resulting equation of the trend line will express y in terms of the coded x's, and we have to account for this when using the equation to make a prediction. For instance, if we are given economic data for the years 1970, 1971, 1972, 1973, and 1974, and these years are coded as $x = -2, -1, 0, 1$, and 2, we must substitute $x = 6$ to obtain a prediction for the year 1978.

a. For the years 1969 through 1975, a company's profit (in millions of dollars) was 4.5, 5.3, 6.1, 8.0, 8.9, 10.1, and 10.7. Fit a least-squares line after suitably coding the seven years, and use it to predict the company's profit in 1977.

b. For the years 1964 through 1971, income from tourism in Arizona (in millions of dollars) totaled 400, 420, 450, 480, 500, 530, 565, and 600. Fit a least-squares line after suitably coding the eight years, and (assuming that the trend will continue, use it to predict Arizona's total income from tourism in 1978.

c. In the years 1968 through 1972, Great Britain's average monthly exports (in millions of pounds) totaled 658, 693, 753, 818, and 930. Fit a least-squares line after suitably coding the years, and (assuming that the trend will continue to follow the same pattern) use it to predict Great Britain's average monthly exports in 1980.

6. Multiply the expressions on both sides of the first of the two *normal* equations on page 298 by $\sum x$, multiply those of the second of the two normal equations by n, eliminate a by subtraction, and then solve for b. Then, substitute the expression obtained for b into the first of the two normal equations, and solve for a.

The coefficient of correlation

Now that we have learned how to fit a least-squares line to paired data, let us see how we might answer the third question raised on page 294:

How well does the line actually fit the given data? Of course, we can get a fair idea by inspection, say, by looking at a diagram like that of Figure 10.2, but to be more objective let us analyze the *total variation* of the observed y's (in Example 2, the variation among the grades in the proficiency test) and look for possible causes, or explanations. As can be seen from the second column of the table on page 293, there are considerable differences among the y's, with the smallest being 48 and the largest being 89. However, it can also be seen from this table or from Figure 10.2 that these differences must be due partly to the fact that all the applicants to the foreign service jobs did not study German for the same number of years. For instance, the grade of 48 was obtained by an applicant who had studied German for two years, while the grade of 89 was obtained by an applicant who had studied German for five years. This raises the following question:

> **How much of the total variation of the observed y's can be attributed to the relationship between the two variables x and y (namely, the extent to which y depends on x), and how much of it can be attributed to all other factors, including chance?**

With reference to our example, we would thus want to know what part of the differences among the grades can be accounted for by the differences in the number of years that the applicants had studied German, and what part can be attributed to all other factors (including the applicants' intelligence, their health on the day they took the test, . . . , and their being familiar *by chance* with the material covered in the test).

A convenient measure of the *total variation* of the observed y's is the *sum of squares* $\sum (y - \bar{y})^2$, which is simply their *variance* (see page 194) multiplied by $n - 1$.

Example 2 (Continued). Since $\sum y = 697$ according to the table on page 298, we find that $\bar{y} = \frac{697}{10} = 69.7$, and, hence, that for the given data

$$\begin{aligned}\sum (y - \bar{y})^2 &= (57 - 69.7)^2 + (78 - 69.7)^2 + (72 - 69.7)^2 \\ &\quad + (58 - 69.7)^2 + (89 - 69.7)^2 + (63 - 69.7)^2 \\ &\quad + (73 - 69.7)^2 + (84 - 69.7)^2 + (75 - 69.7)^2 \\ &\quad + (48 - 69.7)^2 \\ &= 1{,}504.10\end{aligned}$$

This takes care of the *total variation* of the y's, but it remains to be seen how we might measure its two parts attributed, respectively, to *the relationship between the two variables x and y* and to *all other factors.*

To this end, let us point out that if the number of years the applicants studied German had been the *only* thing that affected their grades in the test, the data points would actually have fallen on a straight line (assuming, of course, that the relationship *is* linear). However, that fact that they did *not* fall on a straight line, as is apparent from Figures 10.2 and 10.4, indicates that there are other factors at play. The extent to which the points fluctuate above and below the line, namely, *the variation caused by these other factors*, is usually measured by the sum of the squared deviations from the points to the line (see Figure 10.4 on page 296); that is, by the quantity

$$\sum [y - (a + bx)]^2$$

which we minimized by imposing the least-squares criterion.

Example 2 (Continued). To calculate this sum of squares for our numerical example, we shall first have to determine $a + bx = 31.6 + 10.9x$ (namely, the y-coordinate of the corresponding point on the line) for *each* of the given values of x, and we get $31.6 + 10.9(3) = 64.3$, $31.6 + 10.9(4) = 75.2$, $31.6 + 10.9(4) = 75.2$, $31.6 + 10.9(2) = 53.4$, $31.6 + 10.9(5) = 86.1$, $31.6 + 10.9(3) = 64.3$, $31.6 + 10.9(4) = 75.2$, $31.6 + 10.9(5) = 86.1$, $31.6 + 10.9(3) = 64.3$, and $31.6 + 10.9(2) = 53.4$. Hence,

$$\begin{aligned}\sum [y - (a + bx)]^2 &= (57 - 64.3)^2 + (78 - 75.2)^2 + (72 - 75.2)^2 \\ &\quad + (58 - 53.4)^2 + (89 - 86.1)^2 + (63 - 64.3)^2 \\ &\quad + (73 - 75.2)^2 + (84 - 86.1)^2 + (75 - 64.3)^2 \\ &\quad + (48 - 53.4)^2 \\ &= 255.53\end{aligned}$$

and we can claim that

$$\frac{\sum [y - (a + bx)]^2}{\sum (y - \bar{y})^2} = \frac{255.53}{1{,}504.10} = 0.17$$

is the proportion of the total variation of the grades which can be attributed to *all other factors*, while

$$1 - \frac{\sum [y - (a + bx)]^2}{\sum (y - \bar{y})^2} = 1 - 0.17 = 0.83$$

is the proportion of the total variation of the grades which can be attributed to *differences in the number of years which the applicants had studied German.*

If we take the square root of this last proportion (namely, the proportion of the total variation of the y's that is accounted for by the relationship with x), we obtain the statistical measure, which is called the **coefficient of correlation**. It is generally denoted by the letter r, and its *sign* is always chosen so that it is the same as that of b in the equation of the least-squares line.

Example 2 (Continued). Thus, for our example we get

$$r = \sqrt{0.83} = 0.91$$

rounded to two decimals.

To understand (that is, interpret) a value of r calculated on the basis of a set of paired data, note that

> **If the dependence between x and y is strong, most of the variation of the y's can be attributed to the relationship with x, and r will be close to 1 or $-$ 1.**

Indeed, $r = 1$ or $r = -1$ is indicative of a *perfect fit*; namely, it is indicative of the fact that the observed points actually fall on a straight line. On the other hand,

> **If the dependence between x and y is weak, very little of the variation of the y's can be attributed to the relationship with x, and r will be close to 0.**

When $r = 0$, we say that there is *no correlation*, which means that none of the variation of the y's can be attributed to the relationship with x. (Of course, r cannot take on a value greater than 1 or less than -1, as it was defined as *plus or minus the square root of a proportion*, namely, as *plus or minus the square root of a number from 0 to 1*.)

It should also be observed that r is *positive* when the least-squares line has an *upward slope*, that is, when the relationship between x and y is such that *small* values of y tend to go with *small* values of x and *large* values of y tend to go with *large* values of x. Correspondingly, r is *negative* when the least-squares line has a *downward slope*, that is, when the relationship between x and y is such that *small* values of y tend to go with *large* values of x and *large* values of y tend to go with *small* values of x. Depending on the sign of r, we thus say that there is a **positive correlation** or a **negative correlation**, as is illustrated in Figure 10.6.

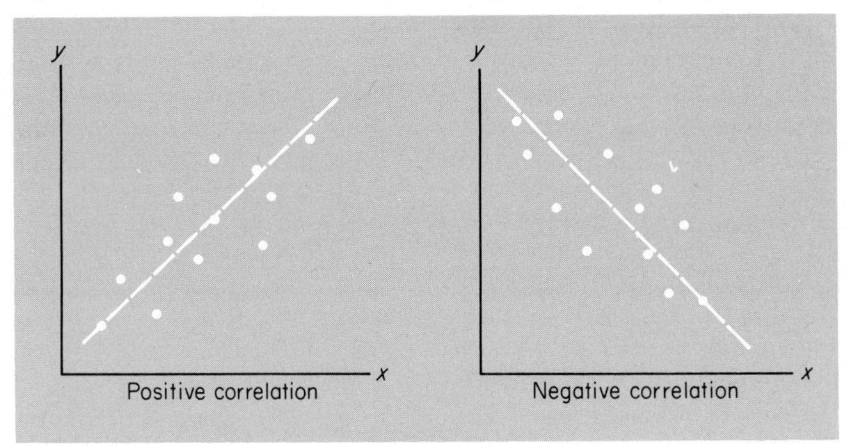

Figure 10.6 Positive and negative correlation.

The correlation coefficient is a very widely used measure of the correlation (relationship, association, or dependence) between two variables. In actual practice, we seldom, if ever, determine r as we did in our example; instead we use the *shortcut computing formula*

$$r = \frac{n(\sum xy) - (\sum x)(\sum y)}{\sqrt{n(\sum x^2) - (\sum x)^2}\sqrt{n(\sum y^2) - (\sum y)^2}}$$

which has the added advantage that it automatically gives r the correct sign. Although this shortcut formula may look rather forbidding, it is actually quite easy to use. All we have to do, really, is substitute the values of n, $\sum x$, $\sum y$, $\sum x^2$, $\sum y^2$, and $\sum xy$ for the paired data with which we happen to be concerned, and it should be observed that, except for $\sum y^2$, these are the same quantities which we needed to fit a straight line by the method of least squares.

Example 2 (Continued). If we square the y's in the table on page 298 we get $\sum y^2 = 50{,}085$, and if we substitute this quantity together with $n = 10$ and the column totals of the table on page 298 into the above formula for r, we get

$$r = \frac{10(2{,}554) - (35)(697)}{\sqrt{10(133) - (35)^2}\sqrt{10(50{,}085) - (697)^2}} = 0.91$$

This agrees, as it should, with the result obtained on page 306.

As a word of caution let us add that the coefficient of correlation is not only a very widely used statistical measure, but that it is also very widely *abused*. For instance, it is often abused in the sense that we forget that our whole discussion has been about *linear relationships* and, hence, *linear correlation*. In other words, r should be calculated only when it is reasonable to assume that the relationship between the two variables under consideration can be described fairly well by means of a straight line. The correlation coefficient is also abused in the sense that values of r close to -1 or 1 are (regrettably often) interpreted as *cause–effect* relationships. For instance, it is easy to find data on the annual consumption of liquor in the United States and teachers' salaries in the same years which will yield a very high value of r (namely, a value of r close to 1). Obviously, though, this is not indicative of a cause–effect relationship; it results from the fact that both variables are *effects of a common cause*—the overall standard of living. (If it is high, greater amounts of money are available for teachers' salaries as well as liquor than when it is low.) Another classical example is the strong positive correlation which has been obtained for the number of storks seen nesting in English villages and the number of childbirths recorded in the same communities. (*In the large villages there are lots of houses with chimneys for storks to nest in, and hence lots of families, storks, and babies; in the smaller villages there are fewer houses with chimneys for storks to nest in, and hence fewer families, fewer storks, and fewer babies.*)

Finally, it is sometimes overlooked that when r is calculated on the basis of sample data, we may get a fairly strong (positive or negative) correlation *purely by chance*, even though there is no relationship whatsoever between the two variables under consideration.

Example 3. To illustrate this point, suppose that we take a pair of dice, one red and one green, roll them together five times, and get the following results:

Red Die x	*Green Die* y
5	4
6	4
1	2
4	6
2	2

If we calculate r for these data, we get the surprisingly high value $r = 0.66$, and this raises the obvious question whether anything is wrong with the idea that there should be no relationship—*after all, one die does not "know" what the other die is doing*. To answer this question, we shall have to see whether this relatively high value of r can be attributed to chance.

To test the null hypothesis of *no correlation* (namely, the hypothesis that there is no linear relationship between two given variables), we shall refer to a special table, Table VI at the end of the book, which is based on the assumption that the x's and y's are values of *independent* (and, hence, *uncorrelated*) random variables having at least approximately normal distributions. With reference to this table

We say the value of r which we get for a set of paired data is significant at the level of significance $\alpha = 0.05$ if it exceeds $r_{0.025}$ or is less than $-r_{0.025}$.

If the value we get for r falls between $-r_{0.025}$ and $r_{0.025}$, we say that there is *no significant correlation*, or that the value of r is *not statistically significant*. Incidentally, Table VI also contains values of $r_{0.005}$, which we would use instead of $r_{0.025}$ when the level of significance is to be $\alpha = 0.01$.

Example 3 (Continued). If we apply this test to the last example, where we rolled the two dice, we find that the result, namely, $r = 0.66$, is *not significant*; it is less than $r_{0.025} = 0.878$, the tabular value for $n = 5$. Thus, we conclude that the apparent relationship between the number of points rolled with the two dice can be attributed to chance.

Example 2 (Continued). If we apply the test to the result which we obtained for the number of years which the applicants for the foreign service jobs had studied German and their grades on the proficiency test, namely, $r - 0.91$, we find that this value of r *is significant* at the level of significance $\alpha = 0.01$ as it *exceeds* $r_{0.005} = 0.765$ for $n = 10$.

Exercises

1. State in each case whether you would expect to obtain a positive correlation, a negative correlation, or no correlation:

a. the ages of husbands and wives;

b. the amount of rubber on a tire and the number of miles it has been driven;

c. a man's shirt size and his sense of humor;

d. income and education;

e. the number of hours which a *distance runner practices* and the number of minutes in which he can run the mile;

f. shoe size and a person's knowledge of foreign affairs;

g. temperature and heating cost;

h. number of persons getting flu shots and number of persons catching the flu;

i. the demand for a product and its total sales;

j. pollen count and the sale of antiallergy drugs.

2. Calculate r for the data of Exercise 2 on page 301 and test its significance at the level of significance $\alpha = 0.05$.

3. Calculate r for the data of Exercise 1 on page 301 and test its significance at the level of significance $\alpha = 0.01$.

4. Calculate r for the data of Exercise 4 on page 302 and test its significance at the level of significance $\alpha = 0.05$.

5. The following data pertain to a study of the effects of environmental pollution on wildlife; in particular, the effect of DDT on the thickness of the eggshells of certain birds:

DDT Residues in Yolk Lipids (parts per million)	*Thickness of Eggshells (millimeters)*
117	0.49
65	0.52
393	0.37
98	0.53
122	0.49

Calculate r for these data and test its significance at the level of significance $\alpha = 0.05$.

6. The following are the grades which 16 students received in final examinations in economics and anthropology:

Economics	*Anthropology*	*Economics*	*Anthropology*
51	74	45	68
68	70	73	87
72	88	93	89
97	93	66	73
55	67	20	33
73	73	91	91
95	99	74	80
74	73	80	86

Calculate r for these data and test its significance at the level of significance $\alpha = 0.01$.

7. If we actually calculated r for each of the following sets of data, should we be surprised to get $r = 1$ and $r = -1$, respectively? Explain.

(*a*)

x	y
19	7
25	11

(*b*)

x	y
13	5
20	2

8. Since r does not depend on the scales of measurement used for x and y, its calculation can often be simplified by adding a suitable positive or negative number to each x, to each y, or to both, or by multiplying each x, each y, or both by a suitable positive constant.

a. Recalculate r for the data of Example 2 on page 293 after having subtracted 3 from each x and 70 from each y.

b. Rework Exercise 5 after having multiplied each of the eggshell thicknesses by 100 and then subtracted 48.

9. RANK CORRELATION Since the calculation of r for large sets of paired data can be fairly tedious, it is sometimes convenient to base r on the ranks of the observations instead of their actual numerical values. Thus, we first rank the x's among themselves, giving Rank 1 to the largest value, Rank 2 to the second largest value, and so forth; then we similarly rank the y's among themselves and calculate the **coefficient of rank correlation** by means of the formula

$$r' = 1 - \frac{6(\sum d^2)}{n(n^2 - 1)}$$

Here n is the number of pairs of observations and the d's are the *differences* between the respective ranks. If there are **ties**, we assign to each of the tied observations the *mean* of the ranks which they jointly occupy; thus, if the third and fourth largest values are identical we assign each the rank $\frac{3 + 4}{2} = 3.5$, and if the fifth, sixth, and seventh largest values are identical we assign each the rank $\frac{5 + 6 + 7}{3} = 6$. To illustrate this technique, let us consider the data in the first two columns of the following table, which pertain to the number of hours which ten students studied for an examination and the grades which they received:

Number of Hours Studied x	*Grade in Examination* y	*Rank of* x	*Rank of* y	d	d^2
8	56	6.5	7	−0.5	0.25
5	44	8.5	9	−0.5	0.25
11	79	4	3	1.0	1.00
13	72	3	4	−1.0	1.00
10	70	5	5	0.0	0.00
5	54	8.5	8	0.5	0.25
18	94	1	1	0.0	0.00
15	85	2	2	0.0	0.00
2	33	10	10	0.0	0.00
8	65	6.5	6	0.5	0.25
					3.00

If we substitute $n = 10$ and $\sum d^2 = 3$ into the formula for r', we get

$$r' = 1 - \frac{6 \cdot 3}{10(10^2 - 1)} = 0.98$$

and it should be noted that if we had calculated r for these data we would have obtained 0.96; thus, the difference is very small.

a. Calculate r' for the data of Exercise 1 on page 301 and compare it with the value of r obtained in Exercise 3 above.

b. Calculate r' for the data of Exercise 4 on page 302 and compare it with the value of r obtained in Exercise 4 above.

c. Calculate r' for the data of Exercise 6 and compare it with the value of r obtained in that exercise.

d. Calculate r' for the data of Example 2 on page 293.

e. Calculate r' for the following attendance figures of the 12 baseball teams in the National League:

	1969	*1973*
Atlanta	1,458,320	800,678
Chicago	1,674,993	1,351,971
Cincinnati	987,991	2,017,601
Houston	1,442,995	1,393,563
Los Angeles	1,784,527	2,136,210
Montreal	1,212,608	1,246,843
New York	2,175,373	1,914,365
Philadelphia	519,414	1,476,733
Pittsburgh	769,368	1,320,915
St. Louis	1,682,783	1,574,276
San Diego	512,970	611,827
San Francisco	873,603	834,193

f. The following table shows how a panel of nutrition experts and a panel of housewives ranked fifteen breakfast foods:

Breakfast Food	*Nutrition Experts*	*Housewives*
I	3	5
II	7	4
III	11	8
IV	9	14
V	1	2
VI	4	6
VII	10	12
VIII	8	7
IX	5	1
X	13	15
XI	12	9
XII	2	3
XIII	15	10
XIV	6	11
XV	14	13

Calculate r' as a measure of the consistency of the two rankings.

10. RANK CORRELATION, Continued If there is no relationship between the two variables under consideration (namely, if the x's and the y's are *randomly matched*), it can be shown that the statistic

$$z = r'\sqrt{n - 1}$$

has approximately the standard normal distribution. Thus, we can test the null hypothesis of no correlation on the basis of the criteria of Figure 8.6 on page 230, and for the illustration of the preceding exercise we get

$$z = 0.98\sqrt{10 - 1} = 2.94$$

Since this exceeds 2.58, the right-hand dividing line of the $\alpha = 0.01$ criterion of Figure 8.6, we conclude (at this level of significance) that the relationship between the number of hours which the students studied for the examination and the grades which they received *is significant.*

a. Check whether the value of r' obtained in part (a) of Exercise 9 is significant at the level of significance $\alpha = 0.05$.

b. Check whether the value of r' obtained in part (b) of Exercise 9 is significant at the level of significance $\alpha = 0.05$.

c. Check whether the value of r' obtained in part (c) of Exercise 9 is significant at the level of significance $\alpha = 0.01$.

11. **CONFIDENCE LIMITS FOR β** As we indicated on page 300, the value we obtain for b by the method of least squares is only an *estimate* of β, the slope of the *true* regression line $y = \alpha + \beta x$. Using r, we can now write 0.95 confidence limits for β as

$$b\left[1 \pm t_{0.025} \cdot \frac{\sqrt{1 - r^2}}{r\sqrt{n - 2}}\right]$$

where the number of degrees of freedom for $t_{0.025}$ is $n - 2$. For instance, for Example 2 we had $n = 10$, $b = 10.9$ (from page 299), $r = 0.91$ (from page 307), and since $t_{0.025} = 2.306$ for $10 - 2 = 8$ degrees of freedom, we get

$$10.9\left[1 \pm 2.306 \cdot \frac{\sqrt{1 - (0.91)^2}}{0.91\sqrt{10 - 2}}\right]$$

which reduces to $10.9(1 \pm 0.37)$, namely, 6.9 and 14.9. Thus, we can write the 0.95 confidence interval for the *slope* of the regression line as

$$6.9 < \beta < 14.9$$

a. Use this method to construct a 0.95 confidence interval for β on the basis of the data of Exercise 1 on page 301, for which r was obtained in Exercise 3 above.

b. Use this method to construct a 0.95 confidence interval for β on the basis of the data of Exercise 4 on page 302, for which r was obtained in Exercise 4 above.

11

Statistical thinking: conclusions

Introduction

As the title of this book suggests, it was written for a first course in statistics, and it does not qualify the reader to call himself a professional statistician even if he has faithfully worked each exercise and studied each example. Indeed, the objectives of this book are fairly modest—it is meant to introduce the reader to many of the basic ideas of modern statistics, and to present him with some of the *standard methods* which we referred to on page 8 as "bread and butter techniques." Unfortunately, however, these *standard methods* apply only to *standard situations*, and these are about as easy to find as a truly "average" housewife or a really "normal" person. To emphasize this point, let us remind the reader of the many "ifs" that preceded each new method which we discussed—"*if* the sample is random," "*if* there are no biases," "*if* the experiment is properly designed," "*if* the trials are independent," "*if* the population from which we are sampling has the shape of a normal distribution," "*if* the probability is the same for each trial," "*if* the samples are independent," "*if* it is reasonable to substitute s for σ," "*if* all the populations have the same standard deviation," "*if* the sample size is at least 30," "*if* none of the expected frequencies is less than 5," and so on. Thus, if the reader decides to *continue* with the study of statistics, he will find that most of his time will be devoted to the question of *what to do about these "ifs."* In the remainder of this chapter we shall mention some of the special topics which provide answers to this question; books dealing with these subjects are listed in the Bibliography immediately following this chapter.

Sampling

All the methods which we discussed in Chapters 7 through 10 assumed that we are dealing with random samples. We did not even consider the possibility that under certain conditions there might be samples which are *better* (say, *cheaper* or *more reliable*) than random samples, and we did not go into any details about the question of what might be done when random sampling is impossible. Thus, let us take a brief look at the following questions:

Can we improve on simple random sampling?
What can we do when it is physically impractical or otherwise impossible to use random sampling?
What can we do when we have little or no choice about the selection of the data?
How about the cost of sampling?

The first of these questions can be answered in the *affirmative* provided we have some special knowledge about the "makeup," or composition, of the population with which we are concerned.

Example 1. Suppose, for instance, that income level is a factor which has an important bearing on public opinion concerning an election, and that it is known that 30% of all eligible voters have low incomes, 60% have medium incomes, and 10% have high incomes. In that case, a pollster who intends to interview 800 voters could improve on simple random sampling (that is, make his prediction of the outcome of the election more reliable) by intentionally selecting a random sample of $0.30(800) = 240$ voters from the low-income group, a random sample of $0.60(800) = 480$ voters from the medium-income group, a random sample of $0.10(800) = 80$ voters from the high-income group, and combining the results.

What we have illustrated here is a form of **stratified random sampling,** namely, a kind of sampling in which definite portions of the total sample are allocated to different parts, or **strata**, of the population. Stratified sampling is widely used, and it can be very effective provided one stratifies with respect to *truly relevant characteristics* of the population; unfortunately, it can be quite difficult to determine what characteristics are really relevant in a given situation.

As we already pointed out on page 185, there are situations where simple random sampling is very impractical or virtually impossible, and the examples which we used pertained among other things to the sampling of trees in the Blue Ridge Mountains. In a situation like this we might use what is called **cluster sampling** by dividing the entire area into sections (say, 5 acres each), randomly selecting a few of these sections, and then measuring the diameter (or whatever happens to be of interest) of each tree, or of a random sample of trees, within each of the chosen sections. Cluster sampling is used very extensively in sample surveys conducted by government agencies and private research organizations, and when the subdivisions are counties, villages, city blocks, . . . , or acreages as in our example, we refer to this kind of sampling also as **area sampling**.

In some instances, the most practical way of choosing a sample is to select, say, every 10th name on a list, every 20th voucher in a file, every 50th piece coming off an assembly line, or every 12th house on one side of a street. This is called **systematic sampling**, and it can be made random (at least in part) by using random numbers or some gambling device to pick the unit with which to start. The danger of this kind of sampling lies in the possible existence of *hidden periodicities*—for instance, if we selected every 12th house on one side of a street, our results would probably be biased if it so happened that each of these houses is a corner house on a double lot. Similarly, our results would be very misleading if we inspected every 50th piece coming off an assembly line, and it so happened that (due to a regularly recurring failure) every 25th piece is defective.

Although results based on cluster samples or systematic samples are generally not as reliable as results based on simple random samples *of the same size*, they are usually more reliable *per unit cost*. For instance, it is much cheaper to interview families living, say, in the same city block or in several city blocks, than it is to interview families selected at random over a wide area, and in a situation like this a cluster sample of size 500 may be *cheaper to obtain and more reliable* than a small random sample of size 50.

As we already indicated on page 186, we often have to keep our fingers crossed that statistical theory intended for random samples can be employed even though we had little or no control over the way in which our data were obtained. For instance, we would really have no choice but to rely on whatever records happen to be available, say, if we wanted to predict a department store's December sales, if we wanted to estimate the mortality rate of a very rare disease, or if we wanted to study traffic patterns at a dangerous intersection. In situations like this we can only hope to make sure that there are no obvious indications that "anything might be wrong"—that there is no trend as in Example 4 on page 186, that there are no sudden *shifts or changes* which might occur, for instance, when different measuring instruments are used for different parts of an experiment, and that there are no *regularly recurring patterns* like those which we might find in the work of a secretary who is very alert and efficient each morning but tends to get tired each afternoon.

In recent years, several methods have been developed which make it possible to judge the randomness of a sample (or the lack of it) on the basis of the *order* in which the observations are actually obtained. For instance, there is the **theory of runs**, which might be used to detect a *trend* in a set of data by denoting each value falling below the median (of the data) with the letter b, each value falling above the median with the letter a, and checking whether there might be a clustering of b's followed by a clustering of a's (indicating an *upward trend*) or a clustering of a's followed by a clustering of b's (indicating a *downward trend*). To be more specific, a **run** is defined as a *succession of identical letters (or other kinds of symbols) which is followed and preceded by different letters or no letters at all.*

Example 2. For instance, if

$$\underbrace{HHHHHHH}\underbrace{DDD}\underbrace{HHHHHH}\underbrace{DDDDD}\underbrace{HHHHHH}\underbrace{DDD}\underbrace{H}$$

is the order in which healthy, H, and diseased, D, piñon trees were observed in a survey conducted by the Forestry Service, there are *relatively few runs*, only seven to be exact, and we might suspect that the diseased trees tend to come in clusters, namely, that the arrangement of H's and D's is *not random.* On the other hand, if

m f m m f m f m f f m f m f m m f f f m f f m m f m f f m

is the order in which men, m, and women, f, pass through a theater turnstile, there seem to be *so many runs*, 21 to be exact, that we might well suspect that the arrangement of m's and f's is *not random.* To make sure of this, we would have to refer to special tables, or otherwise perform a suitable test of significance.

In the methods which we discussed in Chapters 7 through 10 it was always assumed that we were dealing with a sample whose size was determined before any data were actually collected. This is called **single sampling**, and it has the disadvantage that a predetermined sample size can be (in fact, often is) unnecessarily large.

Example 3. For instance, if a public opinion poll wants to predict a gubernatorial election on the basis of a sample of size 4,000 and it so happens that the contest is very one-sided, it may well be that a sample of size 200 or 300 would have sufficed. To economize, it may thus be advantageous to start out with a small sample, and to continue sampling only if necessary, namely, only if it is impossible to reach a decision on the basis of the initial sample.

Example 4. To give another example, the quality control division of a large manufacturer might use the following scheme in the inspection of incoming lots of electronic components: First they take a random sample of size 50 from each lot and accept the lot if the number of defectives is 2 or less, reject the lot if the number of defectives is 6 or more, and take an additional sample, say, of size 100, if the number of defectives in the initial sample is 3, 4, or 5. If the second sample is needed, they combine the data and accept the lot only if there are *altogether* fewer than 7 defectives; otherwise the lot is rejected.

The kind of sampling we have illustrated in Example 4 is called **double sampling**, and it is easy to picture how it can be generalized to **multiple sampling**, in which we may have to take three, four, or more samples before a decision can be reached.

Carrying this idea one step further, we can even take observations *one at a time*, and check after each observation whether the combined data enable us to reach a decision or whether we have to continue sampling. This is called **sequential sampling**, and (as in double sampling and in multiple sampling) the main advantage of this kind of sampling is that it can produce very substantial reductions in cost. This is true especially in **destructive sampling**, where products cannot be used after they have been examined—for instance, when we test the strength of an explosive, subject airplane wings to vibrations until they fall apart, or in taste-testing, where we might sample the quality of foods or wine.

Nonparametric methods

Most of the methods of Chapters 7 through 9 required specific assumptions about the population (or populations) from which the samples were obtained; for instance, we often had to assume that these populations could be approximated closely with normal distributions. Since there are many situations where some, or all, of these requirements cannot be met, statisticians have developed alternative techniques based on less stringent assumptions, which have become known as **nonparametric methods**. (This name is meant to imply that we are *not directly* concerned with specific parameters or specific kinds of populations.) Many non-parametric methods have become quite popular, for they are often easier to understand (or explain) than the "standard" techniques which they replace, and many of them fall under the heading of "quick and easy" or "shortcut" statistics. The main *disadvantage* of nonparametric methods is that they are often *wasteful of information* (as will be apparent from the examples which follow); however, this is offset to some extent by the fact that they can be used under much more general conditions.

Several nonparametric methods have already been discussed in this text—two applications of the *sign test* were given in Exercises 11 and 13 on pages 268 and 270, the *median test* was given in Exercise 8 on page 279, and some *rank correlation techniques* were discussed in Exercises 9 and 10 on pages 311 and 313. Indeed, many nonparametric methods are based on ranks, as in the following **rank-sum** alternative to the "standard" significance test for the difference between two sample means.

Example 5. Suppose that two samples of students were taught French, group A *with* certain audio equipment and group B *without* it, and that after they completed the course they obtained the following scores in an appropriate achievement test:

Group A: 77, 86, 43, 93, 71, 54, 91, 69

Group B: 46, 57, 92, 63, 29, 70, 80, 23

The means of these two samples are, respectively, 73.0 and 57.5, and the problem is to decide whether their fairly large difference is really significant. Upon examining these data carefully, we find that the grades in the second sample vary quite a bit more than those in the first, and this suggests that the method of Exercise 5 on page 245 cannot be used (for it requires that the samples come from populations with *equal standard deviations*). To use a nonparametric alternative based on rank sums, we rank the data *jointly* (as if they were one sample) in an increasing order of magnitude, using the letters A and B to indicate for each value whether it belongs to group A or to group B. Thus, we get

23	29	43	46	54	57	63	69	70	71	77	80	86	91	92	93
B	*B*	*A*	*B*	*A*	*B*	*B*	*A*	*B*	*A*	*A*	*B*	*A*	*A*	*B*	*A*

and if we assign the 16 scores *in this order* the ranks 1, 2, 3, . . . , and 16, we find that the values of the first sample (group A) occupy ranks 3, 5, 8, 10, 11, 13, 14, and 16, while those of the second sample (group B) occupy ranks 1, 2, 4, 6, 7, 9, 12, and 15. (There were no *ties* in this example, but if there had been, we could have proceeded as in

Exercise 9 on page 311 and assigned each of the tied observations the *mean* of the ranks which they jointly occupy.)

The sum of *all* the ranks is $1 + 2 + 3 + \cdots + 15 + 16 = 136$, and since there are equally many values in each sample, it would seem reasonable to conclude that the sum of the ranks for each sample should be about $\frac{136}{2} = 68$ *if the null hypothesis that the two samples came from identical populations is true.* Now, if we actually add the respective ranks, we find that those of group A total

$$3 + 5 + 8 + 10 + 11 + 13 + 14 + 16 = 80$$

while those of group B total

$$1 + 2 + 4 + 6 + 7 + 9 + 12 + 15 = 56$$

which shows that the students in group A got *higher ranks* than we might have expected, while the opposite is true for those in group B. Of course, it still remains to be seen whether the difference between 80 and 68 (or 56 and 68) is significant, and for this purpose we would have to refer to a special table.

The method which we have described in this example is essentially that of the *Mann–Whitney test*, and it can be generalized so that it applies to situations in which there are more than two samples; this would provide a nonparametric alternative to the analysis of variance technique described on pages 247 through 253. Another example of a nonparametric test, one that is of an entirely different nature, is the test of randomness based on the *theory of runs*, which we mentioned on page 318.

Experimental design

As we already emphasized in Chapter 1, statistics must concern itself with *all* questions concerning the collection of data, including the preliminary planning of surveys or experiments. Otherwise, it can easily happen that a survey which is purported to demonstrate one thing leads to results which are entirely irrelevant, or that an experiment which is improperly designed serves no useful purpose at all.

Example 6. Suppose, for instance, that we want to compare the performance of two kinds of gasoline, and that in 12 test runs, 6 each with a gallon of the respective gasolines, the following results were obtained:

Gasoline A: 22, 20, 19, 21, 22, and 22 miles

Gasoline B: 18, 21, 20, 16, 17, and 19 miles

The means of these two samples are, respectively, 21 and 18.5, and it can be shown that their difference is significant at the level of significance $\alpha = 0.05$, but does this really prove that the first gasoline is better than the second? *Is it not possible that the observed difference between the two means might be due to the fact that the gasolines were tested in different cars, by different drivers, or over differently surfaced roads?* Any one of these factors, not to mention numerous other possibilities, can make it very difficult to interpret the result of the experiment in a meaningful way.

There are essentially two ways of handling the problems raised in Example 6. One is to perform a rigorously **controlled experiment** in which all variables except the one with which we are concerned are held fixed.

Example 6 (Continued). Thus, all the test runs could be made with the same car (which is carefully inspected after each run), with the same driver, and over identical routes. In that case, if there is a significant difference in the average mileage yield of the two kinds of gasoline, we know that it is *not* due to differences in cars, drivers, or routes. On the positive side we know that one gasoline performs better than the other *if it is used in a specific kind of car, by a specific driver, and over a specific route,* but this does not tell us very much, for it would be dangerous to infer that similar results would be obtained with different kinds of cars, different drivers, and over different routes (say, freeway driving instead of crawling in heavy city traffic).

The other way of handling problems of this kind is to **design** the experiment in such a way that we cannot only investigate the particular variable with which we are concerned, but at the same time study the effects of other variables.

Example 6 (Continued). To illustrate how this might be done, suppose that the test runs are performed in two cars, a low-priced car L and a high-priced car H, by two drivers, a good driver Mr. G and a poor driver Mr. P, and over two routes, a city route C and a freeway route F. As before, each test run will be performed with a gallon of either kind of gasoline, but there will be 16 test runs, 8 with gasoline A and 8 with gasoline B, and they might be planned as follows:

Test Run	*Gasoline*	*Car*	*Driver*	*Route*
1	*B*	*L*	G	*C*
2	*B*	*H*	P	*C*
3	*A*	*H*	P	*C*
4	*B*	*L*	P	*F*
5	*A*	*H*	G	*C*
6	*B*	*H*	G	*C*
7	*B*	*L*	G	*F*
8	*A*	*L*	P	*F*
9	*A*	*L*	G	*F*
10	*A*	*H*	P	*F*
11	*B*	*L*	P	*C*
12	*A*	*H*	G	*F*
13	*A*	*L*	G	*C*
14	*B*	*H*	G	*F*
15	*A*	*L*	P	*C*
16	*B*	*H*	P	*F*

This means that the first test run is performed with gasoline B in the low-priced car, by the good driver, over the city route; the second test run is performed with gasoline B in the high-priced car, by the poor driver, over the city route; and so forth.

It is customary to refer to the kind of scheme which we gave in the last example as a **complete factorial experiment**—each kind of gasoline is used once with each possible combination of cars, drivers, and routes. This can easily be checked. Another important feature of the above scheme (which may not be immediately apparent) is that we protected ourselves by **randomization**.

Example 6 (Continued). This means that we first wrote down the 16 possible ways in which we can select one of the two gasolines, one of the two cars, one of the two drivers, and one of the two routes, and then we *randomly selected the order in which the test runs are to be performed.* If we did not randomize the experiment in this fashion, extraneous factors might conceivably upset the results. For instance, if we used gasoline A in the first 8 test runs and gasoline B in the others, the results might be affected by the deterioration of equipment, driver fatigue, or differences in traffic conditions along the chosen routes. Similarly, we might have asked for trouble if we had performed the first 8 test runs with car L and the others with car H, if we had performed the first 8 test runs with Mr. G and the others with Mr. P, and so on.

The purpose of Example 6 has been to introduce some of the basic ideas of experimental design. The actual analysis of a **four-factor experi-**

ment such as the one just described requires a generalization of the analysis of variance technique described in Chapter 8, but it is fairly complicated, and this is why it was not discussed in this text. It is important, though, to realize that the above design not only enables us to decide whether there really is a difference in the quality of the two kinds of gasoline, but also whether their performance is affected by differences in cars, differences in drivers, or differences in driving conditions.

Nonlinear and multivariate methods

Two obvious generalizations of the problems studied in Chapter 10 are to consider *paired data for which the underlying relationship is not linear*, and to investigate *situations in which there are more than two variables*, and, hence, *triples, quadruples, etc., instead of pairs.*

Example 7. To illustrate the first kind of generalization, let us consider the following data on the drying time of a varnish and the amount of a certain chemical that has been added:

Amount of Additive (grams) x	*Drying Time (hours)* y
1	7.2
2	6.7
3	4.7
4	3.7
5	4.7
6	4.2
7	5.2
8	5.7

As can be seen from Figure 11.1, where we have plotted the corresponding data points, the y-values *first decrease and then increase*, so that a straight line will not give a good fit. As more advanced methods would show, a **parabola** (see dashed curve of Figure 11.1) could be used to describe this kind of relationship, and if it were fit by the method of least squares, we would get an equation of the form

$$y = a + bx + cx^2$$

Besides parabolas, **exponential curves**, which often arise in problems of *growth* (say, the enrollment of a college, the size of a culture of bacteria, or the spreading of rumors), and **hyperbolic curves**, which apply, among

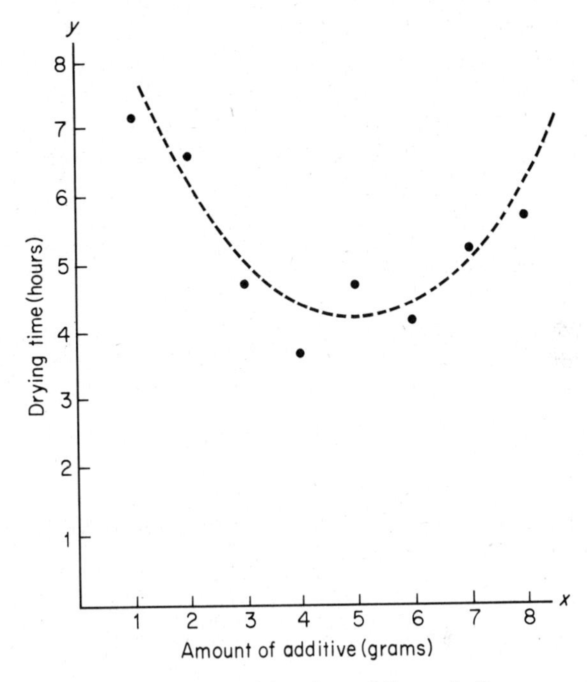

Figure 11.1 Parabola fit to data of Example 7.

other things, to problems involving supply and demand, are often used to describe relationships among observed data.

Although there are many problems in which one variable can be predicted quite accurately in terms of another, it stands to reason that predictions should be improved if one considers additional relevant information. For instance, we should be able to make better predictions of the demand for pork chops if we considered not only their own price, but also that of other kinds of meat; similarly, we should be able to make better predictions of the attendance at a concert if we considered the quality of the performers in addition to the size of the community and, perhaps, what other forms of entertainment are available.

Many mathematical formulas can serve to express relationships among more than two variables, but most widely used in statistics (partly for reasons of simplicity) are linear equations of the form

$$y = b_0 + b_1x_1 + b_2x_2 + \ldots + b_kx_k$$

Here y is the variable which is to be predicted, $x_1, x_2, \ldots,$ and x_k are the k known variables on which the predictions are to be based, and b_0, b_1, $b_2, \ldots,$ and b_k are numerical constants.

Example 8. To give an example, consider the following equation, obtained in a study of the demand for beef and veal:

$$y = 3.489 - 0.090x_1 + 0.064x_2 + 0.019x_3$$

where y stands for the total consumption of federally inspected beef and veal in millions of pounds, x_1 stands for the retail price of beef in cents per pound, x_2 stands for the retail price of pork in cents per pound, and x_3 stands for personal income as measured by a certain payroll index. Once an equation like this has been obtained, it can be used to *forecast* the total consumption of federally inspected beef and veal on the basis of known (or estimated) values of the variables x_1, x_2, and x_3.

The main questions that arise in connection with linear equations in more than two variables are exactly like those of the two-variable case: *We must find the values of the constants* b_0, b_1, b_2, . . . , *and* b_k *to get the best possible fit, and once this is done, we must ask whether the fit is any good.* The first part of this question is usually taken care of by using the method of least squares, and the second part is taken care of by calculating the **multiple correlation coefficient**, which is similar, conceptually, to the ordinary correlation coefficient, whose significance was explained on page 306. However, in the three-or-more-variable case, things are a bit more complicated, for there are other kinds of relationships which may be of interest. For instance, we may want to know whether there is a correlation between two of the variables when all other variables are held fixed (say, whether there is a relationship between family income and church attendance among families belonging to the *same* general age group, having the *same* educational background, and the *same* number of children). This is a problem of **partial correlation**, for which (as the reader may have guessed) appropriate statistical measures have been devised. All the techniques mentioned in this and the two preceding paragraphs come under the general heading of **multivariate analysis**.

Decision theory

This is one aspect of modern statistics which we mentioned briefly in the last section of Chapter 1 and touched upon lightly in Chapters 3 and 4. In many ways, decision theory presents the most challenging problems, as it delves deeply into the philosophical foundations of statistics. In recent years, the concept of *personal probability*, which we introduced on page 83, and the idea that the parameters of populations are things about which we can have different *strengths of beliefs* (and, hence, different personal probabilities) have given considerable impetus to the use of Bayes' rule (see page 120) in problems of inference, and, hence, in statistical decision making. This is true, particularly, in business applications, where it is possible (or at least easier) to put "cash values" on the consequences of one's decisions.

One of the most fascinating developments of decision theory is that it enables us to look upon statistical decision problems as **games**. In fact, it may have occurred to the reader that in Example 28 on page 90 the director of the research laboratory was playing some sort of game with the government as his opponent—the director of the research laboratory had the choice of two different "moves," to expand the facilities right away or to wait at least another year, while the government also had the choice of two different "moves," to give them the contract or not.

Example 9. To illustrate this further, let us refer to Exercise 12 on page 93, which concerned a mining company's decision whether or not to continue an operation at a certain location. Now, if we refer to *nature* (or *fate*), which controls the success or failure of the operation, as player A and the mining company as player B, and to their respective choices as strategies, all the information given in that exercise (namely, the mining company's potential losses or gains) can be presented as in the following table:

		Player A (Nature)	
		Strategy I (Operation Is Successful)	*Strategy II (Operation Is Failure)*
Player B (mining company)	*Strategy 1 (continue)*	1,500,000	−900,000
	Strategy 2 (stop)	−600,000	150,000

For instance, if player A chooses strategy I and player B chooses strategy 1, player B (the mining company) will "win" \$1,500,000; if player A chooses strategy II and player B chooses strategy 1, player B (the mining company) will "lose" \$900,000; and so on.

This table takes care of part (a) of Exercise 12 on page 93, and as the reader may recall, in parts (b), (c), and (d) we were always given the odds for success, namely, the odds whether nature would choose strategy I or strategy II. *This raises the question of what we might do when we have no idea whatsoever what our opponent decides to do.*

One possibility is to use the **minimax criterion** of *minimizing one's maximum losses* (or *maximizing one's minimum gains*, which is the same). For instance, if player B (the mining company) chose strategy 1, the worst that could happen would be a loss of \$900,000, and if player B chose strategy 2, the worst that could happen would be a loss of \$600,000. Using the minimax criterion (according to which a possible maximum loss of \$600,000 is preferable to a possible maximum loss of \$900,000), player B would thus choose strategy 2. Suppose now that player A is a *malevolent opponent* (who is trying to make things as rough as possible for player B); clearly, he would choose strategy I to make sure that his opponent will lose \$600,000 rather than win \$150,000. This would work nicely, unless player B figures that this is precisely what player A intends to do, and that it would, therefore, be smart to switch to strategy 1 and win \$1,500,000. Of course, if player A thinks that player B will try to outsmart him by choosing strategy 1, he can in turn try to outfox player B by switching to strategy II and cause his opponent to lose \$900,000. This argument can be continued *ad infinitum*, and the only way in which a player can avoid being outsmarted by his opponent in this fashion is to *mix up* his strategies in some way, namely, to introduce an element of chance.

Suppose, thus, that player B (the mining company) uses random numbers or some other gambling device which is rigged in such a way that he will choose strategy 1 with the probability p and strategy 2 with the probability $1 - p$. Using the formula for a mathematical expectation on page 88, we thus find that player B can *expect* to win

$$E = 1{,}500{,}000p - 600{,}000(1 - p)$$

dollars *if player A selects strategy I*, and that player B can *expect* to win

$$E = -900{,}000p + 150{,}000(1 - p)$$

dollars *if player A selects strategy II*. Graphically, this situation is pictured in Figure 11.2, where we have plotted the two lines whose equations are

$$E = 1{,}500{,}000p - 600{,}000(1 - p)$$

and

$$E = -900{,}000p + 150{,}000(1 - p)$$

for values of p from 0 to 1. Suppose now that we apply the minimax criterion to the *expected* winnings of player B. As is apparent from

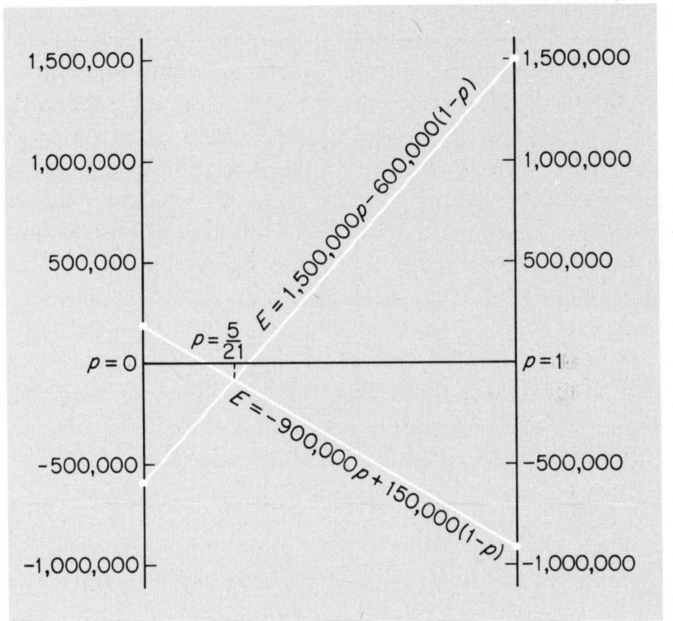

Figure 11.2 Diagram for Example 9.

Figure 11.2, the *least* player B can expect to win (or, better, the smaller of the two values of E for any given value of p) is *greatest* where the two lines intersect, and simple algebra will show that the corresponding value of p is $\frac{5}{21}$. This means that if player B labels five slips of paper "continue the operation" and 16 slips of paper "stop the operation," shuffles them thoroughly, and then acts according to which kind of slip he randomly selects, he will be applying the minimax criterion to his expected winnings. They are

$$1{,}500{,}000(\tfrac{5}{21}) - 600{,}000(\tfrac{16}{21}) = -100{,}000$$

namely, *expected losses* of $100,000. This is too bad for the mining company, but not as bad as the possible losses of $900,000 to which they would be exposed by strategy 1 or the possible losses of $600,000 to which they would be exposed by strategy 2.

The method by which we have analyzed this problem assumes that player B has no idea what his opponent might do, and that nature is, in fact, a *malevolent opponent*—this last assumption is crucial, and whether or not it is "reasonable" in any given situation can only be judged on an individual basis. In any case, it would probably take a lot to convince management and government officials that there are situations in which it is "best" to leave important decisions to the flip of a coin or the roll of a pair of dice. Also, let us not forget that this kind of decision making requires that we can assign "cash values" to the individual **payoffs**, namely, to the consequences of all possible actions; as we already indicated on page 7, this can pose nearly insurmountable difficulties. Nevertheless, the decision-theory approach to statistics has the positive advantage of forcing one to formulate problems clearly, to anticipate the consequences of one's actions, to eliminate the irrelevant, and all this can hardly fail to be of great help.

In case the reader has not been able to fully understand all the material in this chapter, he should not become overly concerned. More advanced books (for instance, those listed in the Bibliography which follows) treat these problems in great detail, and our discussion was intended to be nothing more than a preview of what lies ahead.

Bibliography

A. Statistics for the layman

Federer, W. T., *Statistics and Society*. New York: Marcel Dekker, Inc., 1973.

Levinson, H. C., *Chance, Luck, and Statistics*. New York: Dover Publications, Inc., 1963.

Moroney, M. J., *Facts from Figures*. London: Pelican Books, 1956.

Mosteller, F., Kruskal, W. H., Pieters, R. S., Rising, G. R., and Link, R. F., *Statistics by Example*. Reading, Mass.: Addison-Wesley Publishing Co., Inc., 1973.

Reichman, W. J., *Use and Abuse of Statistics*. New York: Oxford University Press, 1962.

Sielaff, T. J. (ed.), *Statistics in Action*. San Jose, Calif.: Lansford Press, 1963.

Tanur, J. M. (ed.), *Statistics: A Guide to the Unknown*. San Francisco: Holden-Day, Inc., 1972.

B. Some books on the theory of probability and statistics

Freund, J. E., *Introduction to Probability*. Encino, Calif.: Dickenson Publishing Co., Inc., 1973.

Freund, J. E., *Mathematical Statistics*, 2nd ed. Englewood Cliffs, N.J.: Prentice-Hall, Inc., 1971.

Goldberg, S., *Probability—An Introduction*. Englewood Cliffs, N.J.: Prentice-Hall, Inc., 1960.

Hodges, J. L., and Lehmann, E. L., *Elements of Finite Probability*. San Francisco: Holden-Day, Inc., 1965.

Hoel, P., *Introduction to Mathematical Statistics*, 4th ed. New York: John Wiley & Sons., Inc., 1971.

Mosteller, F., Rourke, R. E. K., and Thomas, G. B., *Probability with Statistical Applications*, 2nd ed. Reading, Mass.: Addison-Wesley Publishing Co., Inc., 1970.

C. Some books dealing with special topics

Cochran, W. G., *Sampling Techniques*, 2nd ed. New York: John Wiley & Sons, Inc., 1963.

Dresher, M., *Games of Strategy: Theory and Applications*. Englewood Cliffs, N.J.: Prentice-Hall, Inc., 1961.

Harris, R. J., *A Primer of Multivariate Statistics*. New York: Academic Press, Inc., 1974.

Hicks, C. R., *Fundamental Concepts in the Design of Experiments*. New York: Holt, Rinehart and Winston, Inc., 1964.

Noether, G. E., *Elements of Nonparametric Statistics*. New York: John Wiley & Sons, Inc., 1967.

Stuart, A., *Basic Ideas of Scientific Sampling*. New York: Hafner Press, 1962.

D. Some general reference works and tables

Freund, J. E., and Williams, F. J., *Dictionary/Outline of Basic Statistics*. New York: McGraw-Hill Book Company, 1966.

Hauser, P. M., and Leonard, W. R., *Government Statistics for Business Use*, 2nd ed. New York: John Wiley & Sons, Inc., 1956.

Kendall, M. G., and Buckland, W. R., *A Dictionary of Statistical Terms*, 3rd ed. New York: Hafner Press, 1971.

National Bureau of Standards, *Tables of the Binomial Probability Distribution*. Washington, D.C.: Government Printing Office, 1950.

Owen, D. B., *Handbook of Statistical Tables*. Reading, Mass.: Addison-Wesley Publishing Co., Inc., 1962.

RAND Corporation, *A Million Random Digits with 100,000 Normal Deviates*. New York: Free Press, 1955.

Statistical tables

I. The Standard Normal Distribution

II. The t Distribution

III. The χ^2 Distribution

IV. The F Distribution

V. Confidence Intervals for Proportions

VI. Critical Values of r

VII. Factorials

VIII. Binomial Coefficients

IX. Values of e^{-x}

X. Square Roots

XI. Binomial Probabilities

XII. Random Numbers

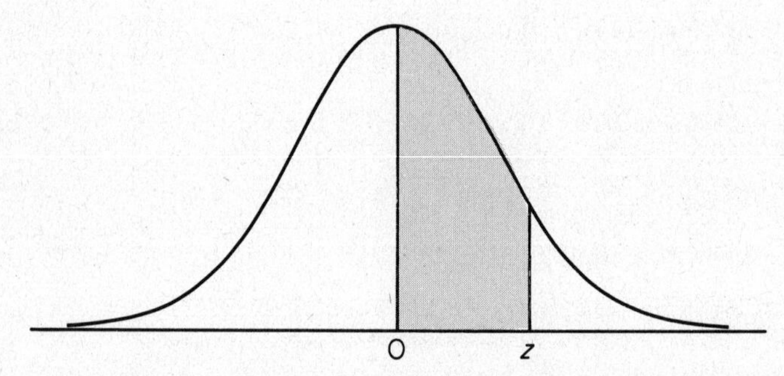

The entries in Table I are the probabilities that a random variable having the standard normal distribution takes on a value between 0 and z; they are given by the area under the curve shaded in the diagram.

Table I

THE STANDARD NORMAL DISTRIBUTION

z	.00	.01	.02	.03	.04	.05	.06	.07	.08	.09
0.0	.0000	.0040	.0080	.0120	.0160	.0199	.0239	.0279	.0319	.0359
0.1	.0398	.0438	.0478	.0517	.0557	.0596	.0636	.0675	.0714	.0753
0.2	.0793	.0832	.0871	.0910	.0948	.0987	.1026	.1064	.1103	.1141
0.3	.1179	.1217	.1255	.1293	.1331	.1368	.1406	.1443	.1480	.1517
0.4	.1554	.1591	.1628	.1664	.1700	.1736	.1772	.1808	.1844	.1879
0.5	.1915	.1950	.1985	.2019	.2054	.2088	.2123	.2157	.2190	.2224
0.6	.2257	.2291	.2324	.2357	.2389	.2422	.2454	.2486	.2517	.2549
0.7	.2580	.2611	.2642	.2673	.2704	.2734	.2764	.2794	.2823	.2852
0.8	.2881	.2910	.2939	.2967	.2995	.3023	.3051	.3078	.3106	.3133
0.9	.3159	.3186	.3212	.3238	.3264	.3289	.3315	.3340	.3365	.3389
1.0	.3413	.3438	.3461	.3485	.3508	.3531	.3554	.3577	.3599	.3621
1.1	.3643	.3665	.3686	.3708	.3729	.3749	.3770	.3790	.3810	.3830
1.2	.3849	.3869	.3888	.3907	.3925	.3944	.3962	.3980	.3997	.4015
1.3	.4032	.4049	.4066	.4082	.4099	.4115	.4131	.4147	.4162	.4177
1.4	.4192	.4207	.4222	.4236	.4251	.4265	.4279	.4292	.4306	.4319
1.5	.4332	.4345	.4357	.4370	.4382	.4394	.4406	.4418	.4429	.4441
1.6	.4452	.4463	.4474	.4484	.4495	.4505	.4515	.4525	.4535	.4545
1.7	.4554	.4564	.4573	.4582	.4591	.4599	.4608	.4616	.4625	.4633
1.8	.4641	.4649	.4656	.4664	.4671	.4678	.4686	.4693	.4699	.4706
1.9	.4713	.4719	.4726	.4732	.4738	.4744	.4750	.4756	.4761	.4767
2.0	.4772	.4778	.4783	.4788	.4793	.4798	.4803	.4808	.4812	.4817
2.1	.4821	.4826	.4830	.4834	.4838	.4842	.4846	.4850	.4854	.4857
2.2	.4861	.4864	.4868	.4871	.4875	.4878	.4881	.4884	.4887	.4890
2.3	.4893	.4896	.4898	.4901	.4904	.4906	.4909	.4911	.4913	.4916
2.4	.4918	.4920	.4922	.4925	.4927	.4929	.4931	.4932	.4934	.4936
2.5	.4938	.4940	.4941	.4943	.4945	.4946	.4948	.4949	.4951	.4952
2.6	.4953	.4955	.4956	.4957	.4959	.4960	.4961	.4962	.4963	.4964
2.7	.4965	.4966	.4967	.4968	.4969	.4970	.4971	.4972	.4973	.4974
2.8	.4974	.4975	.4976	.4977	.4977	.4978	.4979	.4979	.4980	.4981
2.9	.4981	.4982	.4982	.4983	.4984	.4984	.4985	.4985	.4986	.4986
3.0	.4987	.4987	.4987	.4988	.4988	.4989	.4989	.4989	.4990	.4990

Table II

THE t DISTRIBUTION*

d.f.	$t_{.050}$	$t_{.025}$	$t_{.010}$	$t_{.005}$	*d.f.*
1	6.314	12.706	31.821	63.657	**1**
2	2.920	4.303	6.965	9.925	**2**
3	2.353	3.182	4.541	5.841	**3**
4	2.132	2.776	3.747	4.604	**4**
5	2.015	2.571	3.365	4.032	**5**
6	1.943	2.447	3.143	3.707	**6**
7	1.895	2.365	2.998	3.499	**7**
8	1.860	2.306	2.896	3.355	**8**
9	1.833	2.262	2.821	3.250	**9**
10	1.812	2.228	2.764	3.169	**10**
11	1.796	2.201	2.718	3.106	**11**
12	1.782	2.179	2.681	3.055	**12**
13	1.771	2.160	2.650	3.012	**13**
14	1.761	2.145	2.624	2.977	**14**
15	1.753	2.131	2.602	2.947	**15**
16	1.746	2.120	2.583	2.921	**16**
17	1.740	2.110	2.567	2.898	**17**
18	1.734	2.101	2.552	2.878	**18**
19	1.729	2.093	2.539	2.861	**19**
20	1.725	2.086	2.528	2.845	**20**
21	1.721	2.080	2.518	2.831	**21**
22	1.717	2.074	2.508	2.819	**22**
23	1.714	2.069	2.500	2.807	**23**
24	1.711	2.064	2.492	2.797	**24**
25	1.708	2.060	2.485	2.787	**25**
26	1.706	2.056	2.479	2.779	**26**
27	1.703	2.052	2.473	2.771	**27**
28	1.701	2.048	2.467	2.763	**28**
29	1.699	2.045	2.462	2.756	**29**
inf.	1.645	1.960	2.326	2.576	inf.

* This table is abridged from Table IV of R. A. Fisher, *Statistical Methods for Research Workers*, published by Oliver & Boyd, Ltd., Edinburgh, by permission of the author's literary executor and publishers.

Table III

THE χ^2 DISTRIBUTION*

d.f.	$\chi^2_{.05}$	$\chi^2_{.01}$	*d.f.*
1	3.841	6.635	1
2	5.991	9.210	2
3	7.815	11.345	3
4	9.488	13.277	4
5	11.070	15.086	5
6	12.592	16.812	6
7	14.067	18.475	7
8	15.507	20.090	8
9	16.919	21.666	9
10	18.307	23.209	10
11	19.675	24.725	11
12	21.026	26.217	12
13	22.362	27.688	13
14	23.685	29.141	14
15	24.996	30.578	15
16	26.296	32.000	16
17	27.587	33.409	17
18	28.869	34.805	18
19	30.144	36.191	19
20	31.410	37.566	20
21	32.671	38.932	21
22	33.924	40.289	22
23	35.172	41.638	23
24	36.415	42.980	24
25	37.652	44.314	25
26	38.885	45.642	26
27	40.113	46.963	27
28	41.337	48.278	28
29	42.557	49.588	29
30	43.773	50.892	30

* This table is based on Table 8 of *Biometrika Tables for Statisticians*, Vol. I (New York: Cambridge University Press, 1966) by permission of the Biometrika trustees.

Table IVa

THE F DISTRIBUTION (VALUES OF $F_{0.05}$)*

	Degrees of freedom for numerator																		
Degrees of freedom for denominator	1	2	3	4	5	6	7	8	9	10	12	15	20	24	30	40	60	120	∞
1	161	200	216	225	230	234	237	239	241	242	244	246	248	249	250	251	252	253	254
2	18.5	19.0	19.2	19.2	19.3	19.3	19.4	19.4	19.4	19.4	19.4	19.4	19.4	19.5	19.5	19.5	19.5	19.5	19.5
3	10.1	9.55	9.28	9.12	9.01	8.94	8.89	8.85	8.81	8.79	8.74	8.70	8.66	8.64	8.62	8.59	8.57	8.55	8.53
4	7.71	6.94	6.59	6.39	6.26	6.16	6.09	6.04	6.00	5.96	5.91	5.86	5.80	5.77	5.75	5.72	5.69	5.66	5.63
5	6.61	5.79	5.41	5.19	5.05	4.95	4.88	4.82	4.77	4.74	4.68	4.62	4.56	4.53	4.50	4.46	4.43	4.40	4.37
6	5.99	5.14	4.76	4.53	4.39	4.28	4.21	4.15	4.10	4.06	4.00	3.94	3.87	3.84	3.81	3.77	3.74	3.70	3.67
7	5.59	4.74	4.35	4.12	3.97	3.87	3.79	3.73	3.68	3.64	3.57	3.51	3.44	3.41	3.38	3.34	3.30	3.27	3.23
8	5.32	4.46	4.07	3.84	3.69	3.58	3.50	3.44	3.39	3.35	3.28	3.22	3.15	3.12	3.08	3.04	3.01	2.97	2.93
9	5.12	4.26	3.86	3.63	3.48	3.37	3.29	3.23	3.18	3.14	3.07	3.01	2.94	2.90	2.86	2.83	2.79	2.75	2.71
10	4.96	4.10	3.71	3.48	3.33	3.22	3.14	3.07	3.02	2.98	2.91	2.85	2.77	2.74	2.70	2.66	2.62	2.58	2.54
11	4.84	3.98	3.59	3.36	3.20	3.09	3.01	2.95	2.90	2.85	2.79	2.72	2.65	2.61	2.57	2.53	2.49	2.45	2.40
12	4.75	3.89	3.49	3.26	3.11	3.00	2.91	2.85	2.80	2.75	2.69	2.62	2.54	2.51	2.47	2.43	2.38	2.34	2.30
13	4.67	3.81	3.41	3.18	3.03	2.92	2.83	2.77	2.71	2.67	2.60	2.53	2.46	2.42	2.38	2.34	2.30	2.25	2.21
14	4.60	3.74	3.34	3.11	2.96	2.85	2.76	2.70	2.65	2.60	2.53	2.46	2.39	2.35	2.31	2.27	2.22	2.18	2.13
15	4.54	3.68	3.29	3.06	2.90	2.79	2.71	2.64	2.59	2.54	2.48	2.40	2.33	2.29	2.25	2.20	2.16	2.11	2.07
16	4.49	3.63	3.24	3.01	2.85	2.74	2.66	2.59	2.54	2.49	2.42	2.35	2.28	2.24	2.19	2.15	2.11	2.06	2.01
17	4.45	3.59	3.20	2.96	2.81	2.70	2.61	2.55	2.49	2.45	2.38	2.31	2.23	2.29	2.15	2.10	2.06	2.01	1.96
18	4.41	3.55	3.16	2.93	2.77	2.66	2.58	2.51	2.46	2.41	2.34	2.27	2.19	2.15	2.11	2.06	2.02	1.97	1.92
19	4.38	3.52	3.13	2.90	2.74	2.63	2.54	2.48	2.42	2.38	2.31	2.23	2.16	2.11	2.07	2.03	1.98	1.93	1.88
20	4.35	3.49	3.10	2.87	2.71	2.60	2.51	2.45	2.39	2.35	2.28	2.20	2.12	2.08	2.04	1.99	1.95	1.90	1.84
21	4.32	3.47	3.07	2.84	2.68	2.57	2.49	2.42	2.37	2.32	2.25	2.18	2.10	2.05	2.01	1.96	1.92	1.87	1.81
22	4.30	3.44	3.05	2.82	2.66	2.55	2.46	2.40	2.34	2.30	2.23	2.15	2.07	2.03	1.98	1.94	1.89	1.84	1.78
23	4.28	3.42	3.03	2.80	2.64	2.53	2.44	2.37	2.32	2.27	2.20	2.13	2.05	2.01	1.96	1.91	1.86	1.81	1.76
24	4.26	3.40	3.01	2.78	2.62	2.51	2.42	2.36	2.30	2.25	2.18	2.11	2.03	1.98	1.94	1.89	1.84	1.79	1.73
25	4.24	3.39	2.99	2.76	2.60	2.49	2.40	2.34	2.28	2.24	2.16	2.09	2.01	1.96	1.92	1.87	1.82	1.77	1.71
30	4.17	3.32	2.92	2.69	2.53	2.42	2.33	2.27	2.21	2.16	2.09	2.01	1.93	1.89	1.84	1.79	1.74	1.68	1.62
40	4.08	3.23	2.84	2.61	2.45	2.34	2.25	2.18	2.12	2.08	2.00	1.92	1.84	1.79	1.74	1.69	1.64	1.58	1.51
60	4.00	3.15	2.76	2.53	2.37	2.25	2.17	2.10	2.04	1.99	1.92	1.84	1.75	1.70	1.65	1.59	1.53	1.47	1.39
120	3.92	3.07	2.68	2.45	2.29	2.18	2.09	2.02	1.96	1.91	1.83	1.75	1.66	1.61	1.55	1.50	1.43	1.35	1.25
∞	3.84	3.00	2.60	2.37	2.21	2.10	2.01	1.94	1.88	1.83	1.75	1.67	1.57	1.52	1.46	1.39	1.32	1.22	1.00

* This table is reproduced from M. Merrington and C. M. Thompson, "Tables of percentage points of the inverted beta (F) distribution," *Biometrika,* Vol. 33 (1943), by permission of the Biometrika trustees.

Table IVb

THE F DISTRIBUTION (VALUES OF $F_{0.01}$)*

Degrees of freedom for numerator (columns); Degrees of freedom for denominator (rows)

	1	2	3	4	5	6	7	8	9	10	12	15	20	24	30	40	60	120	∞
1	4,052	5,000	5,403	5,625	5,764	5,859	5,928	5,982	6,023	6,056	6,106	6,157	6,209	6,235	6,261	6,287	6,313	6,339	6,366
2	98.5	99.0	99.2	99.2	99.3	99.3	99.4	99.4	99.4	99.4	99.4	99.4	99.4	99.5	99.5	99.5	99.5	99.5	99.5
3	34.1	30.8	29.5	28.7	28.2	27.9	27.7	27.5	27.3	27.2	27.1	26.9	26.7	26.6	26.5	26.4	26.3	26.2	26.1
4	21.2	18.0	16.7	16.0	15.5	15.2	15.0	14.8	14.7	14.5	14.4	14.2	14.0	13.9	13.8	13.7	13.7	13.6	13.5
5	16.3	13.3	12.1	11.4	11.0	10.7	10.5	10.3	10.2	10.1	9.89	9.72	9.55	9.47	9.38	9.29	9.20	9.11	9.02
6	13.7	10.9	9.78	9.15	8.75	8.47	8.26	8.10	7.98	7.87	7.72	7.56	7.40	7.31	7.23	7.14	7.06	6.97	6.88
7	12.2	9.55	8.45	7.85	7.46	7.19	6.99	6.84	6.72	6.62	6.47	6.31	6.16	6.07	5.99	5.91	5.82	5.74	5.65
8	11.3	8.65	7.59	7.01	6.63	6.37	6.18	6.03	5.91	5.81	5.67	5.52	5.36	5.28	5.20	5.12	5.03	4.95	4.86
9	10.6	8.02	6.99	6.42	6.06	5.80	5.61	5.47	5.35	5.26	5.11	4.96	4.81	4.73	4.65	4.57	4.48	4.40	4.31
10	10.0	7.56	6.55	5.99	5.64	5.39	5.20	5.06	4.94	4.85	4.71	4.56	4.41	4.33	4.25	4.17	4.08	4.00	3.91
11	9.65	7.21	6.22	5.67	5.32	5.07	4.89	4.74	4.63	4.54	4.40	4.25	4.10	4.02	3.94	3.86	3.78	3.69	3.60
12	9.33	6.93	5.95	5.41	5.06	4.82	4.64	4.50	4.39	4.30	4.16	4.01	3.86	3.78	3.70	3.62	3.54	3.45	3.36
13	9.07	6.70	5.74	5.21	4.86	4.62	4.44	4.30	4.19	4.10	3.96	3.82	3.66	3.59	3.51	3.43	3.34	3.25	3.17
14	8.86	6.51	5.56	5.04	4.70	4.46	4.28	4.14	4.03	3.94	3.80	3.66	3.51	3.43	3.35	3.27	3.18	3.09	3.00
15	8.68	6.36	5.42	4.89	4.56	4.32	4.14	4.00	3.89	3.80	3.67	3.52	3.37	3.29	3.21	3.13	3.05	2.96	2.87
16	8.53	6.23	5.29	4.77	4.44	4.20	4.03	3.89	3.78	3.69	3.55	3.41	3.26	3.18	3.10	3.02	2.93	2.84	2.75
17	8.40	6.11	5.19	4.67	4.34	4.10	3.93	3.79	3.68	3.59	3.46	3.31	3.16	3.08	3.00	2.92	2.83	2.75	2.65
18	8.29	6.01	5.09	4.58	4.25	4.01	3.84	3.71	3.60	3.51	3.37	3.23	3.08	3.00	2.92	2.84	2.75	2.66	2.57
19	8.19	5.93	5.01	4.50	4.17	3.94	3.77	3.63	3.52	3.43	3.30	3.15	3.00	2.92	2.84	2.76	2.67	2.58	2.49
20	8.10	5.85	4.94	4.43	4.10	3.87	3.70	3.56	3.46	3.37	3.23	3.09	2.94	2.86	2.78	2.69	2.61	2.52	2.42
21	8.02	5.78	4.87	4.37	4.04	3.81	3.64	3.51	3.40	3.31	3.17	3.03	2.88	2.80	2.72	2.64	2.55	2.46	2.36
22	7.95	5.72	4.82	4.31	3.99	3.76	3.59	3.45	3.35	3.26	3.12	2.98	2.83	2.75	2.67	2.58	2.50	2.40	2.31
23	7.88	5.66	4.76	4.26	3.94	3.71	3.54	3.41	3.30	3.21	3.07	2.93	2.78	2.70	2.62	2.54	2.45	2.35	2.26
24	7.82	5.61	4.72	4.22	3.90	3.67	3.50	3.36	3.26	3.17	3.03	2.89	2.74	2.66	2.58	2.49	2.40	2.31	2.21
25	7.77	5.57	4.68	4.18	3.86	3.63	3.46	3.32	3.22	3.13	2.99	2.85	2.70	2.62	2.53	2.45	2.36	2.27	2.17
30	7.56	5.39	4.51	4.02	3.70	3.47	3.30	3.17	3.07	2.98	2.84	2.70	2.55	2.47	2.39	2.30	2.21	2.11	2.01
40	7.31	5.18	4.31	3.83	3.51	3.29	3.12	2.99	2.89	2.80	2.66	2.52	2.37	2.29	2.20	2.11	2.02	1.92	1.80
60	7.08	4.98	4.13	3.65	3.34	3.12	2.95	2.82	2.72	2.63	2.50	2.35	2.20	2.12	2.03	1.94	1.84	1.73	1.60
120	6.85	4.79	3.95	3.48	3.17	2.96	2.79	2.66	2.56	2.47	2.34	2.19	2.03	1.95	1.86	1.76	1.66	1.53	1.38
∞	6.63	4.61	3.78	3.32	3.02	2.80	2.64	2.51	2.41	2.32	2.18	2.04	1.88	1.79	1.70	1.59	1.47	1.32	1.00

* This table is reproduced from M. Merrington and C. M. Thompson, "Tables of percentage points of the inverted beta (F) distribution," *Biometrika*, Vol. 33 (1943), by permission of the Biometrika trustees.

Table Va

0.95 CONFIDENCE INTERVALS FOR PROPORTIONS*

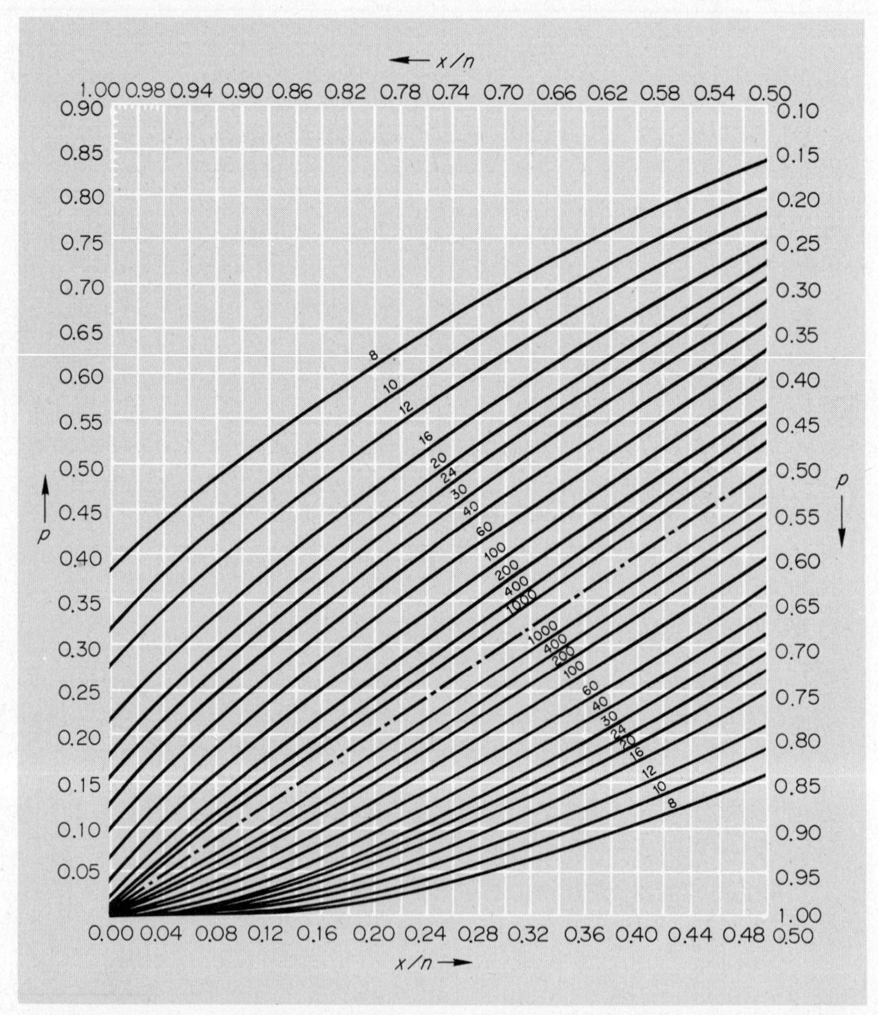

* This table is reproduced from Table 41 of the *Biometrika Tables for Statisticians*, Vol. I (New York: Cambridge University Press, 1966) by permission of the Biometrika trustees.

Table Vb

0.99 CONFIDENCE INTERVALS FOR PROPORTIONS*

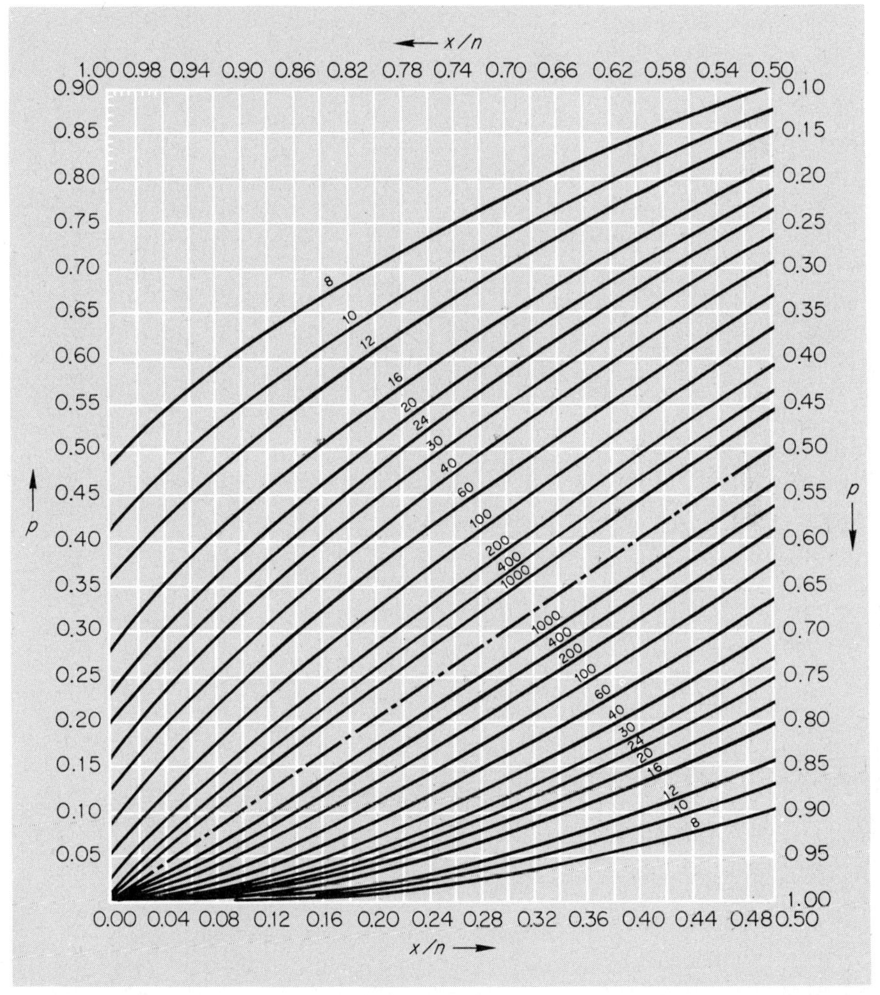

* This table is reproduced from Table 41 of the *Biometrika Tables for Statisticians*, Vol. I (New York: Cambridge University Press, 1966) by permission of the Biometrika trustees.

Table VI

CRITICAL VALUES OF r*

n	$r_{.025}$	$r_{.005}$	n	$r_{.025}$	$r_{.005}$
3	0.997		**18**	0.468	0.590
4	0.950	0.999	**19**	0.456	0.575
5	0.878	0.959	**20**	0.444	0.561
6	0.811	0.917	**21**	0.433	0.549
7	0.754	0.875	**22**	0.423	0.537
8	0.707	0.834	**27**	0.381	0.487
9	0.666	0.798	**32**	0.349	0.449
10	0.632	0.765	**37**	0.325	0.418
11	0.602	0.735	**42**	0.304	0.393
12	0.576	0.708	**47**	0.288	0.372
13	0.553	0.684	**52**	0.273	0.354
14	0.532	0.661	**62**	0.250	0.325
15	0.514	0.641	**72**	0.232	0.302
16	0.497	0.623	**82**	0.217	0.283
17	0.482	0.606	**92**	0.205	0.267

* This table is abridged from Table VI of R. A. Fisher and F. Yates, *Statistical Tables for Biological, Agricultural, and Medical Research*, published by Oliver & Boyd, Ltd., Edinburgh, by permission of the author's literary executor and publishers.

Table VII

FACTORIALS

n	$n!$
0	1
1	1
2	2
3	6
4	24
5	120
6	720
7	5,040
8	40,320
9	362,880
10	3,628,800
11	39,916,800
12	479,001,600
13	6,227,020,800
14	87,178,291,200
15	1,307,674,368,000

Table VIII

BINOMIAL COEFFICIENTS

n	$\binom{n}{0}$	$\binom{n}{1}$	$\binom{n}{2}$	$\binom{n}{3}$	$\binom{n}{4}$	$\binom{n}{5}$	$\binom{n}{6}$	$\binom{n}{7}$	$\binom{n}{8}$	$\binom{n}{9}$	$\binom{n}{10}$
0	1										
1	1	1									
2	1	2	1								
3	1	3	3	1							
4	1	4	6	4	1						
5	1	5	10	10	5	1					
6	1	6	15	20	15	6	1				
7	1	7	21	35	35	21	7	1			
8	1	8	28	56	70	56	28	8	1		
9	1	9	36	84	126	126	84	36	9	1	
10	1	10	45	120	210	252	210	120	45	10	1
11	1	11	55	165	330	462	462	330	165	55	11
12	1	12	66	220	495	792	924	792	495	220	66
13	1	13	78	286	715	1287	1716	1716	1287	715	286
14	1	14	91	364	1001	2002	3003	3432	3003	2002	1001
15	1	15	105	455	1365	3003	5005	6435	6435	5005	3003
16	1	16	120	560	1820	4368	8008	11440	12870	11440	8008
17	1	17	136	680	2380	6188	12376	19448	24310	24310	19448
18	1	18	153	816	3060	8568	18564	31824	43758	48620	43758
19	1	19	171	969	3876	11628	27132	50388	75582	92378	92378
20	1	20	190	1140	4845	15504	38760	77520	125970	167960	184756

If necessary, use the identity $\binom{n}{k} = \binom{n}{n-k}$.

Table IX

VALUES OF e^{-x}

x	e^{-x}	x	e^{-x}	x	e^{-x}	x	e^{-x}
0.0	1.000	2.5	0.082	5.0	0.0067	7.5	0.00055
0.1	0.905	2.6	0.074	5.1	0.0061	7.6	0.00050
0.2	0.819	2.7	0.067	5.2	0.0055	7.7	0.00045
0.3	0.741	2.8	0.061	5.3	0.0050	7.8	0.00041
0.4	0.670	2.9	0.055	5.4	0.0045	7.9	0.00037
0.5	0.607	3.0	0.050	5.5	0.0041	8.0	0.00034
0.6	0.549	3.1	0.045	5.6	0.0037	8.1	0.00030
0.7	0.497	3.2	0.041	5.7	0.0033	8.2	0.00028
0.8	0.449	3.3	0.037	5.8	0.0030	8.3	0.00025
0.9	0.407	3.4	0.033	5.9	0.0027	8.4	0.00023
1.0	0.368	3.5	0.030	6.0	0.0025	8.5	0.00020
1.1	0.333	3.6	0.027	6.1	0.0022	8.6	0.00018
1.2	0.301	3.7	0.025	6.2	0.0020	8.7	0.00017
1.3	0.273	3.8	0.022	6.3	0.0018	8.8	0.00015
1.4	0.247	3.9	0.020	6.4	0.0017	8.9	0.00014
1.5	0.223	4.0	0.018	6.5	0.0015	9.0	0.00012
1.6	0.202	4.1	0.017	6.6	0.0014	9.1	0.00011
1.7	0.183	4.2	0.015	6.7	0.0012	9.2	0.00010
1.8	0.165	4.3	0.014	6.8	0.0011	9.3	0.00009
1.9	0.150	4.4	0.012	6.9	0.0010	9.4	0.00008
2.0	0.135	4.5	0.011	7.0	0.0009	9.5	0.00008
2.1	0.122	4.6	0.010	7.1	0.0008	9.6	0.00007
2.2	0.111	4.7	0.009	7.2	0.0007	9.7	0.00006
2.3	0.100	4.8	0.008	7.3	0.0007	9.8	0.00006
2.4	0.091	4.9	0.007	7.4	0.0006	9.9	0.00005

To find the square root of any positive number rounded to two digits, use the following rule to decide whether to take the entry of the $\sqrt{n}$ or the $\sqrt{10n}$ column:

Move the decimal point an even number of places to the right or to the left until a number greater than or equal to 1 but less than 100 is reached. If the resulting number is less than 10 go to the $\sqrt{n}$ column; if it is 10 or more, go to the $\sqrt{10n}$ column.

Thus, to find the square root of 14,000 or 0.032 we go to the $\sqrt{n}$ column since the decimal point has to be moved, respectively, four places to the left to give 1.4 or two places to the right to give 3.2. Similarly, to find the square root of 2,200 or 0.000016 we go to the $\sqrt{10n}$ column since the decimal point has to be moved, respectively, two places to the left to give 22 or six places to the right to give 16.

Having found the entry in the appropriate column of Table X, the only thing that remains to be done is to put the decimal point in the right place in the result. To this end we use the following rule:

Having previously moved the decimal point an even number of places to the left or to the right to get a number greater than or equal to 1 but less than 100, the decimal point of the appropriate entry of Table X is moved half as many places in the opposite direction.

For example, to find the square root of 14,000 we first note that the decimal point has to be moved *four places to the left* to give 1.4. We thus take the entry of the $\sqrt{n}$ column corresponding to 1.4, move its decimal point *two places to the right*, and get $\sqrt{14{,}000} = 118.32$. Similarly, to find the square root of 0.000016 we note that the decimal point has to be moved *six places to the right* to give 16. We thus take the entry of the $\sqrt{10n}$ column corresponding to 1.6, move the decimal point *three places to the left*, and get $\sqrt{0.000016} = 0.004$.

SQUARE ROOTS

n	$\sqrt{n}$	$\sqrt{10n}$
1.00	1.0000	3.1623
1.01	1.0050	3.1780
1.02	1.0100	3.1937
1.03	1.0149	3.2094
1.04	1.0198	3.2249
1.05	1.0247	3.2404
1.06	1.0296	3.2558
1.07	1.0344	3.2711
1.08	1.0392	3.2863
1.09	1.0440	3.3015
1.10	1.0488	3.3166
1.11	1.0536	3.3317
1.12	1.0583	3.3466
1.13	1.0630	3.3615
1.14	1.0677	3.3764
1.15	1.0724	3.3912
1.16	1.0770	3.4059
1.17	1.0817	3.4205
1.18	1.0863	3.4351
1.19	1.0909	3.4496
1.20	1.0954	3.4641
1.21	1.1000	3.4785
1.22	1.1045	3.4928
1.23	1.1091	3.5071
1.24	1.1136	3.5214
1.25	1.1180	3.5355
1.26	1.1225	3.5496
1.27	1.1269	3.5637
1.28	1.1314	3.5777
1.29	1.1358	3.5917
1.30	1.1402	3.6056
1.31	1.1446	3.6194
1.32	1.1489	3.6332
1.33	1.1533	3.6469
1.34	1.1576	3.6606
1.35	1.1619	3.6742
1.36	1.1662	3.6878
1.37	1.1705	3.7014
1.38	1.1747	3.7148
1.39	1.1790	3.7283
1.40	1.1832	3.7417
1.41	1.1874	3.7550
1.42	1.1916	3.7683
1.43	1.1958	3.7815
1.44	1.2000	3.7947
1.45	1.2042	3.8079
1.46	1.2083	3.8210
1.47	1.2124	3.8341
1.48	1.2166	3.8471
1.49	1.2207	3.8601

n	$\sqrt{n}$	$\sqrt{10n}$
1.50	1.2247	3.8730
1.51	1.2288	3.8859
1.52	1.2329	3.8987
1.53	1.2369	3.9115
1.54	1.2410	3.9243
1.55	1.2450	3.9370
1.56	1.2490	3.9497
1.57	1.2530	3.9623
1.58	1.2570	3.9749
1.59	1.2610	3.9875
1.60	1.2649	4.0000
1.61	1.2689	4.0125
1.62	1.2728	4.0249
1.63	1.2767	4.0373
1.64	1.2806	4.0497
1.65	1.2845	4.0620
1.66	1.2884	4.0743
1.67	1.2923	4.0866
1.68	1.2961	4.0988
1.69	1.3000	4.1110
1.70	1.3038	4.1231
1.71	1.3077	4.1352
1.72	1.3115	4.1473
1.73	1.3153	4.1593
1.74	1.3191	4.1713
1.75	1.3229	4.1833
1.76	1.3266	4.1952
1.77	1.3304	4.2071
1.78	1.3342	4.2190
1.79	1.3379	4.2308
1.80	1.3416	4.2426
1.81	1.3454	4.2544
1.82	1.3491	4.2661
1.83	1.3528	4.2778
1.84	1.3565	4.2895
1.85	1.3601	4.3012
1.86	1.3638	4.3128
1.87	1.3675	4.3243
1.88	1.3711	4.3359
1.89	1.3748	4.3474
1.90	1.3784	4.3589
1.91	1.3820	4.3704
1.92	1.3856	4.3818
1.93	1.3892	4.3932
1.94	1.3928	4.4045
1.95	1.3964	4.4159
1.96	1.4000	4.4272
1.97	1.4036	4.4385
1.98	1.4071	4.4497
1.99	1.4107	4.4609

n	$\sqrt{n}$	$\sqrt{10u}$
2.00	1.4142	4.4721
2.01	1.4177	4.4833
2.02	1.4213	4.4944
2.03	1.4248	4.5056
2.04	1.4283	4.5166
2.05	1.4318	4.5277
2.06	1.4353	4.5387
2.07	1.4387	4.5497
2.08	1.4422	4.5607
2.09	1.4457	4.5717
2.10	1.4491	4.5826
2.11	1.4526	4.5935
2.12	1.4560	4.6043
2.13	1.4595	4.6152
2.14	1.4629	4.6260
2.15	1.4663	4.6368
2.16	1.4697	4.6476
2.17	1.4731	4.6583
2.18	1.4765	4.6690
2.19	1.4799	4.6797
2.20	1.4832	4.6904
2.21	1.4866	4.7011
2.22	1.4900	4.7117
2.23	1.4933	4.7223
2.24	1.4967	4.7329
2.25	1.5000	4.7434
2.26	1.5033	4.7539
2.27	1.5067	4.7645
2.28	1.5100	4.7749
2.29	1.5133	4.7854
2.30	1.5166	4.7958
2.31	1.5199	4.8062
2.32	1.5232	4.8166
2.33	1.5264	4.8270
2.34	1.5297	4.8374
2.35	1.5330	4.8477
2.36	1.5362	4.8580
2.37	1.5395	4.8683
2.38	1.5427	4.8785
2.39	1.5460	4.8888
2.40	1.5492	4.8990
2.41	1.5524	4.9092
2.42	1.5556	4.9193
2.43	1.5588	4.9295
2.44	1.5620	4.9396
2.45	1.5652	4.9497
2.46	1.5684	4.9598
2.47	1.5716	4.9699
2.48	1.5748	4.9800
2.49	1.5780	4.9900

SQUARE ROOTS (Continued)

n	$\sqrt{n}$	$\sqrt{10n}$	n	$\sqrt{n}$	$\sqrt{10n}$	n	$\sqrt{n}$	$\sqrt{10n}$
2.50	1.5811	5.0000	3.00	1.7321	5.4772	3.50	1.8708	5.9161
2.51	1.5843	5.0100	3.01	1.7349	5.4863	3.51	1.8735	5.9245
2.52	1.5875	5.0200	3.02	1.7378	5.4955	3.52	1.8762	5.9330
2.53	1.5906	5.0299	3.03	1.7407	5.5045	3.53	1.8788	5.9414
2.54	1.5937	5.0398	3.04	1.7436	5.5136	3.54	1.8815	5.9498
2.55	1.5969	5.0498	3.05	1.7464	5.5227	3.55	1.8841	5.9582
2.56	1.6000	5.0596	3.06	1.7493	5.5317	3.56	1.8868	5.9666
2.57	1.6031	5.0695	3.07	1.7521	5.5408	3.57	1.8894	5.9749
2.58	1.6062	5.0794	3.08	1.7550	5.5498	3.58	1.8921	5.9833
2.59	1.6093	5.0892	3.09	1.7578	5.5588	3.59	1.8947	5.9917
2.60	1.6125	5.0990	3.10	1.7607	5.5678	3.60	1.8974	6.0000
2.61	1.6155	5.1088	3.11	1.7635	5.5767	3.61	1.9000	6.0083
2.62	1.6186	5.1186	3.12	1.7664	5.5857	3.62	1.9026	6.0166
2.63	1.6217	5.1284	3.13	1.7692	5.5946	3.63	1.9053	6.0249
2.64	1.6248	5.1381	3.14	1.7720	5.6036	3.64	1.9079	6.0332
2.65	1.6279	5.1478	3.15	1.7748	5.6125	3.65	1.9105	6.0415
2.66	1.6310	5.1575	3.16	1.7776	5.6214	3.66	1.9131	6.0498
2.67	1.6340	5.1672	3.17	1.7804	5.6303	3.67	1.9157	6.0581
2.68	1.6371	5.1769	3.18	1.7833	5.6391	3.68	1.9183	6.0663
2.69	1.6401	5.1865	3.19	1.7861	5.6480	3.69	1.9209	6.0745
2.70	1.6432	5.1962	3.20	1.7889	5.6569	3.70	1.9235	6.0828
2.71	1.6462	5.2058	3.21	1.7916	5.6657	3.71	1.9261	6.0910
2.72	1.6492	5.2154	3.22	1.7944	5.6745	3.72	1.9287	6.0992
2.73	1.6523	5.2249	3.23	1.7972	5.6833	3.73	1.9313	6.1074
2.74	1.6553	5.2345	3.24	1.8000	5.6921	3.74	1.9339	6.1156
2.75	1.6583	5.2440	3.25	1.8028	5.7009	3.75	1.9365	6.1237
2.76	1.6613	5.2536	3.26	1.8055	5.7096	3.76	1.9391	6.1319
2.77	1.6643	5.2631	3.27	1.8083	5.7184	3.77	1.9416	6.1400
2.78	1.6673	5.2726	3.28	1.8111	5.7271	3.78	1.9442	6.1482
2.79	1.6703	5.2820	3.29	1.8138	5.7359	3.79	1.9468	6.1563
2.80	1.6733	5.2915	3.30	1.8166	5.7446	3.80	1.9494	6.1644
2.81	1.6763	5.3009	3.31	1.8193	5.7533	3.81	1.9519	6.1725
2.82	1.6793	5.3104	3.32	1.8221	5.7619	3.82	1.9545	6.1806
2.83	1.6823	5.3198	3.33	1.8248	5.7706	3.83	1.9570	6.1887
2.84	1.6852	5.3292	3.34	1.8276	5.7793	3.84	1.9596	6.1968
2.85	1.6882	5.3385	3.35	1.8303	5.7879	3.85	1.9621	6.2048
2.86	1.6912	5.3479	3.36	1.8330	5.7966	3.86	1.9647	6.2129
2.87	1.6941	5.3572	3.37	1.8358	5.8052	3.87	1.9672	6.2209
2.88	1.6971	5.3666	3.38	1.8385	5.8138	3.88	1.9698	6.2290
2.89	1.7000	5.3759	3.39	1.8412	5.8224	3.89	1.9723	6.2370
2.90	1.7029	5.3852	3.40	1.8439	5.8310	3.90	1.9748	6.2450
2.91	1.7059	5.3944	3.41	1.8466	5.8395	3.91	1.9774	6.2530
2.92	1.7088	5.4037	3.42	1.8493	5.8481	3.92	1.9799	6.2610
2.93	1.7117	5.4129	3.43	1.8520	5.8566	3.93	1.9824	6.2690
2.94	1.7146	5.4222	3.44	1.8547	5.8652	3.94	1.9849	6.2769
2.95	1.7176	5.4314	3.45	1.8574	5.8737	3.95	1.9875	6.2849
2.96	1.7205	5.4406	3.46	1.8601	5.8822	3.96	1.9900	6.2929
2.97	1.7234	5.4498	3.47	1.8628	5.8907	3.97	1.9925	6.3008
2.98	1.7263	5.4589	3.48	1.8655	5.8992	3.98	1.9950	6.3087
2.99	1.7292	5.4681	3.49	1.8682	5.9076	3.99	1.9975	6.3166

SQUARE ROOTS (Continued)

n	$\sqrt{n}$	$\sqrt{10n}$
4.00	2.0000	6.3246
4.01	2.0025	6.3325
4.02	2.0050	6.3403
4.03	2.0075	6.3482
4.04	2.0100	6.3561
4.05	2.0125	6.3640
4.06	2.0149	6.3718
4.07	2.0174	6.3797
4.08	2.0199	6.3875
4.09	2.0224	6.3953
4.10	2.0248	6.4031
4.11	2.0273	6.4109
4.12	2.0298	6.4187
4.13	2.0322	6.4265
4.14	2.0347	6.4343
4.15	2.0372	6.4420
4.16	2.0396	6.4498
4.17	2.0421	6.4576
4.18	2.0445	6.4653
4.19	2.0469	6.4730
4.20	2.0494	6.4807
4.21	2.0518	6.4885
4.22	2.0543	6.4962
4.23	2.0567	6.5038
4.24	2.0591	6.5115
4.25	2.0616	6.5192
4.26	2.0640	6.5269
4.27	2.0664	6.5345
4.28	2.0688	6.5422
4.29	2.0712	6.5498
4.30	2.0736	6.5574
4.31	2.0761	6.5651
4.32	2.0785	6.5727
4.33	2.0809	6.5803
4.34	2.0833	6.5879
4.35	2.0857	6.5955
4.36	2.0881	6.6030
4.37	2.0905	6.6106
4.38	2.0928	6.6182
4.39	2.0952	6.6257
4.40	2.0976	6.6332
4.41	2.1000	6.6408
4.42	2.1024	6.6483
4.43	2.1048	6.6558
4.44	2.1071	6.6633
4.45	2.1095	6.6708
4.46	2.1119	6.6783
4.47	2.1142	6.6858
4.48	2.1166	6.6933
4.49	2.1190	6.7007
4.50	2.1213	6.7082
4.51	2.1237	6.7157
4.52	2.1260	6.7231
4.53	2.1284	6.7305
4.54	2.1307	6.7380
4.55	2.1331	6.7454
4.56	2.1354	6.7528
4.57	2.1378	6.7602
4.58	2.1401	6.7676
4.59	2.1424	6.7750
4.60	2.1448	6.7823
4.61	2.1471	6.7897
4.62	2.1494	6.7971
4.63	2.1517	6.8044
4.64	2.1541	6.8118
4.65	2.1564	6.8191
4.66	2.1587	6.8264
4.67	2.1610	6.8337
4.68	2.1633	6.8411
4.69	2.1656	6.8484
4.70	2.1679	6.8557
4.71	2.1703	6.8629
4.72	2.1726	6.8702
4.73	2.1749	6.8775
4.74	2.1772	6.8848
4.75	2.1794	6.8920
4.76	2.1817	6.8993
4.77	2.1840	6.9065
4.78	2.1863	6.9138
4.79	2.1886	6.9210
4.80	2.1909	6.9282
4.81	2.1932	6.9354
4.82	2.1954	6.9426
4.83	2.1977	6.9498
4.84	2.2000	6.9570
4.85	2.2023	6.9642
4.86	2.2045	6.9714
4.87	2.2068	6.9785
4.88	2.2091	6.9857
4.89	2.2113	6.9929
4.90	2.2136	7.0000
4.91	2.2159	7.0071
4.92	2.2181	7.0143
4.93	2.2204	7.0214
4.94	2.2226	7.0285
4.95	2.2249	7.0356
4.96	2.2271	7.0427
4.97	2.2293	7.0498
4.98	2.2316	7.0569
4.99	2.2338	7.0640
5.00	2.2361	7.0711
5.01	2.2383	7.0781
5.02	2.2405	7.0852
5.03	2.2428	7.0922
5.04	2.2450	7.0993
5.05	2.2472	7.1063
5.06	2.2494	7.1134
5.07	2.2517	7.1204
5.08	2.2539	7.1274
5.09	2.2561	7.1344
5.10	2.2583	7.1414
5.11	2.2605	7.1484
5.12	2.2627	7.1554
5.13	2.2650	7.1624
5.14	2.2672	7.1694
5.15	2.2694	7.1764
5.16	2.2716	7.1833
5.17	2.2738	7.1903
5.18	2.2760	7.1972
5.19	2.2782	7.2042
5.20	2.2804	7.2111
5.21	2.2825	7.2180
5.22	2.2847	7.2250
5.23	2.2869	7.2319
5.24	2.2891	7.2388
5.25	2.2913	7.2457
5.26	2.2935	7.2526
5.27	2.2956	7.2595
5.28	2.2978	7.2664
5.29	2.3000	7.2732
5.30	2.3022	7.2801
5.31	2.3043	7.2870
5.32	2.3065	7.2938
5.33	2.3087	7.3007
5.34	2.3108	7.3075
5.35	2.3130	7.3144
5.36	2.3152	7.3212
5.37	2.3173	7.3280
5.38	2.3195	7.3348
5.39	2.3216	7.3417
5.40	2.3238	7.3485
5.41	2.3259	7.3553
5.42	2.3281	7.3621
5.43	2.3302	7.3689
5.44	2.3324	7.3756
5.45	2.3345	7.3824
5.46	2.3367	7.3892
5.47	2.3388	7.3959
5.48	2.3409	7.4027
5.49	2.3431	7.4095

Table X

SQUARE ROOTS (Continued)

n	$\sqrt{n}$	$\sqrt{10n}$	n	$\sqrt{n}$	$\sqrt{10n}$	n	$\sqrt{n}$	$\sqrt{10n}$
5.50	2.3452	7.4162	6.00	2.4495	7.7460	6.50	2.5495	8.0623
5.51	2.3473	7.4229	6.01	2.4515	7.7524	6.51	2.5515	8.0685
5.52	2.3495	7.4297	6.02	2.4536	7.7589	6.52	2.5534	8.0747
5.53	2.3516	7.4364	6.03	2.4556	7.7653	6.53	2.5554	8.0808
5.54	2.3537	7.4431	6.04	2.4576	7.7717	6.54	2.5573	8.0870
5.55	2.3558	7.4498	6.05	2.4597	7.7782	6.55	2.5593	8.0932
5.56	2.3580	7.4565	6.06	2.4617	7.7846	6.56	2.5612	8.0994
5.57	2.3601	7.4632	6.07	2.4637	7.7910	6.57	2.5632	8.1056
5.58	2.3622	7.4699	6.08	2.4658	7.7974	6.58	2.5652	8.1117
5.59	2.3643	7.4766	6.09	2.4678	7.8038	6.59	2.5671	8.1179
5.60	2.3664	7.4833	6.10	2.4698	7.8102	6.60	2.5690	8.1240
5.61	2.3685	7.4900	6.11	2.4718	7.8166	6.61	2.5710	8.1302
5.62	2.3707	7.4967	6.12	2.4739	7.8230	6.62	2.5729	8.1363
5.63	2.3728	7.5033	6.13	2.4759	7.8294	6.63	2.5749	8.1425
5.64	2.3749	7.5100	6.14	2.4779	7.8358	6.64	2.5768	8.1486
5.65	2.3770	7.5166	6.15	2.4799	7.8422	6.65	2.5788	8.1548
5.66	2.3791	7.5233	6.16	2.4819	7.8486	6.66	2.5807	8.1609
5.67	2.3812	7.5299	6.17	2.4839	7.8549	6.67	2.5826	8.1670
5.68	2.3833	7.5366	6.18	2.4860	7.8613	6.68	2.5846	8.1731
5.69	2.3854	7.5432	6.19	2.4880	7.8677	6.69	2.5865	8.1792
5.70	2.3875	7.5498	6.20	2.4900	7.8740	6.70	2.5884	8.1854
5.71	2.3896	7.5565	6.21	2.4920	7.8804	6.71	2.5904	8.1915
5.72	2.3917	7.5631	6.22	2.4940	7.8867	6.72	2.5923	8.1976
5.73	2.3937	7.5697	6.23	2.4960	7.8930	6.73	2.5942	8.2037
5.74	2.3958	7.5763	6.24	2.4980	7.8994	6.74	2.5962	8.2098
5.75	2.3979	7.5829	6.25	2.5000	7.9057	6.75	2.5981	8.2158
5.76	2.4000	7.5895	6.26	2.5020	7.9120	6.76	2.6000	8.2219
5.77	2.4021	7.5961	6.27	2.5040	7.9183	6.77	2.6019	8.2280
5.78	2.4042	7.6026	6.28	2.5060	7.9246	6.78	2.6038	8.2341
5.79	2.4062	7.6092	6.29	2.5080	7.9310	6.79	2.6058	8.2401
5.80	2.4083	7.6158	6.30	2.5100	7.9373	6.80	2.6077	8.2462
5.81	2.4104	7.6223	6.31	2.5120	7.9436	6.81	2.6096	8.2523
5.82	2.4125	7.6289	6.32	2.5140	7.9498	6.82	2.6115	8.2583
5.83	2.4145	7.6354	6.33	2.5159	7.9561	6.83	2.6134	8.2644
5.84	2.4166	7.6420	6.34	2.5179	7.9624	6.84	2.6153	8.2704
5.85	2.4187	7.6485	6.35	2.5199	7.9687	6.85	2.6173	8.2765
5.86	2.4207	7.6551	6.36	2.5219	7.9750	6.86	2.6192	8.2825
5.87	2.4228	7.6616	6.37	2.5239	7.9812	6.87	2.6211	8.2885
5.88	2.4249	7.6681	6.38	2.5259	7.9875	6.88	2.6230	8.2946
5.89	2.4269	7.6746	6.39	2.5278	7.9937	6.89	2.6249	8.3006
5.90	2.4290	7.6811	6.40	2.5298	8.0000	6.90	2.6268	8.3066
5.91	2.4310	7.6877	6.41	2.5318	8.0062	6.91	2.6287	8.3126
5.92	2.4331	7.6942	6.42	2.5338	8.0125	6.92	2.6306	8.3187
5.93	2.4352	7.7006	6.43	2.5357	8.0187	6.93	2.6325	8.3247
5.94	2.4372	7.7071	6.44	2.5377	8.0250	6.94	2.6344	8.3307
5.95	2.4393	7.7136	6.45	2.5397	8.0312	6.95	2.6363	8.3367
5.96	2.4413	7.7201	6.46	2.5417	8.0374	6.96	2.6382	8.3427
5.97	2.4434	7.7266	6.47	2.5436	8.0436	6.97	2.6401	8.3487
5.98	2.4454	7.7330	6.48	2.5456	8.0498	6.98	2.6420	8.3546
5.99	2.4474	7.7395	6.49	2.5475	8.0561	6.99	2.6439	8.3606

SQUARE ROOTS (Continued)

n	$\sqrt{n}$	$\sqrt{10n}$	n	$\sqrt{n}$	$\sqrt{10n}$	n	$\sqrt{n}$	$\sqrt{10n}$
7.00	2.6458	8.3666	7.50	2.7386	8.6603	8.00	2.8284	8.9443
7.01	2.6476	8.3726	7.51	2.7404	8.6660	8.01	2.8302	8.9499
7.02	2.6495	8.3785	7.52	2.7423	8.6718	8.02	2.8320	8.9554
7,03	2.6514	8.3845	7.53	2.7441	8.6776	8.03	2.8337	8.9610
7.04	2.6533	8.3905	7.54	2.7459	8.6833	8.04	2.8355	8.9666
7.05	2.6552	8.3964	7.55	2.7477	8.6891	8.05	2.8373	8.9722
7.06	2.6571	8.4024	7.56	2.7495	8.6948	8.06	2.8390	8.9778
7.07	2.6589	8.4083	7.57	2.7514	8.7006	8.07	2.8408	8.9833
7.08	2.6608	8.4143	7.58	2.7532	8.7063	8.08	2.8425	8.9889
7.09	2.6627	8.4202	7.59	2.7550	8.7121	8.09	2.8443	8.9944
7.10	2.6646	8.4261	7.60	2.7568	8.7178	8.10	2.8460	9.0000
7.11	2.6665	8.4321	7.61	2.7586	8.7235	8.11	2.8478	9.0056
7.12	2.6683	8.4380	7.62	2.7604	8.7293	8.12	2.8496	9.0111
7.13	2.6702	8.4439	7.63	2.7622	8.7350	8.13	2.8513	9.0167
7.14	2.6721	8.4499	7.64	2.7641	8.7407	8.14	2.8531	9.0222
7.15	2.6739	8.4558	7.65	2.7659	8.7464	8.15	2.8548	9.0277
7.16	2.6758	8.4617	7.66	2.7677	8.7521	8.16	2.8566	9.0333
7.17	2.6777	8.4676	7.67	2.7695	8.7579	8.17	2.8583	9.0388
7.18	2.6796	8.4735	7.68	2.7713	8.7636	8.18	2.8601	9.0443
7.19	2.6814	8.4794	7.69	2.7731	8.7693	8.19	2.8618	9.0499
7.20	2.6833	8.4853	7.70	2.7749	8.7750	8.20	2.8636	9.0554
7.21	2.6851	8.4912	7.71	2.7767	8.7807	8.21	2.8653	9.0609
7.22	2.6870	8.4971	7.72	2.7785	8.7864	8.22	2.8671	9.0664
7.23	2.6889	8.5029	7.73	2.7803	8.7920	8.23	2.8688	9.0719
7.24	2.6907	8.5088	7.74	2.7821	8.7977	8.24	2.8705	9.0774
7.25	2.6926	8.5147	7.75	2.7839	8.8034	8.25	2.8723	9.0830
7.26	2.6944	8.5206	7.76	2.7857	8.8091	8.26	2.8740	9.0885
7.27	2.6963	8.5264	7.77	2.7875	8.8148	8.27	2.8758	9.0940
7.28	2.6981	8.5323	7.78	2.7893	8.8204	8.28	2.8775	9.0995
7.29	2.7000	8.5381	7.79	2.7911	8.8261	8.29	2.8792	9.1049
7.30	2.7019	8.5440	7.80	2.7928	8.8318	8.30	2.8810	9.1104
7.31	2.7037	8.5499	7.81	2.7946	8.8374	8.31	2.8827	9.1159
7.32	2.7055	8.5557	7.82	2.7964	8.8431	8.32	2.8844	9.1214
7.33	2.7074	8.5615	7.83	2.7982	8.8487	8.33	2.8862	9.1269
7.34	2.7092	8.5674	7.84	2.8000	8.8544	8.34	2.8879	9.1324
7.35	2.7111	8.5732	7.85	2.8018	8.8600	8.35	2.8896	9.1378
7.36	2.7129	8.5790	7.86	2.8036	8.8657	8.36	2.8914	9.1433
7.37	2.7148	8.5849	7.87	2.8054	8.8713	8.37	2.8931	9.1488
7.38	2.7166	8.5907	7.88	2.8071	8.8769	8.38	2.8948	9.1542
7.39	2.7185	8.5965	7.89	2.8089	8.8826	8.39	2.8965	9.1597
7.40	2.7203	8.6023	7.90	2.8107	8.8882	8.40	2.8983	9.1652
7.41	2.7221	8.6081	7.91	2.8125	8.8938	8.41	2.9000	9.1706
7.42	2.7240	8.6139	7.92	2.8142	8.8994	8.42	2.9017	9.1761
7.43	2.7258	8.6197	7.93	2.8160	8.9051	8.43	2.9034	9.1815
7.44	2.7276	8.6255	7.94	2.8178	8.9107	8.44	2.9052	9.1869
7.45	2.7295	8.6313	7.95	2.8196	8.9163	8.45	2.9069	9.1924
7.46	2.7313	8.6371	7.96	2.8213	8.9219	8.46	2.9086	9.1978
7.47	2.7331	8.6429	7.97	2.8231	8.9275	8.47	2.9103	9.2033
7.48	2.7350	8.6487	7.98	2.8249	8.9331	8.48	2.9120	9.2087
7.49	2.7368	8.6545	7.99	2.8267	8.9387	8.49	2.9138	9.2141

Table X

SQUARE ROOTS (Continued)

n	$\sqrt{n}$	$\sqrt{10n}$	n	$\sqrt{n}$	$\sqrt{10n}$	n	$\sqrt{n}$	$\sqrt{10n}$
8.50	2.9155	9.2195	9.00	3.0000	9.4868	9.50	3.0822	9.7468
8.51	2.9172	9.2250	9.01	3.0017	9.4921	9.51	3.0838	9.7519
8.52	2.9189	9.2304	9.02	3.0033	9.4974	9.52	3.0854	9.7570
8.53	2.9206	9.2358	9.03	3.0050	9.5026	9.53	3.0871	9.7622
8.54	2.9223	9.2412	9.04	3.0067	9.5079	9.54	3.0887	9.7673
8.55	2.9240	9.2466	9.05	3.0083	9.5131	9.55	3.0903	9.7724
8.56	2.9257	9.2520	9.06	3.0100	9.5184	9.56	3.0919	9.7775
8.57	2.9275	9.2574	9.07	3.0116	9.5237	9.57	3.0935	9.7826
8.58	2.9292	9.2628	9.08	3.0133	9.5289	9.58	3.0952	9.7877
8.59	2.9309	9.2682	9.09	3.0150	9.5341	9.59	3.0968	9.7929
8.60	2.9326	9.2736	9.10	3.0166	9.5394	9.60	3.0984	9.7980
8.61	2.9343	9.2790	9.11	3.0183	9.5446	9.61	3.1000	9.8031
8.62	2.9360	9.2844	9.12	3.0199	9.5499	9.62	3.1016	9.8082
8.63	2.9377	9.2898	9.13	3.0216	9.5551	9.63	3.1032	9.8133
8.64	2.9394	9.2952	9.14	3.0232	9.5603	9.64	3.1048	9.8184
8.65	2.9411	9.3005	9.15	3.0249	9.5656	9.65	3.1064	9.8234
8.66	2.9428	9.3059	9.16	3.0265	9.5708	9.66	3.1081	9.8285
8.67	2.9445	9.3113	9.17	3.0282	9.5760	9.67	3.1097	9.8336
8.68	2.9462	9.3167	9.18	3.0299	9.5812	9.68	3.1113	9.8387
8.69	2.9479	9.3220	9.19	3.0315	9.5864	9.69	3.1129	9.8438
8.70	2.9496	9.3274	9.20	3.0332	9.5917	9.70	3.1145	9.8489
8.71	2.9513	9.3327	9.21	3.0348	9.5969	9.71	3.1161	9.8539
8.72	2.9530	9.3381	9.22	3.0364	9.6021	9.72	3.1177	9.8590
8.73	2.9547	9.3434	9.23	3.0381	9.6073	9.73	3.1193	9.8641
8.74	2.9563	9.3488	9.24	3.0397	9.6125	9.74	3.1209	9.8691
8.75	2.9580	9.3541	9.25	3.0414	9.6177	9.75	3.1225	9.8742
8.76	2.9597	9.3595	9.26	3.0430	9.6229	9.76	3.1241	9.8793
8.77	2.9614	9.3648	9.27	3.0447	9.6281	9.77	3.1257	9.8843
8.78	2.9631	9.3702	9.28	3.0463	9.6333	9.78	3.1273	9.8894
8.79	2.9648	9.3755	9.29	3.0480	9.6385	9.79	3.1289	9.8944
8.80	2.9665	9.3808	9.30	3.0496	9.6437	9.80	3.1305	9.8995
8.81	2.9682	9.3862	9.31	3.0512	9.6488	9.81	3.1321	9.9045
8.82	2.9698	9.3915	9.32	3.0529	9.6540	9.82	3.1337	9.9096
8.83	2.9715	9.3968	9.33	3.0545	9.6592	9.83	3.1353	9.9146
8.84	2.9732	9.4021	9.34	3.0561	9.6644	9.84	3.1369	9.9197
8.85	2.9749	9.4074	9.35	3.0578	9.6695	9.85	3.1385	9.9247
8.86	2.9766	9.4128	9.36	3.0594	9.6747	9.86	3.1401	9.9298
8.87	2.9783	9.4181	9.37	3.0610	9.6799	9.87	3.1417	9.9348
8.88	2.9799	9.4234	9.38	3.0627	9.6850	9.88	3.1432	9.9398
8.89	2.9816	9.4287	9.39	3.0643	9.6902	9.89	3.1448	9.9448
8.90	2.9833	9.4340	9.40	3.0659	9.6954	9.90	3.1464	9.9499
8.91	2.9850	9.4393	9.41	3.0676	9.7005	9.91	3.1480	9.9549
8.92	2.9866	9.4446	9.42	3.0692	9.7057	9.92	3.1496	9.9599
8.93	2.9883	9.4499	9.43	3.0708	9.7108	9.93	3.1512	9.9649
8.94	2.9900	9.4552	9.44	3.0725	9.7160	9.94	3.1528	9.9700
8.95	2.9917	9.4604	9.45	3.0741	9.7211	9.95	3.1544	9.9750
8.96	2.9933	9.4657	9.46	3.0757	9.7263	9.96	3.1559	9.9800
8.97	2.9950	9.4710	9.47	3.0773	9.7314	9.97	3.1575	9.9850
8.98	2.9967	9.4763	9.48	3.0790	9.7365	9.98	3.1591	9.9900
8.99	2.9983	9.4816	9.49	3.0806	9.7417	9.99	3.1607	9.9950

Table XI

BINOMIAL PROBABILITIES

n	x	p = 0.05	0.1	0.2	0.3	0.4	0.5	0.6	0.7	0.8	0.9	0.95
2	0	0.902	0.810	0.640	0.490	0.360	0.250	0.160	0.090	0.040	0.010	0.002
	1	0.095	0.180	0.320	0.420	0.480	0.500	0.480	0.420	0.320	0.180	0.095
	2	0.002	0.010	0.040	0.090	0.160	0.250	0.360	0.490	0.640	0.810	0.902
3	0	0.857	0.729	0.512	0.343	0.216	0.125	0.064	0.027	0.008	0.001	
	1	0.135	0.243	0.384	0.441	0.432	0.375	0.288	0.189	0.096	0.027	0.007
	2	0.007	0.027	0.096	0.189	0.288	0.375	0.432	0.441	0.384	0.243	0.135
	3		0.001	0.008	0.027	0.064	0.125	0.216	0.343	0.512	0.729	0.857
4	0	0.815	0.656	0.410	0.240	0.130	0.062	0.026	0.008	0.002		
	1	0.171	0.292	0.410	0.412	0.346	0.250	0.154	0.076	0.026	0.004	
	2	0.014	0.049	0.154	0.265	0.346	0.375	0.346	0.265	0.154	0.049	0.014
	3		0.004	0.026	0.076	0.154	0.250	0.346	0.412	0.410	0.292	0.171
	4			0.002	0.008	0.026	0.062	0.130	0.240	0.410	0.656	0.815
5	0	0.774	0.590	0.328	0.168	0.078	0.031	0.010	0.002			
	1	0.204	0.328	0.410	0.360	0.259	0.156	0.077	0.028	0.006		
	2	0.021	0.073	0.205	0.309	0.346	0.312	0.230	0.132	0.051	0.008	0.001
	3	0.001	0.008	0.051	0.132	0.230	0.312	0.346	0.309	0.205	0.073	0.021
	4			0.006	0.028	0.077	0.156	0.259	0.360	0.410	0.328	0.204
	5				0.002	0.010	0.031	0.078	0.168	0.328	0.590	0.774
6	0	0.735	0.531	0.262	0.118	0.047	0.016	0.004	0.001			
	1	0.232	0.354	0.393	0.303	0.187	0.094	0.037	0.010	0.002		
	2	0.031	0.098	0.246	0.324	0.311	0.234	0.138	0.060	0.015	0.001	
	3	0.002	0.015	0.082	0.185	0.276	0.312	0.276	0.185	0.082	0.015	0.002
	4		0.001	0.015	0.060	0.138	0.234	0.311	0.324	0.246	0.098	0.031
	5			0.002	0.010	0.037	0.094	0.187	0.303	0.393	0.354	0.232
	6				0.001	0.004	0.016	0.047	0.118	0.262	0.531	0.735
7	0	0.698	0.478	0.210	0.082	0.028	0.008	0.002				
	1	0.257	0.372	0.367	0.247	0.131	0.055	0.017	0.004			
	2	0.041	0.124	0.275	0.318	0.261	0.164	0.077	0.025	0.004		
	3	0.004	0.023	0.115	0.227	0.290	0.273	0.194	0.097	0.029	0.003	
	4		0.003	0.029	0.097	0.194	0.273	0.290	0.227	0.115	0.023	0.004

Table XI

BINOMIAL PROBABILITIES (Continued)

		p										
n	*x*	0.05	0.1	0.2	0.3	0.4	0.5	0.6	0.7	0.8	0.9	0.95
7	5			0.004	0.025	0.077	0.164	0.261	0.318	0.275	0.124	0.041
	6				0.004	0.017	0.055	0.131	0.247	0.367	0.372	0.257
	7					0.002	0.008	0.028	0.082	0.210	0.478	0.698
8	0	0.663	0.430	0.168	0.058	0.017	0.004	0.001				
	1	0.279	0.383	0.336	0.198	0.090	0.031	0.008	0.001			
	2	0.051	0.149	0.294	0.296	0.209	0.109	0.041	0.010	0.001		
	3	0.005	0.033	0.147	0.254	0.279	0.219	0.124	0.047	0.009		
	4		0.005	0.046	0.136	0.232	0.273	0.232	0.136	0.046	0.005	
	5			0.009	0.047	0.124	0.219	0.279	0.254	0.147	0.033	0.005
	6			0.001	0.010	0.041	0.109	0.209	0.296	0.294	0.149	0.051
	7				0.001	0.008	0.031	0.090	0.198	0.336	0.383	0.279
	8					0.001	0.004	0.017	0.058	0.168	0.430	0.663
9	0	0.630	0.387	0.134	0.040	0.010	0.002					
	1	0.299	0.387	0.302	0.156	0.060	0.018	0.004				
	2	0.063	0.172	0.302	0.267	0.161	0.070	0.021	0.004			
	3	0.008	0.045	0.176	0.267	0.251	0.164	0.074	0.021	0.003		
	4	0.001	0.007	0.066	0.172	0.251	0.246	0.167	0.074	0.017	0.001	
	5		0.001	0.017	0.074	0.167	0.246	0.251	0.172	0.066	0.007	0.001
	6			0.003	0.021	0.074	0.164	0.251	0.267	0.176	0.045	0.008
	7				0.004	0.021	0.070	0.161	0.267	0.302	0.172	0.063
	8					0.004	0.018	0.060	0.156	0.302	0.387	0.299
	9						0.002	0.010	0.040	0.134	0.387	0.630
10	0	0.599	0.349	0.107	0.028	0.006	0.001					
	1	0.315	0.387	0.268	0.121	0.040	0.010	0.002				
	2	0.075	0.194	0.302	0.233	0.121	0.044	0.011	0.001			
	3	0.010	0.057	0.201	0.267	0.215	0.117	0.042	0.009	0.001		
	4	0.001	0.011	0.088	0.200	0.251	0.205	0.111	0.037	0.006		
	5		0.001	0.026	0.103	0.201	0.246	0.201	0.103	0.026	0.001	
	6			0.006	0.037	0.111	0.205	0.251	0.200	0.088	0.011	0.001
	7			0.001	0.009	0.042	0.117	0.215	0.267	0.201	0.057	0.010
	8				0.001	0.011	0.044	0.121	0.233	0.302	0.194	0.075
	9					0.002	0.010	0.040	0.121	0.268	0.387	0.315
	10						0.001	0.006	0.028	0.107	0.349	0.599

BINOMIAL PROBABILITIES (Continued)

		p										
n	*x*	0.05	0.1	0.2	0.3	0.4	0.5	0.6	0.7	0.8	0.9	0.95
11	0	0.569	0.314	0.086	0.020	0.004						
	1	0.329	0.384	0.236	0.093	0.027	0.005	0.001				
	2	0.087	0.213	0.295	0.200	0.089	0.027	0.005	0.001			
	3	0.014	0.071	0.221	0.257	0.177	0.081	0.023	0.004			
	4	0.001	0.016	0.111	0.220	0.236	0.161	0.070	0.017	0.002		
	5		0.002	0.039	0.132	0.221	0.226	0.147	0.057	0.010		
	6			0.010	0.057	0.147	0.226	0.221	0.132	0.039	0.002	
	7			0.002	0.017	0.070	0.161	0.236	0.220	0.111	0.016	0.001
	8				0.004	0.023	0.081	0.177	0.257	0.221	0.071	0.014
	9				0.001	0.005	0.027	0.089	0.200	0.295	0.213	0.087
	10					0.001	0.005	0.027	0.093	0.236	0.384	0.329
	11							0.004	0.020	0.086	0.314	0.569
12	0	0.540	0.282	0.069	0.014	0.002						
	1	0.341	0.377	0.206	0.071	0.017	0.003					
	2	0.099	0.230	0.283	0.168	0.064	0.016	0.002				
	3	0.017	0.085	0.236	0.240	0.142	0.054	0.012	0.001			
	4	0.002	0.021	0.133	0.231	0.213	0.121	0.042	0.008	0.001		
	5		0.004	0.053	0.158	0.227	0.193	0.101	0.029	0.003		
	6			0.016	0.079	0.177	0.226	0.177	0.079	0.016		
	7			0.003	0.029	0.101	0.193	0.227	0.158	0.053	0.004	
	8			0.001	0.008	0.042	0.121	0.213	0.231	0.133	0.021	0.002
	9				0.001	0.012	0.054	0.142	0.240	0.236	0.085	0.017
	10					0.002	0.016	0.064	0.168	0.283	0.230	0.099
	11						0.003	0.017	0.071	0.206	0.377	0.341
	12							0.002	0.014	0.069	0.282	0.540
13	0	0.513	0.254	0.055	0.010	0.001						
	1	0.351	0.367	0.179	0.054	0.011	0.002					
	2	0.111	0.245	0.268	0.139	0.045	0.010	0.001				
	3	0.021	0.100	0.246	0.218	0.111	0.035	0.006	0.001			
	4	0.003	0.028	0.154	0.234	0.184	0.087	0.024	0.003			
	5		0.006	0.069	0.180	0.221	0.157	0.066	0.014	0.001		
	6		0.001	0.023	0.103	0.197	0.209	0.131	0.044	0.006		
	7			0.006	0.044	0.131	0.209	0.197	0.103	0.023	0.001	

Table XI

BINOMIAL PROBABILITIES (Continued)

		p										
n	x	0.05	0.1	0.2	0.3	0.4	0.5	0.6	0.7	0.8	0.9	0.95
13	8			0.001	0.014	0.066	0.157	0.221	0.180	0.069	0.006	
	9				0.003	0.024	0.087	0.184	0.234	0.154	0.028	0.003
	10				0.001	0.006	0.035	0.111	0.218	0.246	0.100	0.021
	11					0.001	0.010	0.045	0.139	0.268	0.245	0.111
	12						0.002	0.011	0.054	0.179	0.367	0.351
	13							0.001	0.010	0.055	0.254	0.513
14	0	0.488	0.229	0.044	0.007	0.001						
	1	0.359	0.356	0.154	0.041	0.007	0.001					
	2	0.123	0.257	0.250	0.113	0.032	0.006	0.001				
	3	0.026	0.114	0.250	0.194	0.085	0.022	0.003				
	4	0.004	0.035	0.172	0.229	0.155	0.061	0.014	0.001			
	5		0.008	0.086	0.196	0.207	0.122	0.041	0.007			
	6		0.001	0.032	0.126	0.207	0.183	0.092	0.023	0.002		
	7			0.009	0.062	0.157	0.209	0.157	0.062	0.009		
	8			0.002	0.023	0.092	0.183	0.207	0.126	0.032	0.001	
	9				0.007	0.041	0.122	0.207	0.196	0.086	0.008	
	10				0.001	0.014	0.061	0.155	0.229	0.172	0.035	0.004
	11					0.003	0.022	0.085	0.194	0.250	0.114	0.026
	12					0.001	0.006	0.032	0.113	0.250	0.257	0.123
	13						0.001	0.007	0.041	0.154	0.356	0.359
	14							0.001	0.007	0.044	0.229	0.488
15	0	0.463	0.206	0.035	0.005							
	1	0.366	0.343	0.132	0.031	0.005						
	2	0.135	0.267	0.231	0.092	0.022	0.003					
	3	0.031	0.129	0.250	0.170	0.063	0.014	0.002				
	4	0.005	0.043	0.188	0.219	0.127	0.042	0.007	0.001			
	5	0.001	0.010	0.103	0.206	0.186	0.092	0.024	0.003			
	6		0.002	0.043	0.147	0.207	0.153	0.061	0.012	0.001		
	7			0.014	0.081	0.177	0.196	0.118	0.035	0.003		
	8			0.003	0.035	0.118	0.196	0.177	0.081	0.014		
	9			0.001	0.012	0.061	0.153	0.207	0.147	0.043	0.002	
	10				0.003	0.024	0.092	0.186	0.206	0.103	0.010	0.001
	11				0.001	0.007	0.042	0.127	0.219	0.188	0.043	0.005
	12					0.002	0.014	0.063	0.170	0.250	0.129	0.031
	13						0.003	0.022	0.092	0.231	0.267	0.135
	14							0.005	0.031	0.132	0.343	0.366
	15								0.005	0.035	0.206	0.463

Table XII

RANDOM NUMBERS

48611	62866	33963	14045	79451	04934	45576
78812	03509	78673	73181	29973	18664	04555
19472	63971	37271	31445	49019	49405	46925
51266	11569	08697	91120	64156	40365	74297
55806	96275	26130	47949	14877	69594	83041
77527	81360	18180	97421	55541	90275	18213
77680	58788	33016	61173	93049	04694	43534
15404	96554	88265	34537	38526	67924	40474
14045	22917	60718	66487	46346	30949	03173
68376	43918	77653	04127	69930	43283	35766
93385	13421	67957	20384	58731	53396	59723
09858	52104	32014	53115	03727	98624	84616
93307	34116	49516	42148	57740	31198	70336
04794	01534	92058	03157	91758	80611	45357
86265	49096	97021	92582	61422	75890	86442
65943	79232	45702	67055	39024	57383	44424
90038	94209	04055	27393	61517	23002	96560
97283	95943	78363	36498	40662	94188	18202
21913	72958	75637	99936	58715	07943	23748
41161	37341	81838	19389	80336	46346	91895
23777	98392	31417	98547	92058	02277	50315
59973	08144	61070	73094	27059	69181	55623
82690	74099	77885	23813	10054	11900	44653
83854	24715	48866	65745	31131	47636	45137
61980	34997	41825	11623	07320	15003	56774
99915	45821	97702	87125	44488	77613	56823
48293	86847	43186	42951	37804	85129	28993
33225	31280	41232	34750	91097	60752	69783
06846	32828	24425	30249	78801	26977	92074
32671	45587	79620	84831	38156	74211	82752
82096	21913	75544	55228	89796	05694	91552
51666	10433	10945	55306	78562	89630	41230
54044	67942	24145	42294	27427	84875	37022
66738	60184	75679	38120	17640	36242	99357
55064	17427	89180	74018	44865	53197	74810
69599	60264	84549	78007	88450	06488	72274
64756	87759	92354	78694	63638	80939	98644
80817	74533	68407	55862	32476	19326	95558
39847	96884	84657	33697	39578	90197	80532
90401	41700	95510	61166	33757	23279	85523
78227	90110	81378	96659	37008	04050	04228
87240	52716	87697	79433	16336	52862	69149
08486	10951	26832	39763	02485	71688	90936
39338	32169	03713	93510	61244	73774	01245
21188	01850	69689	49426	49128	14660	14143
13287	82531	04388	64693	11934	35051	68576
53609	04001	19648	14053	49623	10840	31915
87900	36194	31567	53506	34304	39910	79630
81641	00496	36058	75899	46620	70024	88753
19512	50277	71508	20116	79520	06269	74173

Table XII

RANDOM NUMBERS (Continued)

24418	23508	91507	76455	54941	72711	39406
57404	73678	08272	62941	02349	71389	45605
77644	98489	86268	73652	98210	44546	27174
68366	65614	01443	07607	11826	91326	29664
64472	72294	95432	53555	96810	17100	35066
88205	37913	98633	81009	81060	33449	68055
98455	78685	71250	10329	56135	80647	51404
48977	36794	56054	59243	57361	65304	93258
93077	72941	92779	23581	24548	56415	61927
84533	26564	91583	83411	66504	02036	02922
11338	12903	14514	27585	45068	05520	56321
23853	68500	92274	87026	99717	01542	72990
94096	74920	25822	98026	05394	61840	83089
83160	82362	09350	98536	38155	42661	02363
97425	47335	69709	01386	74319	04318	99387
83951	11954	24317	20345	18134	90062	10761
93085	35203	05740	03206	92012	42710	34650
33762	83193	58045	89880	78101	44392	53767
49665	85397	85137	30496	23469	42846	94810
37541	82627	80051	72521	35342	56119	97190
22145	85304	35348	82854	55846	18076	12415
27153	08662	61078	52433	22184	33998	87436
00301	49425	66682	25442	83668	66236	79655
43815	43272	73778	63469	50083	70696	13558
14689	86482	74157	46012	97765	27552	49617
16680	55936	82453	19532	49988	13176	94219
86938	60429	01137	86168	78257	86249	46134
33944	29219	73161	46061	30946	22210	79302
16045	67736	18608	18198	19468	76358	69203
37044	52523	25627	63107	30806	80857	84383
61471	45322	35340	35132	42163	69332	98851
47422	21296	16785	66393	39249	51463	95963
24133	39719	14484	58613	88717	29289	77360
67253	67064	10748	16006	16767	57345	42285
62382	76941	01635	35829	77516	98468	51686
98011	16503	09201	03523	87192	66483	55649
37366	24386	20654	85117	74078	64120	04643
73587	83993	54176	05221	94119	20108	78101
33583	68291	50547	96085	62180	27453	18567
02878	33223	39199	49536	56199	05993	71201
91498	41673	17195	33175	04994	09879	70337
91127	19815	30219	55591	21725	43827	78862
12997	55013	18662	81724	24305	37661	18956
96098	13651	15393	69995	14762	69734	89150
97627	17837	10472	18983	28387	99781	52977
40064	47981	31484	76603	54088	91095	00010
16239	68743	71374	55863	22672	91609	51514
58354	24913	20435	30965	17453	65623	93058
52567	65085	60220	84641	18273	49604	47418
06236	29052	91392	07551	83532	68130	56970

Answers to odd-numbered exercises

Page 8

1. (a) Descriptive methods; (b) descriptive methods; (c) inference; (d) descriptive methods; (e) inference.
3. (a) Descriptive methods; (b) inference; (c) descriptive methods; (d) inference; (e) descriptive methods.

Page 22

1. (a) No; (b) 44; (c) no; (d) 63; (e) no; (f) no.
3. There are only five classes; 11 can go into either of two classes; none of the classes include 23; the fifth class covers a wider range of values than the others.
5. (a) 2, 8, 14, 20, 26, and 32; (b) 7, 13, 19, 25, 31, and 37; (c) 1.5, 7.5, 13.5, 19.5, 25.5, 31.5, and 37.5; (d) 4.5, 10.5, 16.5, 22.5, 28.5, and 34.5; (e) 6.
7. (a) 11.95, 12.95, 13.95, 14.95, 15.95, and 16.95; (b) 11.45, 12.45, 13.45, 14.45, 15.45, 16.45, and 17.45; (c) 1.
9. (a) 4.995, 14.995, 24.995, 34.995, 44.995, 54.995, and 64.995; (b) −0.005, 9.995, 19.995, 29.995, 39.995, 49.995, 59.995, and 69.995; (c) 10.00.
11. (a) The respective frequencies are 2, 8, 13, 11, 5, and 1; (b) the respective cumulative frequencies are 40, 38, 30, 17, 6, 1, and 0.
13. The respective frequencies are 1, 3, 7, 13, 13, 7, 2, and 2.

Page 34

1. Yes, the mean is 2.1.
3. (a) The mean is 103.5 and the *average I.Q.* of the jurors is over 100, but it is better to avoid terms such as "average juror"; (b) the answer is the same, and the calculation of the mean can often be simplified by subtracting a suitable number from each value and then adding it to the result.

5. 72.
7. 45.8.
9. 45.0.
11. (a) 5.74; (b) 45.0; (c) 1,079.5, 0.1090, 119.71, or 4.51.
13. (a) 8.2%; if equal amounts are invested at each interest rate; (b) 7.34%.
15. (a) 242 pounds; (b) $14,918; (c) $11,871.67.
17. (a) $40; (b) 360 miles per hour.
19. (a) 110.1%; (b) 136.7% and 137.4%.

Page 50

1. (a) 22; (b) 21; (c) 22.5.
3. (a) 18.375; (b) 17; (c) 24.
5. (a) 52.3; (b) 51; (c) 52.5; (d) 44 and 75; (e) the median; (f) at most $1,110.
7. (a) The medians are 4, 5, 4, 3, 5, 2, 3, 5, 3, 2, 3, 4, and the means are 4, $4\frac{1}{3}$, 4, $3\frac{1}{3}$, 4, 2, $2\frac{2}{3}$, 4, $3\frac{1}{3}$, 3, 3, 4; (b) the distributions are

	Medians	*Means*
1.5–2.5	2	1
2.5–3.5	4	5
3.5–4.5	3	6
4.5–5.5	3	0

9. (a) 5.67; (b) 5–6.
11. The median is 61.8 and the modal class is 60–69.
13. Independent is the modal choice.
15. (a) 28.95 and 58.18; (b) 3.98 and 7.40; (c) 50.39 and 72.72.
17. (a) 16.17 and 75.12; (b) 4.35 and 6.33; (c) 47.27 and 75.51.

Page 64

1. (a) Each expedition will discover the same number of the ancient cities; (b) the first expedition will discover neither of the ancient cities and the second expedition will discover two; (c) both of the ancient cities will be discovered by the expeditions; (d) the first expedition will discover neither of the ancient cities.
3. (b) K consists of (2, 2), (3, 2), and (3, 3); L consists of (1, 0) and (1, 1); M consists of (0, 0), (1, 1), (2, 2), and (3, 3); (c) M' consists of (1, 0), (2, 0), (2, 1), (3, 0), (3, 1), and (3, 2); it represents the event that at least one operative car is not out on a call; $M \cap L$ consists of (1, 1); it represents the event that only one car is operative and it is out on a call; (d) K and L are

mutually exclusive, K and M are not mutually exclusive, and L and M are not mutually exclusive.

5. (c) D is the event that the first shopper prefers Brand A; E is the event that the second shopper prefers Brand B; F is the event that neither shopper has no preference; G is the event that one of the two shoppers prefers Brand A while the other has no preference; (d) F' consists of (1, 3), (2, 3), (3, 1), (3, 2), and (3, 3); it represents the event that at least one of the two shoppers has no preference; $D \cup E$ consists of (1, 1), (1, 2), (1, 3), (2, 2), and (3, 2); it represents the event that either the first shopper prefers Brand A or the second shopper prefers Brand B; $F \cap E$ consists of (1, 2) and (2, 2); it represents the event that the second shopper prefers Brand B and the first shopper prefers Brand A or Brand B; (e) the first shopper prefers Brand B, the second shopper has no preference, and the third shopper prefers Brand A; all three shoppers give the same response; the first two shoppers prefer Brand B.
7. Region 1 represents the event that the novel will get good reviews and be a best-seller; Region 2 represents the event that the novel will get good reviews but not be a best-seller; Region 3 represents the event that the novel will not get good reviews but be a best-seller; Region 4 represents the event that the novel will not get good reviews and not be a best-seller.
9. (a) 2; (b) 6; (c) 1 and 3; (d) 2, 5, and 7.
11. (a) Not mutually exclusive; (b) mutually exclusive; (c) not mutually exclusive; (d) not mutually exclusive; (e) mutually exclusive; (f) not mutually exclusive; (g) mutually exclusive.

Page 76

5. (b) 80; (c) 120; (d) 60.
7. 30.
9. 40.
11. 240.
13. 4,096.
15. Beauty, Cactus, and Princess; Beauty, Princess, and Cactus; Cactus, Beauty, and Princess; Cactus, Princess, and Beauty; Princess, Beauty, and Cactus; Princess, Cactus, and Beauty.
17. (a) 5,040; (b) 210.
19. 120.
21. (a) 1,365; (b) 816; (c) 286; (d) 1,820.
23. 200.

Page 85

1. 0.76.
3. (a) 0.167; (b) 1 to 5.

5. (a) 0.08; (b) 0.92; (c) 2 to 23.
7. $\frac{2}{3}$.
9. 0.64.
11. (a) At least 0.80; (b) at most 0.20.
13. Greater than 0.88 but at most 0.92.

Page 92

1. \$0.16.
3. (a) \$0.05; (b) no.
5. \$12.40.
7. The probability is greater than 0.80.
9. (a) \$7,430; (b) \$1,930.
11. 2.88 complaints.
13. (b) La Jolla; (c) Mission Beach.

Page 104

1. (a) the sum of the probabilities exceeds 1; (b) probabilities cannot be negative; (c) $0.67 + 0.08$ does not equal 0.79; (d) the two events are not mutually exclusive; (e) the sum of the probabilities exceeds 1.
3. (a) 0.41; (b) 0.86; (c) 0.14; events C and D must be mutually exclusive.
5. Odds against winning are 11 to 9.
7. (a) 0.29; (b) 0.73; (c) 0.99.
9. (a) 0.55; (b) 0.40; (c) 0.60; (d) 0.83.
11. (b) 0.4, 0.7, and 0.6.
13. $\frac{1}{16}$, $\frac{4}{16}$, $\frac{6}{16}$, $\frac{4}{16}$, and $\frac{1}{16}$.
15. (a) 0.37; (b) 0.45; (c) 0.80; (d) 0.17; (e) 0.83; (f) 0.20.
17. (a) 0.20; (b) 0.80; (c) 0.03.

Page 113

1. (a) The probability that the fire was put out without too much damage given that the fire alarm was sounded right away; (b) the probability that the fire was not put out without too much damage given that the fire alarm was not sounded right away; (c) the probability that the fire alarm was sounded right away given that the fire was put out without too much damage; (d) the probability that the fire alarm was sounded right away given that the fire was not put out without too much damage.
3. (a) The probability that a very rare ancient coin is in good condition; (b) the probability that an ancient coin which is not a counterfeit is in good

condition; (c) the probability that an ancient coin which is not very rare is a counterfeit; (d) the probability that a very rare ancient coin is in good condition and a counterfeit; (e) the probability that an ancient coin which is not in good condition is very rare; (f) the probability that an ancient coin which is very rare and a counterfeit is in good condition.

5. (a) $\frac{43}{60}$; (b) $\frac{17}{60}$; (c) $\frac{2}{3}$; (d) $\frac{1}{3}$; (e) $\frac{31}{60}$; (f) $\frac{1}{5}$; (g) $\frac{13}{15}$; (h) $\frac{2}{15}$; (i) $\frac{31}{40}$; (j) $\frac{31}{43}$; (k) $\frac{2}{5}$; (l) $\frac{8}{17}$.
7. (a) $\frac{1}{3}$; (b) $\frac{2}{9}$.
9. (a) 0.90; (b) 0.96.
13. (a) $\frac{1}{7}$; (b) $\frac{12}{35}$; (c) $\frac{18}{35}$.
17. 0.4096.
19. (a) $\frac{1}{64}$; (b) $\frac{1}{256}$; (c) 0.32768; (d) 0.195112; (e) 0.0081.

Page 122

1. 0.34.
3. 0.84.
5. The odds are 5 to 4 that he is not staying in Milan.

Page 135

1. (a) Yes; (b) no, the sum of the probabilities is less than 1; (c) no, f(1) is negative; (d) yes.
5. (a) 0.273; (b) 0.039; (c) 0.0197.
7. (a) 0.185; (b) 0.185.
9. (a) 0.991; (b) 0.991.
11. (a) 0.506; (b) 0.253.
13. (a) 0.512; (b) 0.044.
15. 0.15.
17. (a) 0.545; (b) 0.382; (c) 0.255.
21. (a) 0.326; (b) 0.325.
23. (a) 0.140; (b) 0.179; (c) 0.162.

Page 149

1. (a) $\mu = 1$; (b) $\sigma^2 = 1$.
3. (a) $\mu = 1.8$; (b) $\sigma^2 = 1.8$.
5. (a) $\mu = 450$ and $\sigma = 15$; (b) $\mu = 120$ and $\sigma = 10$; (c) $\mu = 7.2$ and $\sigma = 2.4$; (d) $\mu = 280$ and $\sigma = 9.16$.
7. (a) 0.363, 0.484, 0.145, and 0.009; (b) $\mu = 0.801$; the formula yields $\mu = 0.8$.

9. (a) $\mu = 60$ and $\sigma = 5.48$; (b) the probability is at least $\frac{8}{9}$ that the student will get anywhere from 44 to 76 correct answers; (c) the probability is at least $\frac{15}{16}$ that the student will get anywhere from 39 to 81 correct answers.
11. The probability is at least $\frac{63}{64}$ (or approximately 0.984) that anywhere from 207 to 353 of the students will prefer to live in coed dormitories.
13. For the first 100 flips the proportion of heads is 0.30 and for the first 200 flips it is 0.37. Thus, the proportion of heads is getting closer to 0.50, and this actually supports the Law of Large Numbers.

Page 162

1. (a) The probability that the random variable will take on a value on the interval from 4 to 6; 0.25; (b) 0.625; (c) 0.60; (d) 0.30.
3. (a) 0.2939; (b) 0.4505; (c) 0.7291; (d) 0.4602; (e) 0.7088; (f) 0.0681; (g) 0.2589; (h) 0.4155; (i) 0.5898; (j) 0.6318.
5. (a) 1.92; (b) 0.74; (c) −2.14; (d) 1.12; (e) −2.83 or −2.84; (f) 1.44.
7. (a) 1.64 or 1.65; (b) 1.96; (c) 2.05; (d) 2.33; (e) 2.33; (f) 2.57 or 2.58.
9. (a) 0.9772; (b) 0.0401; (c) 0.8664; (d) 0.8822.
11. 73.1.

Page 172

1. (a) 0.9251; (b) 0.3211.
3. (a) 0.5899; (b) 0.0475; (c) 0.0038.
5. (a) 0.0277; (b) 0.1093; (c) 87; (d) 57.
7. (a) 0.0732; (b) 0.6678.

Page 177

1. 0.2128; the tabular value is 0.209.
3. (a) 0.1530; (b) 0.5438; (c) 0.8888.
5. 0.0778.
7. 0.9090.
9. (a) 0.2358; (b) 0.4908; (c) 0.9556.

Page 182

1. (a) It is only a sample if we are interested in the claims filed against the insurance company during the whole year, or the claims filed against all fire insurance companies during the given month; (b) it is a population if we are

interested only in the claims filed against the particular fire insurance company during the given month.

3. (a) The information would constitute only a sample if the congressman used it to make generalizations about the opinions of all his constituents; (b) the information would constitute a population, say, if the congressman wanted to determine what percentage of the letters he received opposed a given piece of legislation.
5. (a) Finite; all scientists eligible for the award; (b) infinite; the hypothetically infinite number of measurements we could make of the weight of the rock; (c) finite; the novels that are on the given bookstore's shelves; (d) infinite; it consists of the integers 0, 1, 2, 3, . . .; (e) finite; the games played by this team during the 1974 season.

Page 187

1. (a) $\frac{1}{56}$; (b) $\frac{1}{220}$.
3. (a) *ab*, *ac*, *ad*, *ae*, *af*, *bc*, *bd*, *be*, *bf*, *cd*, *ce*, *cf*, *de*, *df*, and *ef*; (b) $\frac{1}{2}$; yes.
5. Using the letters G, X, I, S, P, and A to denote the six corporations, the choices are GXIS, GXIP, GXIA, GXSP, GXSA, GXPA, GISP, GISA, GIPA, GSPA, XISP, XISA, XIPA, XSPA, and ISPA; (a) $\frac{2}{3}$; (b) $\frac{2}{5}$; (c) $\frac{1}{5}$.
9. 489, 113, 238, 337, 496, 375, 221, 271, 3, 438, 146, 166, 339, 160, and 370.
13. (b) The mean of the probability distribution is $\mu = 3.02$.

Page 199

1. 12.5.
3. 1.603.
5. 116.3.
7. If we add or subtract the same number to (from) each measurement, this will not affect their standard deviation; to simplify the work as much as possible, it is generally desirable to subtract from each measurement a number close to their mean. (If it is preferable to keep all the numbers positive, the best thing to do is to subtract from each measurement the value of the smallest measurement or a convenient number near it.)
9. 6.12.
11. (a) 5.38; (b) 6.15; (c) 2.24; differences are due to rounding.
13. $\sum (x - \overline{x}) = \sum x - \sum \overline{x} = n \cdot \overline{x} - n \cdot \overline{x} = 0.$
15. (a) Chicken; (b) the student who got 25 is in a relatively better position at the first university; the student who got 50 is in a relatively better position at the second university.
17. (a) 0.055; (b) 0.279.

Page 211

1. (a) It is divided by 2; (b) it is divided by 3; (c) it is multiplied by 5; (d) it is divided by 1.5.
3. (c) 1.58.
5. (a) The probability is at least 0.84; (b) 0.9876.
7. 0.9742; when the population is finite, the sample becomes a greater proportion of the population; thus, there is relatively more information and the probability is greater.
9. 0.9974.
11. (b) 400.

Page 222

1. The probability is 0.95 that the research worker's error is less than 1.568.
3. (a) The probability is 0.95 that the error is less than 1.708; (b) the probability is 0.99 that the error is less than 2.248.
5. (a) 711; (b) 424; (c) 55; (d) 1,232.
7. $15.363 < \mu < 15.837$.
9. (a) $\$452.61 < \mu < \$481.89$; (b) $\$447.98 < \mu < \$486.52$.
11. $59.71 \text{ mph} < \mu < 63.29 \text{ mph}$.
13. (a) $1.728 < \mu < 2.632$; (b) $5.38 < \mu < 7.82$; (c) $135{,}639 < \mu < 157{,}745$; (d) $6.80 < \mu < 17.20$.
15. $0.935 < \sigma < 1.278$.

Page 237

1. $z = -2.31$; reject the null hypothesis that convicted embezzlers spend on the average 12.3 months in jail.
3. $z = 2.89$; reject the null hypothesis and accept the district superintendent's claim.
5. $z = -2.42$; (a) reject the claim that a pound of the given kind of fish yields on the average 2.41 ounces of FPC; (b) cannot reject this claim (but may be reluctant to accept it because the value of z is numerically quite large).
7. $z = 1.84$; reject the null hypothesis and accept the alternative that the true average daily emission exceeds 18 tons.
9. (a) $t = -0.60$; the null hypothesis that the mean breaking strength is 175 pounds cannot be rejected; (b) $t = 4.84$; reject the null hypothesis and accept the alternative that the true average nicotine content of the cigarets is greater than 0.30 milligrams; (c) $t = -2.68$; reject the null hypothesis and, hence, the claim; (d) $t = -1.17$; the null hypothesis that the engine will operate on the average at the desired standard cannot be rejected.

Page 244

1. $z = 1.44$; the difference is not significant and the claim cannot be accepted.
3. $z = 4.76$; the difference is significant; in the given city male file clerks have on the average higher earnings than female file clerks.
5. (a) $t = 3.62$; the difference is significant; (b) $t = -2.20$; the difference is not significant; cannot reject the hypothesis that on the average the trees in the two orchards are equally tall; (c) $t = 0.85$; the difference is not significant; cannot reject the null hypothesis that the true average mileage is the same for both cars.
7. $t = -4.06$; this supports the claim that the program of physical exercise is effective.

Page 253

1. $F = 2.25$; the hypothesis that there is no difference in the true average yield of the four varieties of wheat cannot be rejected.
3. $F = 6.2$; reject the null hypothesis that there is no difference in the true average time it takes to drive along the four different routes.
5. $F = 0.41$; the differences among the average mileages are not significnat.

Page 260

1. $0.26 < p < 0.31$.
3. 47%–72.5%.
5. (a) $0.12 < p < 0.22$; (b) $0.11 < p < 0.21$.
7. $0.253 < p < 0.307$.
9. (a) $0.53 < p < 0.61$; (b) we can assert with a probability of 0.95 that the error is numerically less than 0.04.
11. We can assert with a probability of 0.99 that the size of the error is less than 0.067.
13. (a) 1,068 and 897; (b) 385; (c) 4,161 and 3,994; (d) $\frac{1}{4} - (p - \frac{1}{2})^2 = \frac{1}{4} - (p^2 - p + \frac{1}{4}) = -p^2 + p = p(-p + 1) = p(1 - p)$; the maximum value is $\frac{1}{4}$ because $(p - \frac{1}{2})^2$ cannot be negative.

Page 267

1. The probability of 3 or more successes is 0.184, so that the null hypothesis cannot be rejected.
3. The probability of at most one success is 0.048, so that (a) the null hypothesis cannot be rejected, and (b) the null hypothesis will have to be rejected.

5. $z = -1.06$; the hypothesis (and, hence, the manufacturer's claim) cannot be rejected.
7. $z = -2.33$; reject the null hypothesis and accept the alternative that fewer than 40% of the calls are life-threatening emergencies.
9. $z = 0.77$; the hypothesis that 60% of the students oppose the plan cannot be rejected.
11. (a) The probability of one or fewer successes is 0.020, so that the null hypothesis will have to be rejected; (b) the probability of 11 or more successes is 0.059, so that the null hypothesis cannot be rejected; (c) the probability of nine or more successes is 0.134, so that the null hypothesis cannot be rejected.
13. (a) The probability of three or fewer successes is 0.029, so that the null hypothesis will have to be rejected; (b) the probability of 13 or more successes is 0.003, so that the null hypothesis will have to be rejected; (c) $z = 2.04$; the null hypothesis will have to be rejected; (d) $z = 2.75$; the null hypothesis will have to be rejected.

Page 276

1. $\chi^2 = 34.9$; the difference is significant.
3. $\chi^2 = 4.98$; the difference is not significant.
5. $\chi^2 = 4.74$; the differences among the proportions are not significant.
7. (a) $z = 2.23$; the null hypothesis cannot be rejected; (b) $z = -0.50$; the null hypothesis cannot be rejected; (c) $z = -0.74$; the null hypothesis cannot be rejected.

Page 285

1. $\chi^2 = 17.0$; reject the null hypothesis that in the three cities housewives behave in the same way with regard to the recycling of empty aluminum cans.
3. $\chi^2 = 1.3$; cannot reject the null hypothesis that the three vendors ship products of equal quality.
5. $\chi^2 = 13.2$; the null hypothesis that there is no relationship cannot be rejected.
7. $\chi^2 = 85$; reject the null hypothesis that there is no relationship between school children's intelligence and their standard of clothing.

Page 289

1. $\chi^2 = 7.0$; the null hypothesis cannot be rejected; that is, there is a fairly good fit.

Page 301

1. (a) $a = 2.34$ and $b = 24.93$; (b) $a = 2.34$ and $b = 24.93$; (c) 151.92 words per minute.
3. (a) $y = 54.0 + 0.6x$; (c) 85.2 grams.
5. (a) $y = 7.66 + 1.11x$; 13.21 million dollars; (b) $y = 493.12 + 14.20x$; 791.32 million dollars; (c) $y = 770.4 + 66.9x$; 1,439.4 million pounds.

Page 309

1. (a) Positive correlation; (b) negative correlation; (c) no correlation; (d) positive correlation; (e) negative correlation; (f) no correlation; (g) negative correlation; (h) negative correlation; (i) positive correlation; (j) positive correlation.
3. $r = 0.99$; significant.
5. $r = -0.98$; significant.
9. (a) $r' = 0.99$; (b) $r' = 0.98$; (c) $r' = 0.85$; (d) $r' = 0.87$; (e) $r' = 0.54$; (f) $r' = 0.75$.
11. (a) $21.4 < \beta < 28.4$; (b) $3.5 < \beta < 5.3$.

Index